Medicinal Plants: Microbial Interactions, Molecular Techniques and Therapeutic Trends

Edited by

Zulqurnain Khan

Department of Biotechnology
Institute of Plant Breeding and Biotechnology (IPBB)
Muhammad Nawaz Shareef University of Agriculture
Multan-60000, Pakistan

Azra Yasmin

Department of Biotechnology
Fatima Jinnah Women University
Rawalpindi, Pakistan

&

Naila Safdar

Department of Biotechnology
Fatima Jinnah Women University
Rawalpindi, Pakistan

Medicinal Plants: Microbial Interactions, Molecular Techniques and Therapeutic Trends

Editors: Zulqurnain Khan, Azra Yasmin and Naila Safdar

ISBN (Online): 978-981-5136-83-8

ISBN (Print): 978-981-5136-84-5

ISBN (Paperback): 978-981-5136-85-2

Published by Bentham Science Publishers Pte. Ltd. Singapore. All Rights Reserved.

First published in 2023.

need for a court order if at any point you breach any terms of this License Agreement. In no event will any delay or failure by Bentham Science Publishers in enforcing your compliance with this License Agreement constitute a waiver of any of its rights.

3. You acknowledge that you have read this License Agreement, and agree to be bound by its terms and conditions. To the extent that any other terms and conditions presented on any website of Bentham Science Publishers conflict with, or are inconsistent with, the terms and conditions set out in this License Agreement, you acknowledge that the terms and conditions set out in this License Agreement shall prevail.

Bentham Science Publishers Pte. Ltd.
80 Robinson Road #02-00
Singapore 068898
Singapore
Email: subscriptions@benthamscience.net

BENTHAM SCIENCE

CONTENTS

FOREWORD

Writing forward of this valuable book is a matter of pleasure and honour for me. This book, "Medicinal Plants: Microbial Interactions, Molecular Techniques and Therapeutic Trends" is divided into three sections focusing on medicinal plants and their therapeutic trends and techniques, which can be used to improve medicinal plants. The first section deals with the plant-microbe interaction strategies, the diversity of microbes related to medicinal plants and the role of these microbes in the value addition of medicinal plants. Microbial Phyto-therapeutics and microflora of medicinal plants in hydroponic systems are providing current trends in the respective field. The second section of the book focuses on molecular techniques available for the genetic improvement of medicinal plants. Latest cutting-edge tools such as genome editing and modern breeding tools and techniques are discussed to manipulate medicinal plants for better growth, yield and quality. In the third section of the book, nano strategies and nanoengineering opportunities and challenges are described which are involved in increasing the therapeutic tendencies of medicinal plants. Medicinal plants and their affiliation with the treatment of a variety of diseases along with the production of bioactive clinically administered drugs urge the need for persuasive therapeutic potencies through recent approaches. Detailed chapters on the contribution of nanotechnology in elucidating the pharmacological profile of medicinal plants are explored in this section. Therapeutic enhancement through most employed metallic and carbon nanostructures, nano-elicitors for the production of augmented bioactive secondary metabolites and the role of nanocarriers in achieving the best possible efficacy of the bioactive medicinal ingredients are comprehensively discussed in this section to cover the recent and up-to-date approaches in the nano-manipulations of the medicinal plants.

I am really convinced by the knowledge and expertise of the editors who have done a great job in compiling and formatting a comprehensive and valuable book on the importance, use, and improvement of medicinal plants. I hope that the book will get a broad readership ranging from students to scientists, academia and industry. I appreciate the editors, authors and publishers for their effortless hard work in writing, formatting, compiling and publishing this tremendous and valuable book.

Asif Ali Khan (T.I.)
MNS University of Agriculture
Multan-60000, Pakistan

PREFACE

"Medicinal Plants: Microbial Interactions, Molecular Techniques, and Therapeutic Trends" is one of the efforts and contributions to bring together the advancements in technologies going on with respect to medicinal plants in the search for healthcare agents. This book will look into the ongoing practices and recent innovations involved in enhancing, modulating, and isolating active therapeutic phytodrugs. The importance of medicinal plants can be realized from the fact that more than half of the clinically administered drugs are derived from natural medicinal plants. This book has been divided into three sections, each section focusing on the recent strategies employed for phytochemistry and therapeutic explication. The book will cover modern-day approaches like microbial-plant interaction, genome editing and nano-engineering of medicinal plants for increasing the therapeutic potencies of the plants that have not been assembled in one book yet. Given that, the book will give a comprehensive scenario of the recent scientific approaches explored for increasing the production of phyto-metabolites and therapeutic attributes. Detailed chapters on the contribution of microbial interactions, genetic alterations and nanotechnology in elucidating the pharmacological profile of medicinal plants have been presented. Hope the readers will find it functional and exemplary useful in the subject of medicinal plants.

Zulqurnain Khan
Department of Biotechnology
Institute of Plant Breeding and Biotechnology (IPBB)
Muhammad Nawaz Shareef University of Agriculture
Multan-60000, Pakistan

Azra Yasmin
Department of Biotechnology
Fatima Jinnah Women University
Rawalpindi, Pakistan

&

Naila Safdar
Department of Biotechnology
Fatima Jinnah Women University
Rawalpindi, Pakistan

List of Contributors

Azra Yasmin	Department of Biotechnology, Fatima Jinnah Women University, Rawalpindi, Pakistan
Ayesha Hassan	Institute of Environmental and Agricultural Science, Faculty of Life Sciences, University of Okara, Okara, 56130, Pakistan
Ahmad Mahmood	Department of Soil and Environmental Sciences, Muhammad Nawaz Shareef University of Agriculture, Multan, Pakistan
Ali Hamid	Institute of Plant Breeding and Biotechnology, Muhammad Nawaz Shareef University of Agriculture, Multan, Pakistan
Akash Fatima	Institute of Plant Breeding and Biotechnology, MNS University of Agriculture, Multan, Pakistan
Abdul Qayyum	Department of Agronomy, The University of Haripur, Haripur, 22620, Pakistan
Ajam C. Shaikh	MAEER's MIT College of Railway Engineering and Research, Barshi-413401, Solapur, Maharashtra, India
Ashfaq A. Shah	Department of Life Sciences, Graphic Era (Deemed to be) University, Dehradun-248001, Uttarakhand, India
Amit Gupta	Department of Life Sciences, Graphic Era (Deemed to be) University, Dehradun-248001, Uttarakhand, India
Bushra Hafeez Kiani	Department of Biological Sciences (Female Campus), International Islamic University, Islamabad, 44000, Pakistan
Babar Farid	Institute of Plant Breeding and Biotechnology, MNS University of Agriculture, Multan, Pakistan
Chandra Kishore	Stem Cell Research Centre, Department of Hematology, Sanjay Gandhi Postgraduate Institute of Medical Sciences, Lucknow, Uttar Pradesh, India
Divya Pa	Department of Pharmaceutics, Dr. D.Y. Patil Institute of Pharmaceutical Sciences and Research, Pimpri, Pune-411018, India
Fatima Javeria	Institute of Plant Breeding and Biotechnology, MNS University of Agriculture, Multan, Pakistan
Fouzia Tanvir	Department of Biochemistry, University of Okara, Punjab, Pakistan
Furqan Ahmad	Institute of Plant Breeding and Biotechnology, MNS University of Agriculture, Multan, Pakistan
Hafiz Kamran Yousaf	Institute of Pure and Applied Zoology, Faculty of Life Sciences, Department of Wildlife and Ecology, University of Okara, Okara, 56130, Pakistan
Hijab Fatima	Department of Biotechnology, Fatima Jinnah Women University, Rawalpindi, Pakistan
Hafiz Shahzad Ahmad	Department of Soil and Environmental Sciences, Muhammad Nawaz Shareef University of Agriculture, Multan, Pakistan
Hammad Naeem	Department of Food Science and Technology, Muhammad Nawaz Sharif University of Agriculture, Multan, Pakistan

Iram Fatima	Department of Biotechnology, Fatima Jinnah Women University, Rawalpindi, Pakistan
Irum Shahzadi	Department of Biotechnology, COMSATS University Islamabad, Abbottabad Campus, Pakistan
Jawad Ali	Department of Food Science and Technology, Muhammad Nawaz Sharif University of Agriculture, Multan, Pakistan
Kalakotla Shanker	Department of Pharmacognosy & Phyto-Pharmacy, JSS College of Pharmacy, JSS Academy of Higher Education & Research, Ooty, Nilgiris, Tamil Nadu, India
Kristina Apryatina	Department of High Molecular Compounds and Colloid Chemistry, Faculty of Chemistry, Lobachevsky State University of Nizhny Novgorod (UNN), Moscow, Russia
Kulsoom Zahara	Department of Botany, PMAS-Arid Agriculture University Rawalpindi, Rawalpindi-46300, Pakistan
Krishna Prakash	ICAR-Indian Agricultural Research Institute (IARI), Hazaribagh, Jharkhand, India
Muhammad Insaf	Institute of Plant Breeding and Biotechnology, MNS University of Agriculture, Multan, Pakistan
Muhammad Abu Bakar Saddique	Institute of Plant Breeding and Biotechnology, MNS University of Agriculture, Multan, Pakistan
Muhammad Ali Sher	Institute of Plant Breeding and Biotechnology, MNS University of Agriculture, Multan, Pakistan
Mahmood Alam Khan	Institute of Plant Breeding and Biotechnology, MNS University of Agriculture, Multan, Pakistan
Muhammad Rahil Afzal	Institute of Environmental and Agricultural Science, Faculty of Life Sciences, University of Okara, Okara, 56130, Pakistan
Mustansar Aslam	Institute of Environmental and Agricultural Science, Faculty of Life Sciences, University of Okara, Okara, 56130, Pakistan
Muhammad Saleem	Department of Biochemistry, University of Okara, Punjab, Pakistan
Muhammad Shahbaz	Department of Food Science and Technology, Muhammad Nawaz Sharif University of Agriculture, Multan, Pakistan
Muhammad Ashfaq	Institute of Plant Protection, MNS University of Agriculture, Multan, Pakistan
Maria Siddique	Department of Environmental Science, COMSATS University Islamabad, Abbottabad Campus, Pakistan
Maria Khan	Department of Education, University of Mianwali, Punjab, Pakistan
Muhammad Usman	Department of Food Science and Technology, MNS University of Agriculture, Multan, Pakistan
Nighat Raza	Department of Food Science and Technology, Muhammad Nawaz Sharif University of Agriculture, Multan, Pakistan
Nadia Iqbal	Department of Biotechnology, Woman University Multan, Multan, Punjab, Pakistan

Naila Safdar	Department of Biotechnology, Fatima Jinnah Women University, Rawalpindi, Pakistan
Oğuz Can Turgay	Department of Soil Science and Plant Nutrition, Ankara University, Ankara, Turkey
Rana Binyamin	Institute of Plant Protection, MNS University of Agriculture, Multan, Pakistan
Raza Ullah	Institute of Environmental and Agricultural Science, Faculty of Life Sciences, University of Okara, Okara, 56130, Pakistan
Ryota Kataoka	Department of Environmental Sciences, Faculty of Life & Environmental Sciences, University of Yamanashi, Takeda, Kofu, Yamanashi, Japan
S.P. Dhanabal	Department of Pharmacognosy & Phyto-Pharmacy, JSS College of Pharmacy, JSS Academy of Higher Education & Research, Ooty, Nilgiris, Tamil Nadu, India
Saira Karimi	Department of Biosciences, COMSATS University, Islamabad, Pakistan
Sadia Shabir	Institute of Plant Breeding and Biotechnology, MNS University of Agriculture, Multan, Pakistan
Sadia Arshad	The Govt. Sadiq College Women University, Bahawalpur, Punjab, Pakistan
Samiya Rehman	Department of Biochemistry, University of Okara, Punjab, Pakistan
Saman Zulfiqar	Department of Biochemistry, University of Okara, Punjab, Pakistan
Shamas Murtaza	Department of Food Science and Technology, Muhammad Nawaz Sharif University of Agriculture, Multan, Pakistan
Saeed Rauf	Department of Plant Breeding and Genetics, College of Agriculture, University of Sargodha, Sargodha, Punjab, Pakistan
Saba Yaseen	Institute of Plant Breeding and Biotechnology, MNS University of Agriculture, Multan, Pakistan
Shoaib ur Rehman	Institute of Plant Breeding and Biotechnology, MNS University of Agriculture, Multan, Pakistan
Shaghufta Perveen	Department of Biotechnology, Fatima Jinnah Women University, Rawalpindi, Pakistan
Sobia Nisa	Department of Microbiology, The University of Haripur, Haripur, 22620, Pakistan
Sushil Y. Raut	Department of Pharmaceutics, Dr. D.Y. Patil Institute of Pharmaceutical Sciences and Research, Pimpri, Pune-411018, India
Tahira Tabassum	Department of Biochemistry, University of Okara, Punjab, Pakistan
Tanveer ul Haq	Department of Soil and Environmental Sciences, Muhammad Nawaz Shareef University of Agriculture, Multan, Pakistan
Tamatam Sunilkumar Reddy	Department of Pharmacognosy & Phyto-Pharmacy, JSS College of Pharmacy, JSS Academy of Higher Education & Research, Ooty, Nilgiris, Tamil Nadu, India
Umar Farooq	Department of Food Science and Technology, Muhammad Nawaz Sharif University of Agriculture, Multan, Pakistan

Ummara Waheed — Institute of Plant Breeding and Biotechnology, MNS University of Agriculture, Multan, Pakistan

Umar Akram — Institute of Plant Breeding and Biotechnology, MNS University of Agriculture, Multan, Pakistan

Vaishali Ji — Department of Botany, Patna Science College, Patna, Bihar, India

Waleed Asghar — Department of Environmental Sciences, Faculty of Life & Environmental Sciences, University of Yamanashi, Takeda, Kofu, Yamanashi, Japan

Wasim Akhtar — Department of Botany, University of Azad Jammu and Kashmir, Muzaffarabad, Pakistan

Yamin Bibi — Department of Botany, PMAS-Arid Agriculture University Rawalpindi, Rawalpindi-46300, Pakistan

Zulqurnain Khan — Department of Biotechnology, Institute of Plant Breeding and Biotechnology (IPBB), Muhammad Nawaz Shareef University of Agriculture, Multan-60000, Pakistan

Zaib un Nisa — Department of Biochemistry, University of Okara, Punjab, Pakistan

Zahid Ishaq — Department of Biochemistry, Nishtar Medical University, Multan, Pakistan

Zarmeen Zafar — Institute of Plant Breeding and Biotechnology, MNS University of Agriculture, Multan, Pakistan

Zareena Ali — Department of Biochemistry and Biotechnology, The Women University, Multan, Pakistan

SECTION 1: MICROBES AND MEDICINAL PLANTS

CHAPTER 1

Association of Plants and Microbes: Past, Present and Future

Wasim Akhtar[1], Iram Fatima[2], Azra Yasmin[2,*], Naila Safdar[2] and **Zulqurnain Khan[3]**

[1] *Department of Botany, University of Azad Jammu and Kashmir, Muzaffarabad, Pakistan*

[2] *Department of Biotechnology, Fatima Jinnah Women University, Rawalpindi, Pakistan*

[3] *Department of Biotechnology, Institute of Plant Breeding and Biotechnology (IPBB), Muhammad Nawaz Shareef University of Agriculture, Multan-60000, Pakistan*

Abstract: Changing climatic conditions, biotic and abiotic stresses along with use of synthetic fertilizers have deteriorated soil quality and crop yield. Microorganisms are natural inhabitants of soil and plant surfaces that form a stable dynamic system with the host plants. The plant microbiome assists in plant growth by solubilizing minerals, recycling nutrients and inducing defense responses by mitigating environmental stresses. These plant-associated microorganisms can be used as functional moieties to enhance overall plant productivity and reduce negative impacts on the ecosystem. The plants and microbes are contemplated as natural partners that harmonize various functional traits, however, the magnitude of friendly or hostile consortium depends on the kind of microorganisms involved. Before the scientific advent of advanced technologies, conventional approaches such as culturing on media, microscopic observations and biochemical tests provided awareness of how these two communicate. Later on, contemporary molecular-based tools like polymerase chain reaction (PCR), microarrays, enzyme-linked assays (ELISA), and nucleic acid-based methods (next-generation sequencing, *etc.*) surfaced. This chapter will comprehend different types of aboveground and subsurface microbes associated with the plants, their impact on sustainable agriculture and high-throughput technologies used to investigate the plant-microbe relationship.

Keywords: Aboveground, Agriculture, Associations, High-throughput approaches, Microbes, Subsurface.

INTRODUCTION

Plants are ubiquitously colonized by microbiome including archaea, protists, fungi and bacteria which exhibit pleiotropic effects on plant health. Biotechnological

---

* **Corresponding author Azra Yasmin:** Department of Biotechnology, Fatima Jinnah Women University, Rawalpindi, Pakistan; E-mail: azrayasmin@fjwu.edu.pk

advancements have enabled us to explore plant microbiome structure and their interaction with the host [1]. Owing to different areas of colonization, microbial species are classified into different groups; for instance, organisms that inhabit the external plant parts are known as epiphytes, while those that occur on the inner side are endophytes. Rhizosphere exists in areas closest to the roots while phyllosphere organisms colonize the leaf surfaces [2]. Owing to the plant exudation occurring at various developmental stages, the root region is explicitly teamed with microbes that are associated with plant health [3].

The plants build various kinds of relationships with the associated microbes. In mutualism, both host and microbial specie acquire benefits from each other by augmenting nutrient and mineral accessibility, heightening immunity against pathogens, and susceptibility to stress conditions. Further, the plant exudates such as sugars, vitamins and other growth factors also promote the colonization of microbes and thus, it can also be termed a synergistic relationship [4].

Commensalism states the symbiotic connection in which the commensal gets benefits, whereas the other partner is neither harmed nor benefitted from the relationship. Typically, the plant by-passes a nutrient to the microorganism that is living in its vicinity [5]. Amensalism is a sort of reciprocal action where the effect of one organism's activity negatively influences the other. Parasitism is a life-long, co-existing association in which a parasite favorably feeds on its host. In this relationship, if the host defense mechanism is vigorous then the interaction will favor the host over the parasite, however, in the opposite scenario, the host becomes ill and eventually die [5]. Thus, microbial heterogeneity makes it essential to understand the interactome association of microbes with the root systems. Due to the plant-microbe interaction, these are also known as holobionts or meta organisms [6].

Microorganisms that can adapt to particular stress conditions are beneficial for plants in different ways [3]. Beneficial microorganisms are normally inoculated in the soil, as a substitute for chemical treatments, to stabilize soil structure, control pests and diseases, mitigate the negative effects of agrochemicals and ameliorate agricultural practices. These inputs could be in the form of biocontrol agents, bioherbicides and biofertilizers. Recently, the use of inoculants has been intensified due to the availability of multifunctional strains that can improve the yield at minimal cost. Rhizobia is one of the extensively used microbial inoculants that form a symbiotic association with legumes and fulfil the requirement of plant nitrogen by influencing the process of biological nitrogen fixation [7]. Likewise, plant growth-promoting bacteria (PGPB) regulate phytohormones synthesis, and phosphate accessibility and elicit the plant's defense mechanism against environmental stresses. Furthermore, some microorganisms also influence crop production through pests and disease management [8].

The recent interest in advanced technologies has warranted in-depth analysis and characterization of microbes that colonize plants. Genome sequencing, metagenomics, proteomics and other contemporary techniques have enabled us to have a clear view of the dynamic belowground and aboveground microorganisms [9]. Hence, the current chapter highlights different microbial interactions, their impact on crop productivity and the advancements that have been adopted in this field.

THE TWO-DIMENSIONAL MICROBIAL COMMUNITIES

Aboveground Microbes

Microbes colonize different aerial parts constituting anthosphere microbiome (flowers), phyllosphere microbiome (leaves), spermosphere microbiome (seeds) and cardosphere microbiome (fruits) [2]. Endophytic and epiphytic organisms colonize plant tissues and acquire nutrients through the aerial tissues *i.e.,* flowers, fruits and leaves and so, their growth and development largely depend on the nutrition present within the organs [10]. However, the phyllosphere organisms exhibit unstable environments with variable physicochemical limitations like UV radiation, desiccation and temperature and they act as commensal-like organisms [11].

These organisms can be visualized in different host species. For instance, in a study conducted on grapes, it was noticed that the phyllosphere was dominated by *Pseudomonas, Acinetobacter, Pantoea, Bacillus, Sphingomonas* and *Curtobacterium* while the endophytes were inhabited by *Burkholderia, Bacillus, Ralstonia* and *Mesorhizobium* [12, 13]. In another study, *Devosia* and *Pedobacter* dominated the potato phyllosphere while *Pantoea* and *Bacillus*were observed in the lettuce phyllosphere. *Acinetobacter* was reported as the dominant community in the stems and leaves of tomatoes [13]. *Alternaria* was dominantly observed in the stem while peel samples, calyx, were populated by *Penicillium*. Hence, it can be proposed that the microorganisms existing in a particular plant depend on the nutrients, host tissue and the physicochemical properties of the soil [14].

Epiphytes can develop resistance towards immunological and antimicrobial chemicals produced by plant tissues and competing microbes. The quorum-sensing (QS) signals are also linked with these organisms such as in the tobacco phyllosphere [15]. Additionally, these organisms can also produce exopolysaccharides that increase resistance to desiccation along with different enzymes and phytohormones. In short, the association between the plant and aboveground microbes is host-dependent and is affected by the signalling circuits accompanied by other species, consequently leading to enhanced plant growth and a reduction in disease incidence [16].

Belowground or Subsurface Microorganisms

The rhizosphere, with plant exudates, recruits' microbes in a niche resulting in parasitism, commensalism, saprophytism, amensalism and symbiotic relationship [6]. The thin layer of soil surrounding the roots exhibits higher microbial diversity such as *Copiotrophs, Pseudomonads, Oligotrophs* and *Actinobacteria* are abundantly present near the root surface than the bulk soil [17]. These microorganisms, substantially influence the aboveground activities and their diversity can be evaluated by the plants and the soil environment using neutral or niche-based processes. In the former case, many organisms use most of the soil surface and are confined to the recruitment parameters and distance among plants. On the contrary, environmental changes affect the microbial population in the niche-based mechanism [18].

Plants hold the ability to recruit microbes in the root system that assist in various roles. For example, the orchid's fitness is strongly reliant on the mycorrhizal fungi [19] and the microbial abundance and richness in cereals depend on the rhizosphere area [20]. The recruitment agent *i.e.,* plants exudate releases hormones, amino acids, polyphenols, cutin monomers and nutrients that mediate the plant-microbe interactions [21]. Some chemicals initiate microbial growth in the rhizosphere while others inhibit their growth to enhance plant growth as benzoxazinioids are produced in maize roots to inhibit *Proteobacteria* and *Actinobacteria*. The microbial population gradually expand by recruiting additional microbes and developing a new niche within the population. Afterwards, the plant exudates start producing biofilms in the rhizosphere [21].

some endophytic microbes occur inside roots and are used in agriculture. For instance, *Piriformospora indica* elevate phosphorus (P) uptake in plants and protect against various stresses [22]. Correspondingly, Gill *et al.* [22] stated that *P. indica* overexpressed cyclophilin which protects against salt stress in tobacco plants. Some microorganisms also act in synergy in acquiring nutrients such as *Azotobacter chroococcum* and *P. indica*. However, some endophytes are involved in chemotaxis activities, for example, *Fusarium oxysporum* reduces the frequency of nematodes in tomatoes [23]. Similarly, biochar in tomato plants produces chemotactic signals towards *Ralstonia solanacearum* to hinder their swarming ability [24].

Root-root Interaction

The root exudates vary among different plant species and are associated with several processes such as increased mineralization and they ls also mediate nutrient competition. The chemicals (phosphatases) present in the soil, generally fulfil the plant requirement of ions and affect the nitrogen cycle [25]. In addition, plants are also governed by the exudates of the neighbouring species. For instance, the legumes release carboxylate in intercropping with non-leguminous species and provide phosphorus (P) to the neighbouring species [26]. The P-levels can also be replenished through the exudates produced by arbuscular mycorrhizal fungi (AMF) [27]. Furthermore, citric acids from cucumber roots and fumaric acid from banana roots attract *Bacillus amyloliquefaciens* and *Bacillus subtilis,* eventuating in biofilm production [28].

The interaction among different roots just not only regulates the nutrients but also affects the roots of neighbouring species through allelopathy. In this case, the plant roots release phytotoxins, such as catechin, which alleviate the root growth of neighbours and thereby, reduce the competition for resources [29]. Volatile organic compounds (VOCs) also act as allelochemicals to monitor rhizosphere signals by the mycorrhizal associations. The allelopathic species act specifically on other plant species excluding a few non-allelopathic neighbours [30]. The root exudates are regulated by the rhizobiome and thus, it can be suggested that the root-root interaction is competitive. Plants exhibiting horizontal root systems possess greater competition in comparison to those species having vertical and deep root systems. Moreover, in this kind of interaction, the plant exudates influence each other as well as the microbiota existing in their vicinity [31].

Root-microbe Interaction

The interaction of roots with microbes principally includes symbiotic and parasitic associations like rhizobacteria-legume, actinobacteria-root, mycorrhizal-root and other root-microbe interactions. In rhizobacteria-legume interaction, organisms produce Nod factors which activate receptors in plants that cause nodule formation. Subsequently, the bacterium uses nodules to produce ammonia which is used in protein synthesis. Proteomic analysis disclosed that legumes release flavonoids which stimulate Nod protein to set nodulation genes [32]. Another study conducted on *Medicago trunculata* revealed that protein content changes in the nodule during the formation of leghemoglobin [33]. However, some legumes like *Rhizobium leguminosarum*produce ethylene-responsive proteins [34].

Rhizobacteria exist as Bacteroides in the plant symbiosomes where all nutrients are monitored by the composition of bacteriode membranes. Heat-shock proteins, nodulin's, proteases, transporters and signalling proteins chiefly occur in these membranes, which indicates that the plant's defense mechanism is continuously controlled to allow nodulation. The nodules also consist of enzymes that are needed for N_2 fixation, heme-synthesis, transporters and stress-related proteins [35]. The proteomic studies displayed that Bacteroidesincrease nitrogen metabolism and suppress fatty acid metabolism [36]. Contrarily, transcriptome analysis indicated the presence of aquaporins, ATPases, nutrient transporters and osmoregulation in the nodules [37]. In general, all these are beneficial in regulating homeostasis in the nodule and in facilitating the transport of nutrients.

Actinomycetes also form symbiotic associations with hosts such as angiospermic species belonging to *Alnus* and *Datisca* genera. A proteomic study on *Alnus* sp. revealed that it secretes proteins, that are involved in forming symbiotic associations [38]. Besides these, mycorrhizal fungi also form a symbiotic association with the plants. These fungi colonize the plant roots and establish arbuscules in the cortical cells or hyphae [39]. Like rhizobacteria, arbuscular mycorrhizal fungi (AMF) also do not hinder the nutrient trafficking between the host and fungus. AMF supplies phosphorus to the plant in return for lipids and carbon [40].

Another symbiotic relationship occurs between the free-living organisms and its positive impact on plant-host interaction from disease and increased plant growth. For example, *Trichoderma* possesses the ability to parasitize other fungi and is thus considered an important biological control agent. *Trichoderma harzianum* produces proteases that can degrade fungal cell walls and *Trichoderma asperellum* induces the production of disease-resistance proteins. Furthermore, these organisms increases ethylene biosynthesis, protein folding and energy metabolism. *T. harzianum, T. atroviride* and *T. asperellum* possess chitinases and cyclophilins which provide maximum resistance in plants [41].

Microbe-microbe Interaction

This type of interaction includes communication between pathogenic microbes, the interaction between pathogens and endophytes, succession by microorganisms and changes in the environment. The pathogens variably influence the microbial population occurring on plant surfaces and soil, as the host is open to infection and increases the susceptibility to other non-pathogenic organisms as well. For example, maize when infected with SLB, reduces the residual microbial community [42] and Brassicaceae species exhibiting white rust are more susceptible to mildew pathogens [43]. *Albugo laibachii* increases susceptibility to *Phytophthora infestans* in *Arabidopsis thaliana* [44]. In this type of interaction, each microbial cell is accountable for the colonization and disease progression in the host. Bacterial species forms biofilm to establish favorable interactions. QS mediated by signals further heightened the pathogenicity of microbiota on the host. Such signals are also aaffect plant proteome and transcriptome by adhering to the plant surfaces and ultimately influence the processes occurring inside the plant [45].

The endophytes give rise to toxic chemicals such as polyketide synthase which prevent the growth of pathogens. *Bacillus thuringiensis, Enterobacter asburiae* and *Erwinia carotovora* are inhibited by the AHL lactonase enzyme present in the endophytes [46]. These endophytic enzymes protect plants against biotic and abiotic stresses. For instance, trehalose aids in stabilizing the membranes and enzymes during drought and also induces systemic resistance to disease [47]. Microbes that inhabit host species always compete for nutrients and space. If endophyte is present within the host before the pathogen infiltration, then it will show more resistance as compared to when both pathogen and endophyte arrive simultaneously in the plants [48]. There are also certain cases in which endophytes protect by co-inoculating with the pathogen such as *Ustilago maydis* was found to be inhibited by direct inoculation with *Fusarium verticillioides* and did not express protection when applied before pathogen infection [49].

The microbiota existing in the soil secrete molecules that induce gene expression in the plants. Some VOCs (alcohols, ketones, alkanes, *etc.*) act as signals in the microbial species existing in the root system and promote activities like disease inhibition, mineralization, nutrient acquisition many other processes. Phytohormones such as gibberellins, cytokinins, abscisic acid, salicylic acid and jasmonic acid are also secreted by microbes that activate plant defense mechanisms [50]. Besides systemic resistance, plant defense can also be heightened by the trans-generational phenomenon. In this, immune memory is transmitted to the next plant generation in response to the pathogens. For example, *Psudomonas syringae* applied on *Arabidopsis* prompted the next generation of plants which increased the salicylic acid content and induced greater disease resistance [51].

UNRAVELING THE IMPACT OF MICROBES ON SUSTAINABLE AGRICULTURE

Microorganisms greatly influence the physico-chemical as well as biological activities occurring in the soil that are associated with plant growth and development. Nowadays, researchers are trying to identify microbial species using various techniques so that they could be effectively utilized in the agricultural industry. Some of the important functional aspects of plant microbiota along with the advanced scientific approaches are presented in Fig. (**1**).

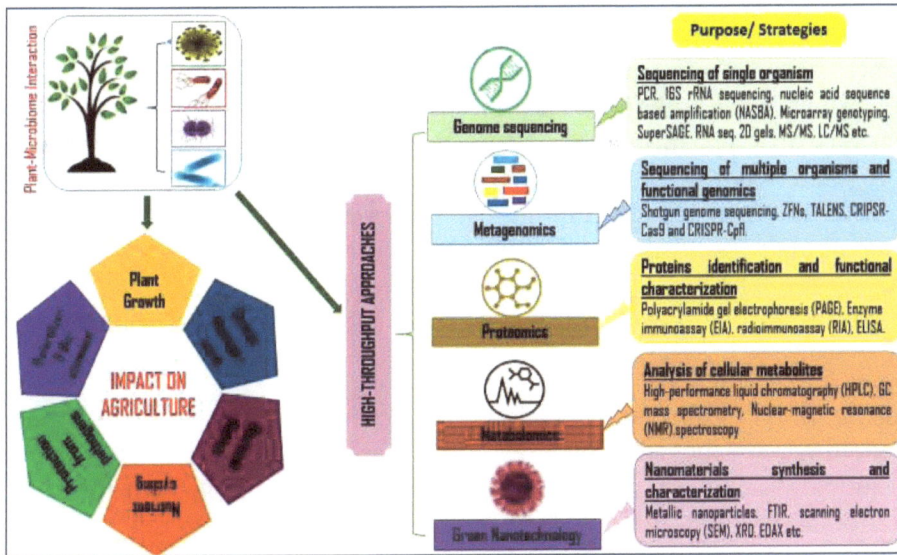

Fig. (1). Schematic overview showing certain benefits of plant-microbe association on sustainable agriculture and various techniques used to investigate host plant and microbial communities.

Growth and Development

Microorganisms use diverse methods to promote plant growth in normal and stressful conditions. These include nitrate reductase activity, phytohormone synthesis, nitrogenase enzyme activity and siderophore generation. Hormones such as auxin, gibberellin, cytokinin, ethylene and abscisic acid are produced by numerous microbes to augment plant growth. Different developmental stages in plants are directly interlinked with the microbial community. For example, microbial composition in *A. thaliana* was examined at four different stages which showed that *Actinobacteria* was abundantly present in seedling and vegetative stages in comparison to the bolting and flowering stages. *Bacteroidetes* and *Cyanobacteria* also increased abundantly with plant growth and reached their peak at the flowering stage [52].

Researchers have genetically modified plant growth-promoting bacteria (PGPB) (or *rhizobacteria*) to increase the synthesis of stress-tolerant hormones, antifreeze proteins, antibiotics, trehalose and lytic enzymes which assist plants in development and surviving stress periods. PGPB needs to develop and maintain a biologically active population in order to compete with indigenous organisms [53]. Microorganisms modified through genetic engineering have developed certain traits such as bioremediation, increased stress tolerance capacity and phytohormone production. The modified strain remains active and survives well in a hostile environment.

Stress Management

Plants are sessile organisms and thus, are subjected to various abiotic and biotic stresses. Plants have to cope with drought, metal toxicity, submergence, salinity and various diseases throughout the seasons. Microorganisms are regarded as prime regulators of stress owing to the presence of biomolecules that are released in the form of hormones and antibiotics. Microbes tend to aggregate and produce biofilms through extracellular polymeric substances (EPS) which help the plant to resist desiccation and other environmental stresses. In this way, plants can combat extreme conditions and get protection against developmental and genetic damages [54].

Hubbard *et al.* [55] observed that fungal endophytes in wheat plants resulted in better yield when exposed to heat. *Enterobacter, Klebsiella* and *Bacillus* produce auxin indole-3-acetic acid (IAA) which enhances the growth of maize in cadmium-stressed conditions [56]. These microbes were also detected from the bitter gourd rhizosphere. Similarly, *Pseudomonas* and *Bacillus* sp. produced IAA during salt stress and increased the growth of *Sulla carnosa* [57].

Bioremediation

Industrialization and advanced farming techniques are increasing the detrimental effects on agricultural land and water by generating hazardous waste, organic contaminants and heavy metals. Some heavy metals are beneficial to plants in trace amounts *viz.* lead (Pb), cadmium (Cd), chromium (Cr), uranium (Ur), zinc (Zn), gold (Au), selenium (Se) and arsenic (As), while their excessive amount alleviates plant growth by impeding photosynthesis and other enzyme activities. Bioremediation is a long-term, cost-effective and sustainable process that utilizes bacteria, fungi, algae or plants to remove heavy metals from polluted environment and eventually restore the natural state of the environment [53].

Different microbial strains are usually inoculated in the soil to detoxify heavy metals. The microorganisms generally use adsorption, biosorption, vacuolar compartmentalization and metal binding approaches to reduce the concentration of metals. *Brevundimonas diminuta* produces ACC deaminase that reduces arsenic toxicity and ethylene concentration under heavy metal stress [58]. It was observed that *Enterobacter asburiae* ameliorates cadmium toxicity in mung beans (*Vigna radiata*) to promote root growth in cadmium-stressed seedlings [59].

Nutrient Cycling

Nitrogen, phosphorus and various other elements are the most limiting nutrients in crop production. Microorganisms consistently recycle these nutrients to get essential components required for DNA, RNA and protein synthesis which are essential for all living beings. The mycorrhizal fungi and nitrogen-fixing bacteria increase the phosphate uptake and transport minerals to the plants by extending surface area for absorption [60]. *Azospirillum* is a free-living organism found in *Zea mays*, sugar cane and wheat that promote plant growth by increasing nutrient uptake through roots [61].

Phosphorus, iron, nitrogen and many other nutrients are usually acquired by PGPB. *Rhizobium* and *Bradyrhizobium* supply nitrogen to the plants by nodulation. However, some microbial species like *Pseudomonas* forms siderophores that are iron-chelators and provides iron to the host species [62]. The microbial community present in leaves also enhances the nutrient accessibility to increase plant growth. It is important to identify and monitor the microbial density in each cycle. The key genes like Nod factors need to be genetically modified to boost nitrogen-fixing abilities. Similar ways can be adopted to improve other nutrient cycling processes [53].

Protection From Pathogens

Rhizospheric microbes specifically augment plant growth and protect against pathogenic organisms. *Rhizobacteria* release metabolites that alleviate the activity of pathogenic organisms, whereas rhizospheric fungi possess the ability to produce antibiotics [63]. *Trichoderma* sp. produces gliovirin, peptaibols, gliotoxin, harzianic acid, massoilactone, tricholin and viridian to counteract pathogenic species [64]. *Agrobacterium radiobacter* produces Agrocin 84 (bacteriocin) that exhibits antibacterial activity [65]. The secondary metabolites along with different genes, such as glucanases and chitinases, effectively increase plant resistance toward pathogens [66].

The microbes present in flowers protect the host species from fire blight and other diseases caused by *Pseudomonas, Erwinia amylovora* and *Pantoea* [67]. The microbes of Fabaceae fruits also possess antibacterial and antifungal activity [68]. The pathogens can easily enter the plant through stomata, hydathodes and wounds. These organisms can be controlled by phyllosphere microbes which induce a plant's immune response through competition between pathogenic and non-pathogenic organisms and by producing antibiotics. In *A. thaliana*, the plant pathogen *P. syringae* was competitively excluded by *Sphingomonas* sp. by competing for sugar, fructose and glucose [69].

Biofertilizers, Biostimulants and Biocontrol Agents

In the absence of pathogens, commensals increase the growth of host plants by producing hormones and VOCs. For instance, radish seeds inoculated with

Pseudomonas fluorescens enhance tryptophan content resulting in increased root biomass [70]. *B. subtilis* increases the photosynthetic ability of *A. thaliana* by modulating endogenous abscisic acid and glucose signalling [71]. Thus, these species can be effectively utilized as biofertilizers, biostimulants and biocontrol agents. Yet, the transition of these microbes from laboratory to market is slow and requires the optimization of a mixture, frequency of application, determination of concentration, selection of carriers and appropriate funding.

Few species have been introduced into the markets as biofertilizers such as Nitragin Golg containing *Rhizobia* (USA) [72], Nodulest 10 with *Bradyrhizobium japonicum* (Argentina) and Phosphobacterin with *Bacillus megaterium* (Russia). Few species are acquainted as biostimulants such as Amase containing *Pseudomonas azotoformans* (Sweden) and biocontrol agents like Cerall with *Pseudomonas chlororaphis* (Sweden) [73, 74]. Another factor that has hampered the utilization of microbes is their potency which is paralleled to the chemical drugs that produce reliable effects on plants and microbes.

TARGETED APPROACHES USED FOR THE ISOLATION AND HIGH-THROUGHPUT CUSTOMIZATION OF PLANT MICROBIOMES

In the beginning, conventional methods like culture studies on media were followed which were extremely laborious, time-consuming and required skilled personnel for morphological and biochemical analysis. Later on, contemporary methods based on electrophoresis, electron microscopy and serological techniques were developed. At the end of the 20th century, ELISA, PCR and many other new tools further revolutionized phytopathogen detection and were routinely used by researchers [75]. The advanced methods have enabled us to compile large amounts of data available on plants and microbes easily through genome sequencing. Previously, Delmont's [76] surveyed Park Grass (2009-2012) and employed six techniques of DNA extraction to present soil microbiome data precisely. It is now possible to scrutinize data using multiple approaches such as genomics, proteomics, metabolomics and transcriptomics platforms. Altogether, these are useful in deciphering the benefits of plant-microbe interactions, enhancing plant growth and ameliorating disease resistance among plants [77].

Genome and Amplicon Sequencing

The genome sequencing platform is frequently used to perform genome-wide annotation of proteins and genes [78]. The first bacterial (*Xylella fastidiosa*) genome was acquired in 1995 and then, *Arabidopsis* in 2000 paved the way for genomic scale annotations to envision the entire organism. Genome sequencing has enabled us to interpret signalling pathways that monitor the defense mechanism in plants. We could have a clear view of the negative and positive effects of microbial interactions by merging their metabolic pathways with different plants [79]. Initially, studies were conducted by using one gene or one protein at a time. Subsequently, more holistic approaches that can identify thousands of taxa at the same time were used to explore any organism and pathogen simultaneously. Contemporary tools include real-time quantitative PCR (RT-qPCR), multiplex PCR and droplet digital PCR (ddPCR). These are also used to ascertain the relative and absolute abundances of pathogenic organisms [80, 81].

Woese and Fox [82] proposed that the 16S rRNA gene can be used for amplification, sequencing and identification of targeted bacterial species. In amplicon sequencing, the universal primers amplify genes by pairing them to highly conserved sequences within the hypervariable region of 16S rRNA. The amplified product is then sequenced and the information obtained is used for the phylogenetic studies of the organisms [83]. This data is further used to deduce the taxonomic information. 18S rRNA and ITS are available for fungal species, while ITS are the most commonly used sequences as they display higher levels of variance and have an extensive reference database. However, amplicon sequencing is unable to identify closely related species with 18S rRNA primers, so ITS region is usually recommended [84].

To obtain more precise relative abundance data of fungal species, non-ITS-based targets must also be included along with ITS sequences [85]. Similar sequences derived from amplicon sequencing are then assigned to the same taxa by the operational taxonomic unit (OUT) having 97% sequence identity with bacterial taxa [86]. The 16S rRNA gene sequencing enables large-scale microbial analysis at relatively moderate expenses and 16S genes are abundantly available, thus facilitating sequencing and determining relationships across taxa [87]. Notwithstanding, sequencing mistakes and chimaeras may occur during DNA amplification and the differences in amplification will also affect the conclusion based on the relative abundance of OUT [88]. Statistical analysis is done using Deblur or Dada2 methods that can be implemented in QIIME2, a software for 16S rRNA analysis [89].

Furthermore, nucleic acid sequence-based amplification (NASBA) was developed for the amplification of RNA sequences containing viruses while loop-mediated isothermal amplification is used to detect plant viruses like plum pox virus [90, 91]. In the early 21st century, cDNA array and SuperSAGE were used to determine gene expression in rice-*Magnaporthe oryzae* and *A. thaliana*-*Pseudomonas syringe* interactions [40, 92]. In microarray, RNA of interest is isolated and reverse transcribed to its complementary DNA. Afterward, they are hybridized with probes on the chip and then detected for signals. Later on,

RNAseq technology was developed to elucidate plant-pathogen profiles more precisely. Over time, 2D gels, MS/MS, LC/MS, GC/MS and iTRAQ techniques were established combining the transcriptome and proteome data of plants and microbes. By using high-throughput technologies, all possible interactions between plants and microbes can be addressed [93].

Metagenomics

Metagenomic sequencing includes the random sequencing of sheared DNA fragments which are re-aligned and provide a more comprehensive functional profile of the microbes. The shotgun genome sequencing approach of metagenomics provides more information on bacteria, viruses, archaea and fungi and requires higher information to distinguish species as compared to the 16S rRNA method [94]. This method requires much more sequencing reads and the information generated often reaches several terabases per study, which increases its costs and bioinformatics demand on sequence assembly and mapping [95]. The online bioinformatic tools assist in analyzing data by comparing reads with the reference databases and/or after de novo assembly [96]. The most common statistical tool used to interpret metagenomics data is a phylogenetic investigation of communities by the reconstruction of unobserved states (PICRUST). It is used to reconstruct the algorithm of the ancestral state which predicts gene families that are present in the sample and then estimates the metagenome [86].

The shotgun metagenome sequencing provides higher efficiency by allowing us to identify up to the strain level in comparison to the amplicons which provide characterization to the taxonomic levels [97]. In metagenomics, the gene sequences may be functionally annotated where the protein-coding sequences are identified and then matched to a protein function [98]. Besides using sequencing platforms, metagenomics screening has some disadvantages like high cost and the risk of missing previously uncharacterized species. So, the bioinformatics tools are updated regularly to improve the quality of reads and information obtained [99]. It can be suggested that a combination of methods should be exploited to acquire maximum information on the total microbial diversity.

Since 2012, the gene editing tool clustered regularly-interspaced short palindromic repeats (CRISPR) has revolutionized life sciences. This technique is used to develop model cell lines, improve gene therapies, identify disease targets, and develop transgenic organisms and transcriptional modulation. Zinc finger nucleases (ZFNs) and transcription activator-like effector nucleases (TALENS) are the predecessors of CRISPR technology [100, 101]. CRISPR reagents are introduced into the plant cells using different vehicles such as physical (microinjection, electroporation), viral (adenovirus and lentivirus) and non-viral (liposomes, gold particles) delivery methods. CRISPR reagents are entered in the form of DNA, RNA or protein-RNA which assemble into site-directed nuclease (SDN) and cleave specific DNA sequences to produce double-strand breaks (DSBs). Plant cells repair these breaks either by annealing DNA ends (SDN-1 editing) which causes few changes in sequences and generate gene knockout, or by integrating piece of DNA at DSB to produce sequence replacements of less than 20 nucleotides (SDN-2 editing) or longer sequence changes (SDN-3 editing) [102].

CRISPR-Cas9 and CRISPR-Cpf1 are the most commonly used CRISPR tools in crop plants including corn, wheat and tomato [103, 104]. The CRISPR field is evolving incredibly, with the number of peer-reviewed publications increasing by 1,435% since 2011. Many Cas9 proteins have been identified from species and used for gene editing such as *Staphylococcus aureus* (SaCas9), *Neisseria meningitides* (NmCas9) and *Streptococcus thermophiles* (StCas9). Its therapeutic potential will further increase with the evolution of technology [105 - 107].

Proteomics

Soil possesses the ability to retain extracellular proteins in different ways, which directly influences the biological systems. Proteomics is used to investigate the protein profile of microbial communities, along with their functions. Species composition and ecological activities can be assessed by unravelling the protein's structural complexities. Polyacrylamide gel electrophoresis (PAGE) is usually used to compare protein profiles, depending on the electric charge and physical size attributes [108]. Different varieties inhabiting ecological niches can be identified using amino acid sequence analysis. However, some issues are concerned with this approach. Firstly, contamination by pollutants sometimes impedes protein extraction. Secondly, extraction techniques also influence the detected metaproteomics which could be circumvented by using extraction methods simultaneously [109]. Thirdly, protein identification is hindered due to an insufficient protein database [110]. In-house libraries are, however, built depending on the data obtained from comparable settings in earlier studies [111].

Furthermore, immunoassays and nucleic acid-based assays are categorized into two groups *i.e.*, direct and labelled methods. The former includes immunoprecipitation, immunodiffusion and immune agglutination while the latter includes enzyme immunoassay (EIA), immunofluorescence and radioimmunoassay (RIA). Earlier, Tempel [112] introduced a gel diffusion test to identify and differentiate formal speciales of *F. oxysporum*. Later on, RIA and ELISA kits were developed to detect *Botrytis cinerea, Rhizoctonia solani* and many other species. ELISA uses epitopes which bind with targeted antibodies conjugated to an enzyme. In the case of irreversible binding, a colour reaction is produced by adding a suitable substrate. This binding specificity is further enhanced by using monoclonal or recombinant antibodies [113]. ELISA assay shas also been used in combination with TaqMan PCR to detect *Phytophthora ramorum* at the species level [114]. ELISA and PCR tests also revealed *Ganoderma lucidum* which causes Ganoderma disease in coconuts [115].

Recently, metagenomics has been integrated with metaproteomics due to the greater efficiency with minimal cost. Next-generation sequencing (NGS) enables us to get maximum reads in less time which helps in establishing optimized databases for protein identification [109]. All-inclusive, metaproteomics is the best platform used to ascertain the biological functions of a microbial community and to compare functional and taxonomic soil makeup in the ecosystem [116]. The soil protein assessment will provide information on the biogeochemical capacity of the soil and pollutants and predict soil health and restoration [117]. This will help us to comprehend contaminants, organic compounds degradation, nutrient cycles and plant-microbial interactions at the molecular level.

Metabolomics

Microorganisms produce and consume metabolites that are linked with microbial functions and host physiology. Metabolomics aids in identifying and quantifying metabolites of plants and microbial species. Gas chromatography-mass spectrometry (GCMS), high-performance liquid chromatography (HPLC) and nuclear magnetic resonance (NMR) spectroscopy are major high-throughput techniques used in this field. Targeted as well as non-targeted metabolite profiling can be addressed easily. In targeted analysis, the specific class of compounds like amino acids, fatty acids, carbohydrates and lipids which are allied with the plant-microbe interactions are determined. However, the non-targeted approach offers a broad overview of metabolic differences by interpreting quantitative or functional variations between microbial species, leading to the discovery of unexpected biomarkers or therapeutic agents [118].

Through advanced sequencing and metabolomics, Hu *et al.* [119] revealed that 6-methoxy-benzoxazolin-2-one could alter the root-associated flora. Furthermore, the relationship between different strains and metabolites can also be investigated using mathematical models such as the MelonnPan model. For instance, the process of recruiting beneficial microbes by the roots can be explored by devices such as an olfactory system that will help in evaluating the ability of root VOCs to recruit soil microbes [120].

Green Nanotechnology

The interception of two domains namely 'green chemistry' and 'nanotechnology' paved the way to a novel nanoscale-oriented field termed 'green technology' (GN). The great contributions of Paul T. Anastas and Richard Philips Feynman in this field are noteworthy [121]. The green synthesis of nanoparticles (NP) using plants and microorganisms offers the advantages of enhanced biocompatibility due to the capping behaviour and is thus, exploited in plant-defense activities and pharmaceutical industries [122]. Their potential application in agriculture has also declined the use of synthetic fertilizers, which in turn help in reducing greenhouse gases such as nitrous oxide [123]. However, the size and shape of NPs play a decisive role in determining their compatibility. For biomedical applications, NPs should be preferably less than 100 nm in size. The instrumental analysis including Fourier-transformed infrared spectroscopy (FTIR) and energy-dispersive X-ray spectroscopy (EDAX) also helps in concluding the capping nature of plant molecules [124].

In general, the nano approach is an economical way to use plants and microbial species and their efficiency can be monitored *via* the biological route of nano-tailoring. Comparatively, the success rate of plants is very high than the microbial-mediated NP. Extensive research has been conducted with plants of different taxonomic groups, revealing their capability of green synthesis. Numerous angiosperms (*Camellia sinensis, Azadirachta indica, Aloe vera* and *Centella asiatica*), gymnosperms (*Ginkgo biloba, Cycas circinalis, torreya nucifera* and *Thuja orientalis*) and pteridophytes (*Nephrolepis exaltata* and *Azolla microphylla*) are the frontline examples of plants that have been explored for medicinal values. The metallic nanoparticles of these species have been experimented with using gold (Au), copper oxide (CuO), palladium (Pd), silver (Ag), iron (Fe_2O_3), zinc oxide (ZnO), magnesium oxide (MgO) and nickel oxide (NiO) [125].

The unicellular and multicellular microorganisms produce intracellular and/or extracellular inorganic materials that can be manipulated for size and shape by controlling the culture parameters. The intracellular and extracellular process of NPs formation varies among different microbes. Gram-positive and negative bacteria exhibit an intercellular process in which positive metal ions interact with negative ions during transport into the cell wall and the cell wall enzymes reduce the metal ions into NPs, which later on, diffuse across the cell walls of bacteria. Conversely, the extracellular production of NPs in fungal species takes place by nitrate reductase-mediated synthesis by using nitrate enzyme, which reduces metal ions into NPs and thus, exhibits greater commercial viability [126]. In comparison, engineered viruses possess self-assembled semiconductor surfaces with highly oriented quantum dots structure along with mono-disperse shape and size of NPs. The genetically engineered tobacco mosaic virus (TMV) produces NPs of certain lengths which can modify inorganic nano-crystals in 3D layered materials [127]. The fabricated viral films can be stored for 7 months for diverse medicinal purposes [128].

CONCLUSION

The green revolution has promoted agricultural value in terms of farm yield with improved plant varieties. However, the wide use of agrochemicals and increased environmental stresses stance a deleterious effect on the environment as well as living organisms. This necessitates the use of microbial inoculations as the plant microbiome communicates beneficial roles to the host with a positive impact on plant productivity and immunity. The high-throughput technologies are extensively used to study the community composition, abundance, genomic profile, phylogeny, sustainability and coherence of microbial communities. Further, CRISPR gene-editing tool is now used to synthesize microbial species to study the relationship between gene clusters and plant phenotypes. Despite the plethora of information on concerned studies, the implementation of these techniques is still in the infancy stage. Thus, there is a need to integrate available data to articulate the ecological principles that could convert this information into better crop yield for sustainable agriculture and food security.

REFERENCES

[1]   Song C, Zhu F, Carrión VJ, Cordovez V. Beyond plant microbiome composition: Exploiting microbial functions and plant traits *via* integrated approaches. Front Bioeng Biotechnol 2020; 8: 896.
        [http://dx.doi.org/10.3389/fbioe.2020.00896] [PMID: 32850744]

[2]   Sivakumar N, Sathishkumar R, Selvakumar G, Shyamkumar R, Arjunekumar K. Phyllospheric Microbiomes: Diversity, ecological significance, and biotechnological applications.Pla Microb Sustain Agricul Sustain DevelopBiodiv. Cham, Switzerland: Springer International Publishing 2020; pp. 113-72.
        [http://dx.doi.org/10.1007/978-3-030-38453-1_5]

[3]   Nadarajah KK. Rhizosphere interactions: Life below ground.Plant-Microbe Interaction: An Approach to Sustainable Agriculture. Singapore: Springer 2016; pp. 3-23.
        [http://dx.doi.org/10.1007/978-981-10-2854-0_1]

[4]   Madigan TM, Martinko MJ, Stahl AD, Clark PD. Brock Biology of Microorganisms. 13th., London: Pearson 2012.

[5]   Willey MJ, Sherwood ML, Woolverton JC. Prescott's Principles of Microbiology. 1st., New York: Mc-Graw-Hill 2009.

[6]   Pascale A, Proietti S, Pantelides IS, Stringlis IA. Modulation of the root microbiome by plant molecules: The basis for targeted disease suppression and plant growth promotion. Front Plant Sci 2020; 10: 1741.
        [http://dx.doi.org/10.3389/fpls.2019.01741] [PMID: 32038698]

[7]   Santos MS, Nogueira MA, Hungria M. Microbial inoculants: Reviewing the past, discussing the present and previewing an outstanding future for the use of beneficial bacteria in agriculture. AMB

Express 2019; 9(1): 205.
[http://dx.doi.org/10.1186/s13568-019-0932-0] [PMID: 31865554]

[8]   Berg G, Köberl M, Rybakova D, Müller H, Grosch R, Smalla K. Plant microbial diversity is suggested as the key to future biocontrol and health trends. FEMS Microbiol Ecol 2017; 93(5).
[http://dx.doi.org/10.1093/femsec/fix050]

[9]   Nadarajah K, Abdul Rahman NSN. Plant–microbe interaction: Aboveground to belowground, from the good to the bad. Int J Mol Sci 2021; 22(19): 10388.
[http://dx.doi.org/10.3390/ijms221910388] [PMID: 34638728]

[10]   Kandel S, Joubert P, Doty S. Bacterial endophyte colonization and distribution within plants. Microorganisms 2017; 5(4): 77.
[http://dx.doi.org/10.3390/microorganisms5040077] [PMID: 29186821]

[11]   Bringel F, Couée I. Pivotal roles of phyllosphere microorganisms at the interface between plant functioning and atmospheric trace gas dynamics. Front Microbiol 2015; 6: 486.
[http://dx.doi.org/10.3389/fmicb.2015.00486] [PMID: 26052316]

[12]   Campisano A, Antonielli L, Pancher M, Yousaf S, Pindo M, Pertot I. Bacterial endophytic communities in the grapevine depend on pest management. PLoS One 2014; 9(11): e112763.
[http://dx.doi.org/10.1371/journal.pone.0112763] [PMID: 25387008]

[13]   Dong CJ, Wang LL, Li Q, Shang QM. Bacterial communities in the rhizosphere, phyllosphere and endosphere of tomato plants. PLoS One 2019; 14(11): e0223847.
[http://dx.doi.org/10.1371/journal.pone.0223847] [PMID: 31703074]

[14]   Abdelfattah A, Wisniewski M, Droby S, Schena L. Spatial and compositional variation in the fungal communities of organic and conventionally grown apple fruit at the consumer point-of-purchase. Hortic Res 2016; 3(1): 16047.
[http://dx.doi.org/10.1038/hortres.2016.47] [PMID: 27766161]

[15]   Ma A, Lv D, Zhuang X, Zhuang G. Quorum quenching in culturable phyllosphere bacteria from tobacco. Int J Mol Sci 2013; 14(7): 14607-19.
[http://dx.doi.org/10.3390/ijms140714607] [PMID: 23857057]

[16]   Vorholt JA. Microbial life in the phyllosphere. Nat Rev Microbiol 2012; 10(12): 828-40.
[http://dx.doi.org/10.1038/nrmicro2910] [PMID: 23154261]

[17]   Donn S, Kirkegaard JA, Perera G, Richardson AE, Watt M. Evolution of bacterial communities in the wheat crop rhizosphere. Environ Microbiol 2015; 17(3): 610-21.
[http://dx.doi.org/10.1111/1462-2920.12452] [PMID: 24628845]

[18]   Mendes LW, Kuramae EE, Navarrete AA, van Veen JA, Tsai SM. Taxonomical and functional microbial community selection in soybean rhizosphere. ISME J 2014; 8(8): 1577-87.
[http://dx.doi.org/10.1038/ismej.2014.17] [PMID: 24553468]

[19]   Li T, Wu S, Yang W, Selosse MA, Gao J. How mycorrhizal associations influence orchid distribution and population dynamics. Front Plant Sci 2021; 12: 647114.
[http://dx.doi.org/10.3389/fpls.2021.647114] [PMID: 34025695]

[20]   Igiehon NO, Babalola OO. Below-ground-above-ground plant-microbial interactions: Focusing on soybean, rhizobacteria and mycorrhizal fungi. Open Microbiol J 2018; 12(1): 261-79.
[http://dx.doi.org/10.2174/1874285801812010261] [PMID: 30197700]

[21]   Santoyo G, Guzmán-Guzmán P, Parra-Cota FI, Santos-Villalobos S, Orozco-Mosqueda MC, Glick BR. Plant growth stimulation by microbial consortia. Agronomy 2021; 11(2): 219.
[http://dx.doi.org/10.3390/agronomy11020219]

[22]   Gill SS, Gill R, Trivedi DK, *et al. Piriformospora indica*: Potential and significance in plant stress tolerance. Front Microbiol 2016; 7: 332.
[http://dx.doi.org/10.3389/fmicb.2016.00332] [PMID: 27047458]

[23]   Eschweiler J, van Holstein-Saj R, Kruidhof HM, Schouten A, Messelink GJ. Tomato inoculation with a non-pathogenic strain of *Fusarium oxysporum* enhances pest control by changing the feeding preference of an omnivorous predator. Front Ecol Evol 2019; 7: 213.
[http://dx.doi.org/10.3389/fevo.2019.00213]

[24]   Gao Y, Lu Y, Lin W, Tian J, Cai K. Microorganisms biochar suppresses bacterial wilt of tomato by improving soil chemical properties and shifting soil microbial community. Microorganisms 2019; 7(12): 676.
[http://dx.doi.org/10.3390/microorganisms7120676] [PMID: 31835630]

[25]   Canarini A, Kaiser C, Merchant A, Richter A, Wanek W. Root Exudation of primary metabolites: Mechanisms and their roles in plant responses to environmental stimuli. Front Plant Sci 2019; 10: 157.
[http://dx.doi.org/10.3389/fpls.2019.00157] [PMID: 30881364]

[26]   Dai J, Qiu W, Wang N, Wang T, Nakanishi H, Zuo Y. From Leguminosae/Gramineae intercropping systems to see benefits of intercropping on iron nutrition. Front Plant Sci 2019; 10: 605.
[http://dx.doi.org/10.3389/fpls.2019.00605] [PMID: 31139203]

[27]   Zemunik G, Lambers H, Turner BL, Laliberté E, Oliveira RS. High abundance of non-mycorrhizal plant species in severely phosphorus-impoverished Brazilian campos rupestres. Plant Soil 2018; 424(1-2): 255-71.
[http://dx.doi.org/10.1007/s11104-017-3503-7]

[28]   Zhang N, Wang D, Liu Y, Li S, Shen Q, Zhang R. Effects of different plant root exudates and their organic acid components on chemotaxis, biofilm formation and colonization by beneficial rhizosphere-associated bacterial strains. Plant Soil 2014; 374(1-2): 689-700.
[http://dx.doi.org/10.1007/s11104-013-1915-6]

[29]   Tharayil N. To survive or to slay. Plant Signal Behav 2009; 4(7): 580-3.
[http://dx.doi.org/10.4161/psb.4.7.8915] [PMID: 19820349]

[30]   Kong CH, Xuan TD, Khanh TD, Tran HD, Trung NT. Allelochemicals and signaling chemicals in plants. Molecules 2019; 24(15): 2737.
[http://dx.doi.org/10.3390/molecules24152737] [PMID: 31357670]

[31]   Montazeaud G, Violle C, Fréville H, *et al.* Crop mixtures: Does niche complementarity hold for belowground resources? An experimental test using rice genotypic pairs. Plant Soil 2018; 424(1-2): 187-202.
[http://dx.doi.org/10.1007/s11104-017-3496-2]

[32]   Lorite MJ, Estrella MJ, Escaray FJ, *et al.* The rhizobia-lotus symbioses: Deeply specific and widely diverse. Front Microbiol 2018; 9: 2055.
[http://dx.doi.org/10.3389/fmicb.2018.02055] [PMID: 30258414]

[33]   Trouvelot S, Héloir MC, Poinssot B, *et al.* Carbohydrates in plant immunity and plant protection: roles and potential application as foliar sprays. Front Plant Sci 2014; 5: 592.
[http://dx.doi.org/10.3389/fpls.2014.00592] [PMID: 25408694]

[34]   Wan J, Torres M, Ganapathy A, *et al.* Proteomic analysis of soybean root hairs after infection by *Bradyrhizobium japonicum.* Mol Plant Microbe Interact 2005; 18(5): 458-67.
[http://dx.doi.org/10.1094/MPMI-18-0458] [PMID: 15915644]

[35]   Davidson AL, Dassa E, Orelle C, Chen J. Structure, function, and evolution of bacterial ATP-binding cassette systems. Microbiol Mol Biol Rev 2008; 72(2): 317-64.
[http://dx.doi.org/10.1128/MMBR.00031-07] [PMID: 18535149]

[36]   Baslam M, Mitsui T, Sueyoshi K, Ohyama T. Recent advances in carbon and nitrogen metabolism in C3 plants. Int J Mol Sci 2020; 22(1): 318.
[http://dx.doi.org/10.3390/ijms22010318] [PMID: 33396811]

[37]   Wang M, Ding L, Gao L, Li Y, Shen Q, Guo S. The interactions of aquaporins and mineral nutrients in higher plants. Int J Mol Sci 2016; 17(8): 1229.

[http://dx.doi.org/10.3390/ijms17081229] [PMID: 27483251]

[38]   Santi C, Bogusz D, Franche C. Biological nitrogen fixation in non-legume plants. Ann Bot 2013; 111(5): 743-67.
[http://dx.doi.org/10.1093/aob/mct048] [PMID: 23478942]

[39]   Nouioui I, Cortés-albayay C, Carro L, *et al.* Genomic insights into plant-growth-promoting potentialities of the genus frankia. Front Microbiol 2019; 10: 1457.
[http://dx.doi.org/10.3389/fmicb.2019.01457] [PMID: 31333602]

[40]   Nadarajah K, Kumar IS. Molecular microbial biodiversity assessment in the mycorrhizosphere.Varma A, Choudhary D. Mycorrhizosphere and Pedogenesis. Singapore: Springer 2019.
[http://dx.doi.org/10.1007/978-981-13-6480-8_23]

[41]   Morán-Diez E, Rubio B, Domínguez S, Hermosa R, Monte E, Nicolás C. Transcriptomic response of Arabidopsis thaliana after 24h incubation with the biocontrol fungus Trichoderma harzianum. J Plant Physiol 2012; 169(6): 614-20.
[http://dx.doi.org/10.1016/j.jplph.2011.12.016] [PMID: 22317785]

[42]   Manching HC, Balint-Kurti PJ, Stapleton AE. Southern leaf blight disease severity is correlated with decreased maize leaf epiphytic bacterial species richness and the phyllosphere bacterial diversity decline is enhanced by nitrogen fertilization. Front Plant Sci 2014; 5: 403.
[http://dx.doi.org/10.3389/fpls.2014.00403] [PMID: 25177328]

[43]   Singh KP, Kumari P, Rai PK. Current status of the disease-resistant gene(s)/QTLs, and strategies for improvement in Brassica juncea. Front Plant Sci 2021; 12: 617405.
[http://dx.doi.org/10.3389/fpls.2021.617405] [PMID: 33747001]

[44]   Prince DC, Rallapalli G, Xu D, *et al.* Albugo-imposed changes to tryptophan-derived antimicrobial metabolite biosynthesis may contribute to suppression of non-host resistance to *Phytophthora infestans* in *Arabidopsis thaliana*. BMC Biol 2017; 15(1): 20.
[http://dx.doi.org/10.1186/s12915-017-0360-z] [PMID: 28320402]

[45]   Chaudhry V, Runge P, Sengupta P, Doehlemann G, Parker JE, Kemen E. Shaping the leaf microbiota: Plant–microbe–microbe interactions. J Exp Bot 2021; 72(1): 36-56.
[http://dx.doi.org/10.1093/jxb/eraa417] [PMID: 32910810]

[46]   Rajesh PS, Ravishankar Rai V. Quorum quenching activity in cell-free lysate of endophytic bacteria isolated from Pterocarpus santalinus Linn., and its effect on quorum sensing regulated biofilm in Pseudomonas aeruginosa PAO1. Microbiol Res 2014; 169(7-8): 561-9.
[http://dx.doi.org/10.1016/j.micres.2013.10.005] [PMID: 24268182]

[47]   Kloepper JW, Ryu C-M. Bacterial endophytes as elicitors of induced systemic resistance. Soc Biol 2006; 9: 33-52.
[http://dx.doi.org/10.1007/3-540-33526-9_3]

[48]   Adame-Álvarez RM, Mendiola-Soto J, Heil M. Order of arrival shifts endophyte-pathogen interactions in bean from resistance induction to disease facilitation. FEMS Microbiol Lett 2014; 355(2): 100-7.
[http://dx.doi.org/10.1111/1574-6968.12454] [PMID: 24801140]

[49]   Pan JJ, Baumgarten AM, May G. Effects of host plant environment and *Ustilago maydis* infection on the fungal endophyte community of maize (*Zea mays*). New Phytol 2008; 178(1): 147-56.
[http://dx.doi.org/10.1111/j.1469-8137.2007.02350.x] [PMID: 18194146]

[50]   Maruri-López I, Aviles-Baltazar NY, Buchala A, Serrano M. Intra and extracellular journey of the phytohormone salicylic acid. Front Plant Sci 2019; 10: 423.
[http://dx.doi.org/10.3389/fpls.2019.00423] [PMID: 31057566]

[51]   Mhlongo MI, Piater LA, Madala NE, Labuschagne N, Dubery IA. The chemistry of plant-microbe interactions in the rhizosphere and the potential for metabolomics to reveal signaling related to defense priming and induced systemic resistance. Front Plant Sci 2018; 9: 112.
[http://dx.doi.org/10.3389/fpls.2018.00112] [PMID: 29479360]

[52]   Chaparro JM, Badri DV, Vivanco JM. Rhizosphere microbiome assemblage is affected by plant development. ISME J 2014; 8(4): 790-803.
       [http://dx.doi.org/10.1038/ismej.2013.196] [PMID: 24196324]

[53]   Rashid MH, Kamruzzaman M, Haque ANA, Krehenbrink M. Soil microbes for sustainable agriculture In sustainable management of soil and environment. Singapore: Springer 2019; pp. 339-82.
       [http://dx.doi.org/10.1007/978-981-13-8832-3_10]

[54]   Morris CE, Barnes MB, McLean RJC. Biofilms on leaf surfaces: Implications for the biology, ecology and management of populations of epiphytic bacteria. Phyllosphere Microbiol 2002; pp. 139-55.

[55]   Hubbard M, Germida JJ, Vujanovic V. Fungal endophytes enhance wheat heat and drought tolerance in terms of grain yield and second-generation seed *via*bility. J Appl Microbiol 2014; 116(1): 109-22.
       [http://dx.doi.org/10.1111/jam.12311] [PMID: 23889706]

[56]   Ahmad I, Akhtar MJ, Asghar HN, Ghafoor U, Shahid M. Differential effects of plant growth-promoting rhizobacteria on maize growth and cadmium uptake. J Plant Growth Regul 2016; 35(2): 303-15.
       [http://dx.doi.org/10.1007/s00344-015-9534-5]

[57]   Hidri R, Barea JM, Mahmoud OMB, Abdelly C, Azcón R. Impact of microbial inoculation on biomass accumulation by Sulla carnosa provenances, and in regulating nutrition, physiological and antioxidant activities of this species under non-saline and saline conditions. J Plant Physiol 2016; 201: 28-41.
       [http://dx.doi.org/10.1016/j.jplph.2016.06.013] [PMID: 27393918]

[58]   Singh N, Marwa N, Mishra S, *et al. Brevundimonas diminuta* mediated alleviation of arsenic toxicity and plant growth promotion in *Oryza sativa* L. Ecotoxicol Environ Saf 2016; 125: 25-34.
       [http://dx.doi.org/10.1016/j.ecoenv.2015.11.020] [PMID: 26650422]

[59]   Kavita B, Shukla S, Naresh Kumar G, Archana G. Amelioration of phytotoxic effects of Cd on mung bean seedlings by gluconic acid secreting rhizobacterium *Enterobacter asburiae* PSI3 and implication of role of organic acid. World J Microbiol Biotechnol 2008; 24(12): 2965-72.
       [http://dx.doi.org/10.1007/s11274-008-9838-8]

[60]   Johnson NC, Graham JH. The continuum concept remains a useful framework for studying mycorrhizal functioning. Plant Soil 2013; 363(1-2): 411-9.
       [http://dx.doi.org/10.1007/s11104-012-1406-1]

[61]   Sharma P, Kumar T, Yadav M, Gill SS, Chauhan NS. Plant-microbe interactions for the sustainable agriculture and food security. Plant Gene 2021; 28: 100325.
       [http://dx.doi.org/10.1016/j.plgene.2021.100325]

[62]   Scavino AF, Pedraza RO. The role of siderophores in plant growth-promoting bacteria.Bacteria in agrobiology: crop productivity. Berlin, Heidelberg: Springer 2013; pp. 265-85.
       [http://dx.doi.org/10.1007/978-3-642-37241-4_11]

[63]   Geetanjali Jain P. Antibiotic production by rhizospheric soil microflora : A review. Int J Pharm Sci Res 2016; 7: 4304-14.

[64]   Druzhinina IS, Seidl-Seiboth V, Herrera-Estrella A, *et al.* Trichoderma: The genomics of opportunistic success. Nat Rev Microbiol 2011; 9(10): 749-59.
       [http://dx.doi.org/10.1038/nrmicro2637] [PMID: 21921934]

[65]   Moore LW, Warren G. Agrobacterium radiobacter strain 84 and biological control of crown gall. Annu Rev Phytopathol 1979; 17(1): 163-79.
       [http://dx.doi.org/10.1146/annurev.py.17.090179.001115]

[66]   Glick BR. Soil microbes and sustainable agriculture. Pedosphere 2018; 28(2): 167-9.
       [http://dx.doi.org/10.1016/S1002-0160(18)60020-7]

[67]   Pusey PL. Biological control agents for fire blight of apple compared under conditions limiting natural dispersal. Plant Dis 2002; 86(6): 639-44.

[http://dx.doi.org/10.1094/PDIS.2002.86.6.639] [PMID: 30823238]

[68]   Chanda S, Dudhatra S, Kaneria M. Antioxidative and antibacterial effects of seeds and fruit rind of nutraceutical plants belonging to the Fabaceae family. Food Funct 2010; 1(3): 308-15.
[http://dx.doi.org/10.1039/c0fo00028k] [PMID: 21776481]

[69]   Innerebner G, Knief C, Vorholt JA. Protection of *Arabidopsis thaliana* against leaf-pathogenic *Pseudomonas syringae* by *Sphingomonas* strains in a controlled model system. Appl Environ Microbiol 2011; 77(10): 3202-10.
[http://dx.doi.org/10.1128/AEM.00133-11] [PMID: 21421777]

[70]   Kamilova F, Kravchenko LV, Shaposhnikov AI, Makarova N, Lugtenberg B. Effects of the tomato pathogen *Fusarium oxysporum* f. sp. radicis-lycopersici and of the biocontrol bacterium *Pseudomonas fluorescens* WCS365 on the composition of organic acids and sugars in tomato root exudate. Mol Plant Microbe Interact 2006; 19(10): 1121-6.
[http://dx.doi.org/10.1094/MPMI-19-1121] [PMID: 17022176]

[71]   Zhang H, Xie X, Kim MS, Kornyeyev DA, Holaday S, Paré PW. Soil bacteria augment arabidopsis photosynthesis by decreasing glucose sensing and abscisic acid levels *in planta*. Plant J 2008; 56(2): 264-73.
[http://dx.doi.org/10.1111/j.1365-313X.2008.03593.x] [PMID: 18573192]

[72]   García-Fraile P, Menéndez E, Celador-Lera L, *et al.* Bacterial probiotics: A truly green revolution.Probiotics and Plant Health. Singapore: Springer 2017; pp. 131-62.
[http://dx.doi.org/10.1007/978-981-10-3473-2_6]

[73]   Adeleke RA, Raimi AR, Roopnarain A, Mokubedi SM. Status and prospects of bacterial inoculants for sustainable management of agroecosystems.Biofertilizers for Sustainable Agriculture and Environment. Cham, Switzerland: Springer 2019; pp. 137-72.
[http://dx.doi.org/10.1007/978-3-030-18933-4_7]

[74]   Basu A, Prasad P, Das SN, *et al.* Plant growth promoting rhizobacteria (pgpr) as green bioinoculants: Recent developments, constraints, and prospects. Sustainability 2021; 13(3): 1140.
[http://dx.doi.org/10.3390/su13031140]

[75]   López MM, Bertolini E, Olmos A, *et al.* Innovative tools for detection of plant pathogenic viruses and bacteria. Int Microbiol 2003; 6(4): 233-43.
[http://dx.doi.org/10.1007/s10123-003-0143-y] [PMID: 13680391]

[76]   Delmont TO, Prestat E, Keegan KP, *et al.* Structure, fluctuation and magnitude of a natural grassland soil metagenome. ISME J 2012; 6(9): 1677-87.
[http://dx.doi.org/10.1038/ismej.2011.197] [PMID: 22297556]

[77]   Gamalero E, Bona E, Glick BR. Current techniques to study beneficial plant-microbe interactions. Microorganisms 2022; 10(7): 1380.
[http://dx.doi.org/10.3390/microorganisms10071380] [PMID: 35889099]

[78]   Duan G, Christian N, Schwachtje J, Walther D, Ebenhöh O. The metabolic interplay between plants and phytopathogens. Metabolites 2013; 3(1): 1-23.
[http://dx.doi.org/10.3390/metabo3010001] [PMID: 24957887]

[79]   Seaver SMD, Henry CS, Hanson AD. Frontiers in metabolic reconstruction and modeling of plant genomes. J Exp Bot 2012; 63(6): 2247-58.
[http://dx.doi.org/10.1093/jxb/err371] [PMID: 22238452]

[80]   Carneiro GA, Matić S, Ortu G, Garibaldi A, Spadaro D, Gullino ML. Development and validation of a TaqMan real-time PCR assay for the specific detection and quantification of *Fusarium fujikuroi* in rice plants and seeds. Phytopathology 2017; 107(7): 885-92.
[http://dx.doi.org/10.1094/PHYTO-10-16-0371-R] [PMID: 28398878]

[81]   Gong S, Zhang X, Jiang S, Chen C, Ma H, Nie Y. A new species of *Ophiognomonia* from Northern China inhabiting the lesions of chestnut leaves infected with *Diaporthe eres*. Mycol Prog 2017; 16(1):

83-91.
[http://dx.doi.org/10.1007/s11557-016-1255-z]

[82]   Woese CR, Fox GE. Phylogenetic structure of the prokaryotic domain: The primary kingdoms. PNAS 1977; 74(11): 5088-90.
[http://dx.doi.org/10.1073/pnas.74.11.5088]

[83]   D'Amore R, Ijaz UZ, Schirmer M, *et al.* A comprehensive benchmarking study of protocols and sequencing platforms for 16S rRNA community profiling. BMC Genomics 2016; 17(1): 55.
[http://dx.doi.org/10.1186/s12864-015-2194-9] [PMID: 26763898]

[84]   Schoch CL, Seifert KA, Huhndorf S, *et al.* Nuclear ribosomal internal transcribed spacer (ITS) region as a universal DNA barcode marker for *Fungi.* Proc Natl Acad Sci 2012; 109(16): 6241-6.
[http://dx.doi.org/10.1073/pnas.1117018109] [PMID: 22454494]

[85]   De Filippis F, Laiola M, Blaiotta G, Ercolini D. Different amplicon targets for sequencing-based studies of fungal diversity. Appl Environ Microbiol 2017; 83(17): e00905-17.
[http://dx.doi.org/10.1128/AEM.00905-17] [PMID: 28625991]

[86]   Langille MGI, Zaneveld J, Caporaso JG, *et al.* Predictive functional profiling of microbial communities using 16S rRNA marker gene sequences. Nat Biotechnol 2013; 31(9): 814-21.
[http://dx.doi.org/10.1038/nbt.2676] [PMID: 23975157]

[87]   Fricker AM, Podlesny D, Fricke WF. What is new and relevant for sequencing-based microbiome research? A mini-review. J Adv Res 2019; 19: 105e12.

[88]   Huse SM, Welch DM, Morrison HG, Sogin ML. Ironing out the wrinkles in the rare biosphere through improved OTU clustering. Environ Microbiol 2010; 12(7): 1889-98.
[http://dx.doi.org/10.1111/j.1462-2920.2010.02193.x] [PMID: 20236171]

[89]   Amir A, McDonald D, Navas-Molina JA, *et al.* Deblur rapidly resolves single-nucleotide community sequence patterns. mSystems 2017; 2(2): e00191-16.
[http://dx.doi.org/10.1128/mSystems.00191-16] [PMID: 28289731]

[90]   Olmos A, Bertolini E, Gil M, Cambra M. Real-time assay for quantitative detection of non-persistently transmitted Plum pox virus RNA targets in single aphids. J Virol Methods 2005; 128(1-2): 151-5.
[http://dx.doi.org/10.1016/j.jviromet.2005.05.011] [PMID: 15964640]

[91]   Varga A, James D. Use of reverse transcription loop-mediated isothermal amplification for the detection of Plum pox virus. J Virol Methods 2006; 138(1-2): 184-90.
[http://dx.doi.org/10.1016/j.jviromet.2006.08.014] [PMID: 17011051]

[92]   de Torres-Zabala M, Truman W, Bennett MH, *et al. Pseudomonas syringae* pv. tomato hijacks the *Arabidopsis* abscisic acid signalling pathway to cause disease. EMBO J 2007; 26(5): 1434-43.
[http://dx.doi.org/10.1038/sj.emboj.7601575] [PMID: 17304219]

[93]   Seaver SMD, Henry CS, Hanson AD. Frontiers in metabolic reconstruction and modeling of plant genomes. J Exp Bot 2012; 63(6): 2247-58.
[http://dx.doi.org/10.1093/jxb/err371] [PMID: 22238452]

[94]   Brumfield KD, Huq A, Colwell RR, Olds JL, Leddy MB. Microbial resolution of whole genome shotgun and 16S amplicon metagenomic sequencing using publicly available NEON data. PLoS One 2020; 15(2): e0228899.
[http://dx.doi.org/10.1371/journal.pone.0228899] [PMID: 32053657]

[95]   Panek M, Čipčić Paljetak H, Barešić A, *et al.* Methodology challenges in studying human gut microbiota : Effects of collection, storage, DNA extraction and next generation sequencing technologies. Sci Rep 2018; 8(1): 5143.
[http://dx.doi.org/10.1038/s41598-018-23296-4] [PMID: 29572539]

[96]   Nadarajah K. Metagenomics for improving soil fertility.Soil Nitrogen Ecology. Cham, Switzerland: Springer 2021; pp. 267-82.
[http://dx.doi.org/10.1007/978-3-030-71206-8_13]

[97]   Snipen L, Angell IL, Rognes T, Rudi K. Reduced metagenome sequencing for strain-resolution taxonomic profiles. Microbiome 2021; 9(1): 79.
[http://dx.doi.org/10.1186/s40168-021-01019-8] [PMID: 33781324]

[98]   Sharifi F, Ye Y. From gene annotation to function prediction for metagenomics. Methods Mol Biol 2017; 1611: 27-34.
[http://dx.doi.org/10.1007/978-1-4939-7015-5_3] [PMID: 28451969]

[99]   Bharti R, Grimm DG. Current challenges and best-practice protocols for microbiome analysis. Brief Bioinform 2021; 22(1): 178-93.
[http://dx.doi.org/10.1093/bib/bbz155] [PMID: 31848574]

[100]   Bibikova M, Beumer K, Trautman JK, Carroll D. Enhancing gene targeting with designed zinc finger nucleases. Science 2003; 300(5620): 764.
[http://dx.doi.org/10.1126/science.1079512] [PMID: 12730594]

[101]   Zhang Z, Zhang S, Huang X, Orwig KE, Sheng Y. Rapid assembly of customized TALENs into multiple delivery systems. PLoS One 2013; 8(11): e80281.
[http://dx.doi.org/10.1371/journal.pone.0080281] [PMID: 24244669]

[102]   Yin K, Gao C, Qiu JL. Progress and prospects in plant genome editing. Nat Plants 2017; 3(8): 17107.
[http://dx.doi.org/10.1038/nplants.2017.107] [PMID: 28758991]

[103]   Jinek M, Chylinski K, Fonfara I, Hauer M, Doudna JA, Charpentier E. A programmable dual-RN-
-guided DNA endonuclease in adaptive bacterial immunity. Science 2012; 337(6096): 816-21.
[http://dx.doi.org/10.1126/science.1225829] [PMID: 22745249]

[104]   Zetsche B, Gootenberg JS, Abudayyeh OO, *et al.* Cpf1 is a single RNA-guided endonuclease of a class 2 CRISPR-Cas system. Cell 2015; 163(3): 759-71.
[http://dx.doi.org/10.1016/j.cell.2015.09.038] [PMID: 26422227]

[105]   Ran FA, Hsu PD, Lin CY, *et al.* Double nicking by RNA-guided CRISPR Cas9 for enhanced genome editing specificity. Cell 2013; 154(6): 1380-9.
[http://dx.doi.org/10.1016/j.cell.2013.08.021] [PMID: 23992846]

[106]   Hou Z, Zhang Y, Propson NE, *et al.* Efficient genome engineering in human pluripotent stem cells using Cas9 from *Neisseria meningitidis.* Proc Natl Acad Sci 2013; 110(39): 15644-9.
[http://dx.doi.org/10.1073/pnas.1313587110] [PMID: 23940360]

[107]   Kleinstiver BP, Prew MS, Tsai SQ, *et al.* Engineered CRISPR-Cas9 nucleases with altered PAM specificities. Nature 2015; 523(7561): 481-5.
[http://dx.doi.org/10.1038/nature14592] [PMID: 26098369]

[108]   Ogunseitan OA. Soil proteomics: Extraction and analysis of proteins from soils In nucleic acid and protein in soil. Berlin, Germany: Springer 2006; Vol. 8: pp. 95-115.
[http://dx.doi.org/10.1007/3-540-29449-X_5]

[109]   Chiapello M, Zampieri E, Mello A. A small effort for researchers, a big gain for soil metaproteomics. Front Microbiol 2020; 11: 88.
[http://dx.doi.org/10.3389/fmicb.2020.00088] [PMID: 32117118]

[110]   Starke R, Jehmlich N, Bastida F. Using proteins to study how microbes contribute to soil ecosystem services: The current state and future perspectives of soil metaproteomics. J Proteomics 2019; 198: 50-8.
[http://dx.doi.org/10.1016/j.jprot.2018.11.011] [PMID: 30445181]

[111]   Mattarozzi M, Manfredi M, Montanini B, *et al.* A metaproteomic approach dissecting major bacterial functions in the rhizosphere of plants living in serpentine soil. Anal Bioanal Chem 2017; 409(9): 2327-39.
[http://dx.doi.org/10.1007/s00216-016-0175-8] [PMID: 28083663]

[112]   Tempel A. A green nano-synthesis to explore the plant microbe interactions.In New and future

developments in microbial biotechnology and bioengineering. Elsevier 1959; pp. 85-105.

[113] Holzloehner P, Schliebs E, Maier N, Füner J, Micheel B, Heilmann K. Production of monoclonal camelid antibodies by means of hybridoma technology (P3376). J Immunol 2013; 190(1_Supplement): 135.14.
[http://dx.doi.org/10.4049/jimmunol.190.Supp.135.14]

[114] Kox LFF, Brouwershaven IR, Vossenberg BTLH, Beld HE, Bonants PJM, Gruyter J. Diagnostic values and utility of immunological, morphological and molecular methods for in planta detection of *Phytophthora ramorum.* Phytopathology 2007; 97(9): 1119-29.
[http://dx.doi.org/10.1094/PHYTO-97-9-1119] [PMID: 18944177]

[115] Karthikeyan M, Radhika K, Bhaskaran R, Mathiyazhagan S, Samiyappan R, Velazhahan R. Rapid detection of Ganoderma disease of coconut and assessment of inhibition effect of various control measures by immunoassay and PCR. Plant Prot Sci 2006; 42(2): 49-57.
[http://dx.doi.org/10.17221/2771-PPS]

[116] Heyer R, Schallert K, Zoun R, Becher B, Saake G, Benndorf D. Challenges and perspectives of metaproteomic data analysis. J Biotechnol 2017; 261: 24-36.
[http://dx.doi.org/10.1016/j.jbiotec.2017.06.1201] [PMID: 28663049]

[117] Bastida F, Jehmlich N, Martínez-Navarro J, Bayona V, García C, Moreno JL. The effects of struvite and sewage sludge on plant yield and the microbial community of a semiarid Mediterranean soil. Geoderma 2019; 337: 1051-7.
[http://dx.doi.org/10.1016/j.geoderma.2018.10.046]

[118] Smirnov KS, Maier TV, Walker A, *et al.* Challenges of metabolomics in human gut microbiota research. Int J Med Microbiol 2016; 306(5): 266-79.
[http://dx.doi.org/10.1016/j.ijmm.2016.03.006]

[119] Hu L, Robert CAM, Cadot S, *et al.* Root exudate metabolites drive plant-soil feedbacks on growth and defense by shaping the rhizosphere microbiota. Nat Commun 2018; 9(1): 2738.
[http://dx.doi.org/10.1038/s41467-018-05122-7] [PMID: 30013066]

[120] Schulz-Bohm K, Gerards S, Hundscheid M, Melenhorst J, de Boer W, Garbeva P. Calling from distance: attraction of soil bacteria by plant root volatiles. ISME J 2018; 12(5): 1252-62.
[http://dx.doi.org/10.1038/s41396-017-0035-3] [PMID: 29358736]

[121] Anastas PT, Perosa A, Selva M. Handbook of green chemistry, green processes, green nanoscience. Hoboken: Wiley 2014; 8.

[122] Pandey S, Mishra A, Giri VP, Kumari M, Soni S. A green nano-synthesis to explore the plant microbe interactions.New and future developments in microbial biotechnology and bioengineering. Elsevier 2019; pp. 85-105.
[http://dx.doi.org/10.1016/B978-0-444-64191-5.00007-9]

[123] Srilatha. Nanotechnology in agriculture. J Nanomed Nanotechnol 2011; 7: 1-2.

[124] Alkilany AM, Murphy CJ. Toxicity and cellular uptake of gold nanoparticles: What we have learned so far? J Nanopart Res 2010; 12(7): 2313-33.
[http://dx.doi.org/10.1007/s11051-010-9911-8] [PMID: 21170131]

[125] Das RK, Pachapur VL, Lonappan L, *et al.* Biological synthesis of metallic nanoparticles: Plants, animals and microbial aspects. Nanotech Environ Eng 2017; 2(1): 18.
[http://dx.doi.org/10.1007/s41204-017-0029-4]

[126] Hulkoti NI, Taranath TC. Biosynthesis of nanoparticles using microbes : A review. Colloids Surf B Biointerfaces 2014; 121: 474-83.
[http://dx.doi.org/10.1016/j.colsurfb.2014.05.027] [PMID: 25001188]

[127] Shenton W, Douglas T, Young M, Stubbs G, Mann S. Inorganic–organic nanotube composites from template mineralization of tobacco mosaic virus. Adv Mater 1999; 11(3): 253-6.
[http://dx.doi.org/10.1002/(SICI)1521-4095(199903)11:3<253::AID-ADMA253>3.0.CO;2-7]

[128]   Mao C, Flynn CE, Hayhurst A, *et al.* Viral assembly of oriented quantum dot nanowires. Proc Natl Acad Sci 2003; 100(12): 6946-51.
[http://dx.doi.org/10.1073/pnas.0832310100] [PMID: 12777631]

CHAPTER 2

Microbiomes in Phytotherapeutics: Pros and Cons

Hijab Fatima[1,*] and **Azra Yasmin**[1]

[1] *Department of Biotechnology, Fatima Jinnah Women University, Rawalpindi, Pakistan*

Abstract: This chapter highlights the significance of microbiomes especially plant microbiomes in the field of therapeutics. The Plant microbiome comprises epiphytes and endophytes inhabiting the surface as well as inside of the tissues of the host. These microbial communities occupy a well-defined habitat and perform various activities developing certain interactions with the host such as commensalism, mutualism, and parasitism. For the establishment and functioning of the plant microbiome, plant root releases exudate according to the nutritional requirement of particular microbial species. In response to the stimulus, microbes chemotactically move towards the roots, colonize and move to other parts of the plant. Microbes also adopt certain mechanisms not only to colonize and multiply in specific hosts but also to promote the growth of the host by secreting various plant growth hormones and exopolysaccharides. The numerous compounds produced by microbes make plants tolerant of biotic and abiotic stresses. The microbial communities in plant microbiome have an active role in maintaining the health, ecology and physiology of the host. As a major portion of the world's population is dependent on phytotherapeutic medicines according to the World Health Organization, the pharmacological characteristics of major medicinal plants such as *Aesculus hippocastanum* and *Ginkgo biloba* are described in detail. This chapter highlights the significance of the core role of the microbiome associated with plants in the synthesis of various medicinal compounds. The phytotherapeutic potential of plant microbiome revealed that endophytes and epiphytes isolated from various plant species showed great potential for the production of antimicrobial as well as anti-inflammatory substances. The medicinally rich compounds such as antibacterial proteins, phenols, saponin glycosides, flavonoids, terpenoids, carbohydrates and fatty acids isolated from plant-associated microbes have various applications in the treatment of fetal diseases and also exhibit anti-inflammatory action. Certain public concerns are raised about the side effects of medicinal plants used in phytotherapeutics. A relevant case study about public concerns along with preventative measures such as rigorous testing is provided in this chapter.

Keywords: Epiphytes, Endophytes, Microbiota, Phytomicrobiomes, Phytotherapeutic potential.

----
* **Corresponding author Hijab Fatima:**Department of Biotechnology, Fatima Jinnah Women University, Rawalpindi, Pakistan; E-mail: hijabfatima72@gmail.com

Zulqurnain Khan, Azra Yasmin & Naila Safdar (Eds.)
All rights reserved-© 2023 Bentham Science Publishers

INTRODUCTION TO MICROBIOME

The word microbiome is derived from two Greek words, *mikros* which means small and *bios* which means life. The microbiome is defined in terms of the microbial community that occupies a well-defined habitat and performs multiple activities. These microbial communities' dwell on the host surface, colonize in various tissues, inhabit both inter and intra-cellular host organisms and have an effect on the health and functions of the host [1]. Nowadays, an engineering approach named host-mediated microbiome selection is applied to improve the health and functionalities of the host. According to this approach, the host is artificially inoculated with beneficial microbes to improve the microbiome [2].

BRIEF HISTORY OF MICROBIOME

The history of microbiomes in the field of microbiology dates back to the seventieth century. According to Whipps, a microbiome is a distinguishing microbial community inhabiting a definite habitat that has distinctive physical and chemical properties. According to this definition, the term microbiome covers microorganisms and their diverse activities [3].

From the historic perspective, the field of microbiome research has derived from the field of microbial ecology. Microbiome research is an interdisciplinary field that links multiple fields such as biotechnology, bioinformatics, bioeconomy, plant sciences, food sciences, medicine and mathematics [4]. The advanced field of microbiome research explains the concept of holobiont theory which is about host-microbial interactions. According to this theory, microbes have a symbiotic relationship with their host and these are transmitted from one generation to the other. The - host-microbial interactions have an impact on the health of holobiont in a particular environment and any mutations in hologenome result variations in microbiota or host genome [5, 6].

MICROBIOME VS MICROBIOTA

The two terms microbiomes and microbiota are often used interchangeably for microbial communities but they are distinct from each other. The term microbiota refers to living microorganisms such as pathogenic, commensal and symbiotic microorganisms living in a particular habitat. Generally, the microbiota includes viruses, archaea, bacteria, fungi and protists [7]. Microbiome involves the collective genome of the microorganisms that occupy a specific habitat. The term microbiome covers all the biotic and abiotic factors and the theater of activities such as the production of metabolites by the microorganisms while inhabiting a particular environmental niche. The structural elements, extracellular DNA and all mobile genetic elements such as phages viroid and prions that are considered non-living microorganisms are included in the microbiome but not in the microbiota [8].

DIVERSE ARRAY OF HOST AND MICROBES INTERACTIONS IN THE MICROBIOME

The microbes inhabiting a specific microbiome enjoying a symbiotic relationship with the host. Symbiotic microbes are defined as microbes that live in symbiosis with other organisms and these are further classified as mutualists, commensals and parasites. Mutualism is a relationship in which both organisms are benefited. In commensalism, one organism is benefited and the other one is neither benefited nor harmed. In parasitism, one organism is benefited and in return, it causes harm to its host [9]. The microbes belonging to a microbiome are both potentially useful and harmful, for example, the human gut microbiome is mostly consisting of symbiotic or commensal communities while few species become pathogenic under specific conditions [10]. The symbiotic microbes are further categorized into two types endosymbionts and ectosymbionts. Endosymbionts are the microbes that live inside the tissues and cells of their host while ectosymbionts reside on the surface of the host [11].

DIFFERENT TYPES OF MICROBIOMES

Marine Microbiome

There are different types of microbiomes such as animal, plant, human, terrestrial and marine microbiomes. Marine animals share water bodies with an enormous number of microbes forming a marine microbiome. Scientists are studying the symbiotic relationship of marine organisms for years but the advancements in technology have provided a new paradigm to discover the correlation between diverse microbial life and sea animals. In marine microbiomes single host symbiont systems are majorly studied. The Advanced marine microbiome studies involve the association between diverse microbes and a variety of marine animals as hosts.

Microorganisms dwelling on or within marine animals are playing an immense role in the production of oxygen, nutrient recycling, and degradation of organic matter [12]. An excellent example of a marine microbiome is a coral reef. The coral polyp lives in symbiotic association with the algae. The coral polys provide shelter to the algae and in return algae gives color to the coral colonies. Algae perform photosynthesis and are the source of oxygen, organic products and host nutrition [13].

Microbial life is improving sea animal behavior, health and ecology through numerous genetic and biochemical interactions [14]. Fig. (**1**) mentioned below is illustrating the host microbiome's relation which is usually in a symbiotic state. Various marine animals and environmental factors are responsible for natural variations that either promote the healthy functioning of the microbiome or result in dysbiosis *i.e.*, halting of the interactions under stress conditions.

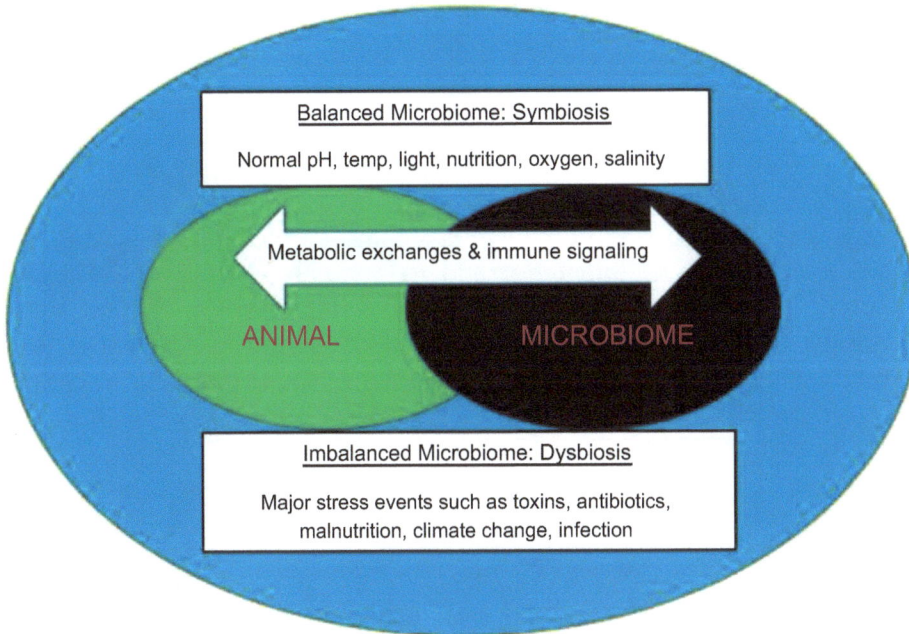

Fig. (1). Model of host-microbiome relationship.

Animal Microbiome

The animal microbiome is a very vast microbiome that covers all animals from tiny insects to large blue whales. It covers different parts of animals such as the gut, skin, oral, *etc*. An immense number of bacterial species residing in the gastrointestinal (GI) tract of animals are discovered and studied by scientists from the beginning of microbiology. With the advancements in the fields of microbiology and microbial genetics, scientists have come to know that animal genetics, diet, and other environmental factors play a great role in the growth, development and regulation of the gut microbiota [15].

Gastro-intestinal tract of mammals is composed of probiotics *i.e.*, the beneficial bacteria which play an essential role in improving the animal's health and modulating the other gut microflora. An animal gut, the perfect example of a complex ecosystem consists of beneficial microbiota such as *Bifidobacterium, Lactobacillus* and *Lactococcus species.* The healthy microbiota performs the function of immunomodulation to inhibit various degenerative diseases such as cancer, liver, cardiovascular and inflammatory bowel diseases [16]. Chen and his coworkers described the essential role of gut microbiome in silkworm growth and development. The microbiome of domesticated as well as improved hybrid species of silkworm is involved in multiple metabolic activities. These metabolic activities involve the transport and metabolism of primary or secondary metabolites and energy production and conservation [17].

Human Microbiome

All the microorganisms residing on or within the human body including skin, glands, tissues, organs, body fluids and GI tract are collectively making a human microbiome. The study of the human microbiome has become easy due to the latest advancements in sequencing and analytical techniques. There are approximately ten to hundred trillion symbiotic cells present in the human body for example bacterial cells of the human gut and to comprehend the function of these symbionts and their impacts on the human body, the human microbiome project was initiated worldwide in 2007. The work on the human microbiota started earlier in the 1680s when Antonie van Leuwenhoek analyzed oral and faecal microbiota. He compared the microbiota of diseased and healthy patients and observed the differences among the microbes residing in two different habitats [18].

The human microbiome is the collection of genomes of all the microorganisms residing on or within the body [19]. The genome of the microbes associated with humans is analyzed and sequenced by high-throughput sequencing techniques. The sequenced genomes helped the scientific community to develop the connection between the human microbiome and various diseases such as autism, obesity, arthritis and inflammatory bowel disease . In healthy and diseased people, the structure and role of the microbiome can certainly be studied by the analysis of metabolomes, transcriptomes, immunomes and proteomes [20].

Human gut microbiome plays an essential role in keeping a body healthy. Any change in the composition and activities of the gut microbiome and impaired gut barrier result in microbiome dysbiosis. Microbiome dysbiosis leads to the production of uremic toxins at an excessive rate. Due to impaired intestinal barrier, toxins are released into the circulatory system and lead to chronic kidney disease and cardiovascular disorders [21].

Plant Microbiome

Phytomicrobiome has a key role in maintaining the health as well as the productivity of plants [22]. Plants and microbes form a close symbiotic relationship. The microbes that dwell inside the plant tissues are named endophytes while the microbes that live on the surface of the plant are named epiphytes.

DIVERSE INTERACTION OF PLANTS ASSOCIATED MICROBES

Microbes inhabiting either the endosphere or episphere of plants develop miscellaneous interactions with the host such as commensalism, mutualism, and parasitism. These plants associated microbes have a dynamic role in the healthy, physiology, and ecology of the host [23]. Microbial species colonizing the aerial parts of plants dominated by leaves are known as phyllobacteria. Phyllosphere such as fruit, stem, flower and leaf surface are nutritionally not as rich as endosphere or rhizosphere but have more variable environment. Phyllosphere microbiome composition and physiology change with various seasonal fluctuations such as temperature, radiation, wind, and water content [24]. Microbial communities of phyllosphere adapt to various strategies to colonize in particular ecological niches. These strategies include modifying habitat by augmenting local nutrient access and secreting extracellular polysaccharides as well as enzymes. After habitat modification, aggregation, ingression and egression take place [25].

Plants-associated microbes are richly found in the rhizoplane and rhizosphere. The microbes such as bacteria living on the roots of plants are classified as rhizoplane bacteria while the bacteria dwelling in thin soil regions (approx. 10 mm) around the plant roots are classified as rhizosphere bacteria. Rhizoplane bacteria stick to the root surface by protein appendages known as pilli and flagella [26]. Beneficial rhizosphere and rhizoplane bacteria also known as plant growth-promoting rhizobacteria (PGPR) promote plant growth by secreting various substances, acting as biocontrol agents and biofertilizers [27].

The rhizosphere and phyllosphere offer specific environments to plant-associated bacteria that are capable of fixing Nitrogen (N_2). Rhizobium is one of the common examples of nitrogen-fixing bacteria. Atmospheric Nitrogen gas is of no use to plants until and unless it is fixed by nitrogen-fixing bacteria such as *Rhizobium* living on the root nodules of the plants. These bacteria convert molecular Nitrogen occurring in the air into products like ammonia, nitrates and nitrites which are consumed by plants. The denitrifying bacteria such as *Pseudomonas* present in the rhizosphere reduce nitrates into nitrous oxide and gaseous Nitrogen [28].

Endophytic bacteria similar to phyllosphere and rhizobacteria play a keen role in plant growth and environment cleaning technologies for example phytoremediation [29]. Endophytes are microorganisms such as bacteria and fungi that inhabit the tissues of the host *i.e.*, roots, stem, flower and leaves. The rudimentary characteristic that differentiates endophytes from phytopathogens is that they do not cause harm to the host in which they dwell [30]. Endophytic bacteria's symbiotic association with plants is contributing to the growth and protection of plants against phytopathogens [31].

ENDOPHYTES IN PLANT MICROBIOME

In the 19th century, the term endophytic bacteria was coined by Bary for the first time and from 1970 onwards, these have become a center of attraction for researchers. Endophytic bacteria have gained immense importance in the field of research because of their ability to produce various pharmacological substances; for example quercin an antifungal agent, Taxol an anticancer agent, various toxins as biocontrol agents and the mass production of cost-effective enzymes [32]. The enzymes and proteins extracted from endophytes are sensitive to genetic manipulation and have numerous biotechnological applications in the field of medicine, industries and molecular biology [33 - 35].

Importance of Symbiotic Association of Endophytes in Phytomicrobiome

Being a part of phytomicrobiome, endophytes establishe a symbiotic association with the plant and deliver a range of advantages to the host. Endophytes are involved in producing plant growth hormones like gibberellins and auxins; fixing nitrogen for plants that act as a substitute to harmful agricultural fertilizers, making plants resistant towards different biotic as well as abiotic stresses, secreting toxins and antibacterial proteins that function against phytopathogens like bacteria, protozoan and fungi [36, 37]. Endophytes obtain nutrients and shelter from the host to grow and multiply [38].

In a phytomicrobiome, roots of different plants produce exudates and in response, endophytes move towards the plant roots. This movement is known as a chemotactic movement. The response of endophytes may vary from specie to specie. The carbon metabolites are the exudates to attract endophytes and promote their growth and activities [39]. For example, exudates of tomato plants attract *Pseudomonas fluorescens* and carbohydrates produced by rice plant roots attract *Corynebacterium flavescens* and *Bacillus pumilus*. Each bacterial species responds to the release of specific exudate based on its nutritional requirement [40].

Endophytic microorganisms also yield certain molecules to colonize and reproduce inside the host tissues. Endophytic fungi and bacteria are demonstrated as brilliant exopolysaccharides (EPS) producers. These exopolysaccharides are either linear homopolysaccharides or greatly branched heteropolysaccharides. According to the case study conducted on the pytomicrobiome of the rice plant, the endophytic and nitrogen-fixing bacterium *Gluconacetobacter diazotrophicus* produces exopolysaccharides that are essential for its attachment to plant and colonization. The exopolysaccharides armour the bacterial cells from oxidative damage, which is necessary for its colonization [41]. The feedback interactions of plants and microbes are indispensable for the functioning of the plant microbiome.

Mechanisms Adopted by Endophytes for Plant Growth

Various mechanisms are adopted by endophytes to boost the growth of plants and to shield the plant against certain biotic and abiotic stress. The stresses usually encountered by plants are fluctuations in temperature, pH, water scarcity, radiations, heavy metals contamination, variation in salts concentration, organic pollutants, and attack of various pathogens [42, 43]. The flow chart given below Fig. (2) demonstrates several mechanisms followed by endophytes for plant growth and protection.

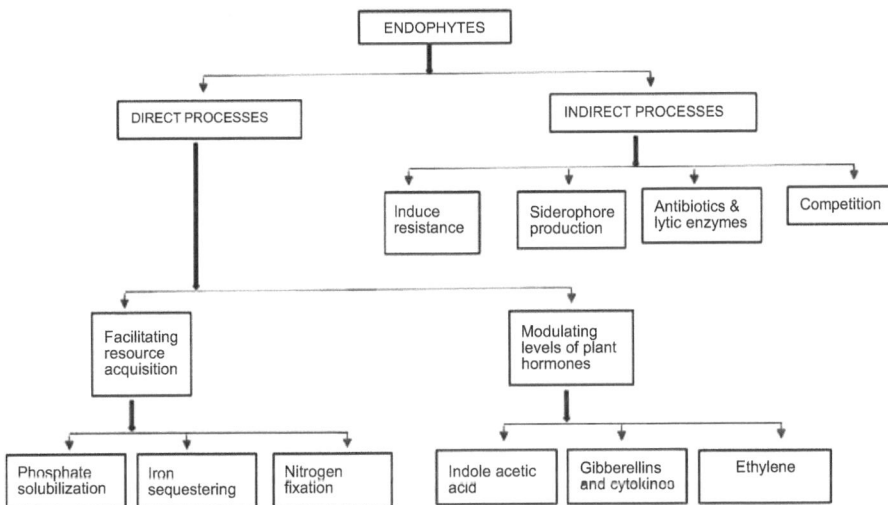

Fig. (2). Different plant growth-promoting mechanisms followed by the endophytes.

Phytotherapeutic Potential of Endophytes in Plant Microbiome

The scientists exploiting the phytotherapeutic potential of plant microbiome revealed that endophytic fungi isolated from *Zanthoxylum simulans* showed great potential for the production of antimicrobial as well as anti-inflammatory substances. *Zanthoxylum simulans* is a medicinal plant and is used as a spice in Chinese dishes. It is commonly named Szechuan pepper and Chinese pepper cultivated on Kinmen Island in Taiwan. The scientists isolated a total of 113 strains of endophytic fungi from the stem and leaves of *Zanthoxylum simulans* during the summer and winter seasons. Among the 113 strains of endophytic fungi, 23 showed excellent antimicrobial potential. The results of the study revealed that diverse species of endophytic fungi are residing in *Zanthoxylum simulans* and their composition differs seasonally. The endophytes yield antimicrobial metabolites to protect the plant against the attack of pathogens and are likely the source of natural antibiotics against *E. coli, Streptococcus agalactiae, Streptococcus aureus, Cryptococcus neoformans, Pseudomonas anguilliseptica* and *Candida albicans* [44].

EPIPHYTES IN PLANT MICROBIOME

Epiphytic microbes residing above and below the ground include phyllosphere and rhizosphere communities. The epiphytic communities inhabit the plant surface and boost the growth and development of their host. The epiphytic microbes have an essential role in enhancing resilience, increasing stress tolerance, improving nutrition and boosting the defense system of the plant [45]. The epiphytic microbes stimulate plant growth by secreting various plant growth regulators such as phytohormones, converting zinc, phosphorous and potassium into a soluble form, biologically fixing nitrogen and secreting siderophore and secondary metabolites that perform inhibition action against pathogens.

Role of Epiphytic Microbes in Plants' Disease Resistance

The epiphytic microbial communities play an essential role in protecting the plant against harmful pathogens so they are the contributor to the plant defense mechanism. Three grape wine cultivars were selected by the scientists to isolate the epiphytic and endophytic bacterial isolates from the leaves sample. The epiphytic microbes showed excellent inhibitory activity against the fungus *Botrytis cinerea* which causes grey mold and oomycete *Phytophthora infestans* which causes downy mildew. According to the results, the epiphytic microbes showed more diversity than endophytic microbes in all three cultivars of grapevine *i.e.*, Pinot Noir, Solaris and Chasselas. All epiphytic microbes were producers of anti-oomycete volatiles. The objective of the study was to discover the function of the phyllosphere microbiome in disease resistance [46].

Variations in Epiphytic Microbial Diversity Throughout the Plant Life Cycle

A lot of diversity is observed in the microbiota inhabiting plants and rhizosphere and this diversity continues throughout the life cycle of the plant. The diversity of the microbial community is influenced by four processes *i.e.*, selection, speciation, drift and dispersal [47]. For example, in the life cycle of the seed plant, the dispersal of seeds is an essential ecological phenomenon, as a result, the microbial species associated with the seed also move across space. Selection of the microbial community depends on the competition among the species occupying the same habitat and survival of the fittest. Drift signifies the profound change in the microbial species abundance and speciation results [48, 49].

Various epiphytic microbial communities are allied to the seed. During plant colonization, seed epiphytes have a potential advantage over the soil microbial community. During the stages of seed germination, the seed-born epiphytic community has a competitive edge over the microbial communities colonizing after the seed germination. After the plant development, the microbial species from the surrounding soil environment might get a chance to colonize new plant habitats. Thus, throughout the life cycle of the plant, the microbial diversity greatly varies.

ROLE OF PLANT GROWTH PROMOTING BACTERIA (PGPB) AGAINST STRESS FACTORS

Plants encounter multiple biotic and abiotic stress throughout their life cycle. The relationship of plant growth-promoting bacteria (PGPB) safeguards the host from various abiotic stresses such as cold, heat, water stress and salinity. For example, Maize plants kept at low temperatures showed poor root and shoot growth because of extreme levels of oxidative damage. The cold-exposed maize plant was treated simply by inoculating it with the *Bacillus amyloliquefaciens*, *Pseudomonas* sp. DSMZ 13134 and *Bacillus simplex* strain R41 with micronutrients (Zn/Mn). The seaweed extract enriched with the microbial community also serves as an effective cold stress protectant [50].

The research was carried out on green gram plants to cope with the water scarcity strains. To prevent the plant from drying out in water stress conditions, green gram plants were inoculated with the *P. fluorescens* Pf1 and the results were compared with the untreated plant. The water stress resistance was increased in plants treated with *P. fluorescens* Pf1. It was observed that the catalase was produced in the treated plant under stress conditions. The catalase enzyme is responsible for detoxifying the compounds stored in green grams during water scarcity conditions [51].

Biotic stress faced by the sessile plants in plant microbiome includes harm caused by viruses, fungi, bacteria, protists, nematodes and insects. Plant growth-promoting bacteria (PGPB) are the beneficial bacteria that promote plant growth directly by producing various plant hormones and uptake of minerals and nutrients or indirectly by combating damaging pathogens. For example, *Fusarium solani* is a filamentous fungus that damages the plant by causing root rot. *F. solani mycelia* is lysed by a Gram-negative, rod-shaped soil bacterium Pseudomonas stutzeri. P. stutzeri secretes extracellular enzymes such as *laminarinase* and *chitinase* that cause the lysis of the cell wall of *F. solani mycelia*.

INTRODUCTION TO PHYTOTHERAPY

According to the latest research, the microbiota of a phytobiome plays a key role in the production of multiple metabolites such as enzymes, peptides, proteins and antibiotics that act against various pathogenic organisms so the scientific community developed a keen interest in studying the importance of microbiomes in phytotherapeutics. Phytotherapy is defined as the use of medicines derived from plants to prevent and treat diseases. Phytotherapy is different from medical herbalism as in a later approach, controlled clinical trials and extensive biomedical research are missing. Phytotherapy is a science that is based on medical practices and this distinguishes it from traditional medicinal herbalism.

BRIEF HISTORY OF PHYTOTHERAPY

The Ebers Papyrus is an orthodox (1600 B.C.) medicine treaty of Egypt. According to this treaty, the plants should be used for therapeutic purposes. The plant named *Achillea millefolium* was used to heal the wounds of soldiers in wars since 1200 A.C. The treaty "De Materia Medica" given by Pedanius Dioscorides in 100 B.C. was also about the therapeutic use of medicinal plants [52].

In 1913, a French physician Henri Leclerc was a person who used the term phytotherapy for the very first time. In 1922, Henri Leclerc published multiple editions of "Handbook of Phytotherapy". In the English language, the definition of the term "Phytotherapy" was given by Eric Frederick William Powell in 1934. Eric Frederick William Powell was an herbalist who practices homeopathy. In 1987 Fred Evans who was a British pharmacognosist edited the journal *"Phytotherapy Research"* which is a milestone in the history of phytotherapy. In the year 1997 Varro Tyler an American pharmacognosist published *"Rational Phytotherapy"*. This was an English translation of a German book written in 1996 and the author was Volker Schulz and Rudolf Hansel [53].

PHYTOTHERAPY AND NATIONAL HEALTHCARE SYSTEMS

The exercise of phytotherapy varies throughout the world. Few countries of the world for example in Japan and South Korea, tested and approved phytotherapy products are incorporated into health insurance coverage while Nepal, China and India provide exceptional health care coverage for traditional herbal medicines. In most of the countries of the world, Phytotherapy products are not incorporated into the health care program rather it is the private concern of the patients, however, the health care practitioners usually prescribe these products to the patients. Thus, phytotherapy is practical science in the medical field. The pharmacologically active drugs analogous to orthodox pharmaceutical drugs are the products of phytotherapy. Phytotherapy does not branch out from complementary and alternative medicine (CAM) which differs principally [53].

Nowadays, due to stochastic advancements in the field of pharmacognosy practitioners of conventional medicine and medical health, professionals have augmented the use of medicinal plants. There are a lot of applications of phytotherapeutic medicine and medicinal plants as phytotherapeutics are hauled out from medicinal plants to treat fetal diseases. The plant derivatives are contributing up to 25% revenue gain of pharmaceutical industries in Brazil. The latest research on the multiple components of conventional medicinal plants, physicochemical and biological functionalities, pharmaco-toxicological characteristics and their biological effects on the patient have enhanced the therapeutic use of medicinal plants.

POPULAR USE OF PHYTOTHERAPEUTICS AND ITS IMPLICATIONS

The importance of phytotherapeutics in current times is highlighted after cases of synthetic medicine resistance have come into the picture. The effective solution to synthetic drug resistance is phytotherapeutics. As we derive a lot of primary and secondary metabolites from the plants such as plant defense system boosters, phytohormones and plant growth regulators *etc.* so that is why plants are named "living factories". But frequent testing and close monitoring are required in the case of phytotherapeutic drugs. It must be prevented after chemotherapy or surgery until and unless full information about its side effects in such cases is obtained [54].

According to World Health Organization (WHO) survey, 85% of the world population is consuming phytotherapeutic medicines. The increasing trend towards the use of the medicinal plant is also because it is economical and readily accessible [55].

PHARMACOLOGICAL CHARACTERISTICS OF MAIN MEDICINAL PLANTS

Among the few most important medicinal plants used in phytotherapeutics Medicinal *Aesculus hippocastanum* is on the top of the list and is native to South East Europe. The local name of the plant *Aesculus hippocastanum* is Horse chestnut. It is used as a remedy for various diseases. The essential metabolites extracted from *Aesculus hippocastanum* are phenols, saponin glycosides, flavonoids, terpenoids, epicatechin, kaempferol, fraxin, tannins, fatty acids, purine bases and carbohydrates. As the plant contains numerous medicinally rich compounds so it has various applications in the treatment of fetal diseases such as chronic venous insufficiency, hemorrhoids, arthritis, diarrhea, varicose veins, phlebitis, post-operative edema and bruises. It has anti-inflammatory action and plays an important role in boosting the antioxidant defense system [56].

Ginkgo biloba is a marvelous medicinal plant because it contains multiple compounds important from a medicinal point of view. For example, biotechnological production of biflavones, ginkgolic acids, acylated flavonol glycosides, ginkgotides and terpene tri lactones is carried out from *Ginkgo biloba*. The leafy extracts are used to cure cardiovascular diseases. Presently clinical trials are being performed on many of the *Ginkgo biloba* products. Each year *Ginkgo biloba* products sale is about 10 billion US dollars. Extraction of compounds with high purity by using traditional methods is difficult because of multiple environmental factors so new strategies must be devised to obtain compounds with maximum purity to meet the current market demand [57].

THERAPEUTIC POTENTIAL OF MEDICINAL PLANTS ASSOCIATED MICROBES

As multidrug-resistant microbes are evolving day by day so it's an urgent need of time to develop medicines that are novel, cheaper and effective against diseases. A lot of microbial communities inhabit on the plants which increase their significance from a medicinal point of view. According to previous records, approximately 80% of the world's antibiotics are derived from microbes and these microbes are mostly originated from the soil and plants are grown in the soil. Scientists have screened new soils and novel plant habitats for the production of the innovative drug. Epiphytes and endophytes develop parasitic, symbiotic and mutualistic relations with the plant and are the source of antimicrobial and anti-cancer compounds, antimycotics, and many essential metabolites [58, 59].

Role of Endophytes Associated with *Taxus Brevifolia* in Phytotherapeutics

The Pacific yew plant, *Taxus brevifolia* is a source of numerous pharmaceuticals, anti-cancer drugs, chemical substances and consumer products. *Taxomyces andreanae* is an endophytic fungus isolated from *Taxus brevifolia*. This endophytic fungus is responsible for the production of the chemotherapy drug paclitaxel (Taxol), which costs billions of dollars. This drug is used for the treatment of various cancers [58, 59].

Role of Endophytes Associated with *Pleurostylia Opposite* in Phytotherapeutics

Pleurostylia opposite is a medicinal shrub found in different parts of the world such as China, India, Sri Lanka, Australia, Malaysia *etc*. According to the latest research, six species of endophytic fungi were isolated from *Pleurostylia opposite*. The isolated fungi were *Doliomyces mysorensis* MGTMMP011, *Mycelia sterilia* MGTMMP012, *Alternaria* sp. MGTMMP013, *Mycelia sterilia* MGTMMP014, *Mycelia sterilia* MGTMMP015 and *Doliomyces mysorensis* MGTMMP016. Among the six isolated strains *Mycelia sterilia* MGTMMP012 showed excellent inhibitory activity against pathogenic *E. coli*, *C.diphtheriae* and *S. typhi* and *Mycelia sterilia* MGTMMP015 showed inhibition zones against *E. coli* and *S.aureus* [60].

Role of Endophytes Associated with *Azadirachta Indica,* in Phytotherapeutics

The neem tree, *Azadirachta indica*, is a traditional medicinal plant which is grown in different parts of Asia such as Pakistan and India. It is used to kill the mosquito that causes malaria. The endophytic fungal specie *Clostridium* is isolated from the Neem plant and it produces Javanicin an antibacterial naphthoquinone. Javanicin showed excellent antibacterial activity against humans and plant pathogens Pseudomonas spp [61, 62].

Role of Epiphytes Associated with *Sea Weeds* in Phytotherapeutics

Seaweeds are an enriching habitat for epiphytic microbes. A lot of antimicrobial compounds are isolated from these surface-associated microbes. In the latest research, scientists have isolated 77 epiphytes from the thallus of eight seaweeds. The seaweeds grow in the Little Andaman Islands in the Bay of Bengal. The epiphytic isolates were subjected to an agar well diffusion assay to determine their antibacterial activity. A total of twelve out of 77 epiphytic bacterial isolates showed positive inhibition zones against *Staphylococcus aureus, Aeromonas hydrophila, Klebsiella pneumonia, Escherichia coli, Shigella flexneri, Shigella dysenteriae, Vibrio cholera* and *Shigella boydii*. Out of these twelve epiphytic bacterial isolates, six showed outstanding inhibitory zones having a size in the range of 10 to 30 mm [63].

Side Effects of the Use of Plants and Associated Microbes in Phytotherapeutics

Despite the fact that medicinal plants are used in phytotherapeutics it is important to know that they have multiple side effects. Rigorous testing is required before their use for the treatment of ailments because severe side effects are also associated with some medicinal plant species. Uncontrolled use of medicinal plants in phytotherapy results in teratogenesis, mutations, carcinogenesis, intoxication and various other medication interactions. Improper regulations and lack of strict policies about the use of these medicines results in the intoxication of human and animals body with traditional drugs, herbicides, pesticides and heavy metals contamination.

Type VI Secretion Systems in Plant-associated Bacteria

Among the microbiota inhabiting the plants, few microbes consist of type VI secretion systems (T6SS). These systems occur frequently in Gram-negative bacteria. Type VI secretion systems play a prominent role as virulence factors. Type VI secretion systems are complex bacterial structures that deliver distinctive virulence mechanism to plant pathogens. The plant pathogens having type VI secretion systems can insert bacterial effectors *i.e.*, proteins directly into the cytoplasm of plant cells while avoiding all the extracellular milieu. These proteins cross the bacteria as well as host cell membranes and entered directly into the host cell cytoplasm where they can easily change the functioning of the cell [64, 65].

Role of Type VI Secretion Systems of Amylovora in Fire Blight Disease

The genome of the Gram-negative bacteria *amylovora* that causes fire blight in pears and apples is being studied by scientists. Three gene clusters of type VI secretion systems (T6SS) *i.e.*, T6SS-1, T6SS-2 and T6SS-3 are identified in *amylovora*. Out of three identified type VI secretion systems two *i.e.*, T6SS-1 and T6SS-3 contained full T6SS gene content while T6SS-2 has reduced T6SS experiments machinery.

Scientists have performed in which they checked the response of the host when in the presence and absence of type VI secretion systems. There was a profound decrease in bacterial virulence in plants in the absence of T6SS [66].

Role of Type VI Secretion Systems of Pectobacterium Wasabiae in Potato Infection

During potato infection, the virulence was caused by two types VI secretion systems present in *Pectobacterium wasabiae* species. The exact mechanism of virulence induction by type VI secretion systems is still unknown. It might be due to the direct insertion of effector protein in the host cell cytoplasm or it provides the fitness advantage to *Pectobacterium wasabiae* species over the other microbiota during the host colonization. As *Pectobacterium wasabiae* consists of eleven *vgrG* and *hcp* genes which are linked to toxic effector protein genes so it can overwhelm the competitors included in the microbiota [67].

Role of Type VI Secretion Systems of Pectobacterium Ananatis in Onion Virulence

Similarly, the absence of type VI secretion systems in *Pectobacterium ananatis* results in loss of virulence in *Allium cepa* common name is an onion that is used for lowering blood pressure, preventing edema, decreasing blood lipid levels and clumping of platelets The protein effectors of *Pectobacterium ananatis* exhibits great inhibitory activity against the Gram-negative bacteria inhabiting in the same host [68].

Major Public Concerns about the Use of Plants Derived Medicines

Medicinal plants derived medicines are popularly used worldwide because of the beneficial metabolites produced by the plant itself and the beneficial microbes inhabiting and on the surface of the plant. Public health and safety concerns about the plant-derived medicines are also increased with time. Chlorinated pesticides, insecticides, heavy metals especially copper, lead and cadmium, and tainting with prescription medicines have led to the contamination of the herbal medicines [69].

CASE STUDY AND THE PREVENTATIVE MEASURES

According to recently collected hospital and survey data about the use of herbal medicines in Ekiti state of Southwest Nigeria, it was figured out that the 10 most commonly used drugs for phytotherapeutic purposes contained heavy metals and were contaminated by prescription medicines adulteration. Cadmium, zinc and copper were found in the sample above the permissible limit described by World Health Organization (WHO) [70].

The samples were analyzed by performing Inductive Coupled Plasma-Optical Emission Spectroscopy (ICP-OES) and Gas Chromatography-Mass Spectrometry (GC-MS). Approximately 85% of the Ekiti state population used these herbal medicines in two years and 57% public showed safety concerns about the uncertified herbal medicines. Stillbirth was reported in patients using herbal medicines and among the total 18089 births, 668 were stillbirths and the fatality rate was 26% [71]. Exposure to heavy metals above the threshold limit resulted in many medical issues and even exposure to heavy metals below the permissible limits resulted in the accumulation of toxic metals in patients. Thus, proper legislative measures, multiple clinical trials and certification of herbal medicines are the solution to the rising public health and safety concerns about the use of medicinal plants for phytotherapeutic purposes

CONCLUSION

The microbiome comprises microbiota and all biotic and abiotic factors occurring in a particular ecological niche. Among multiple terrestrial and marine microbiomes, phytomicrobiomes hold a significant position from the medicinal point of view. Phytomicrobiomes are enriched with microbiota that has symbiotic or pathogenic relations with the host. The diverse array of plants associated microbes includes mutualists, parasites and commensals that live on and inside the tissues of the plants. The microbes inhabiting the surface of the plants are categorized as epiphytes while the microbes dwelling inside the tissues of the plants are known as endophytes. Plants, in order to colonize the beneficial microbes, secrete exudates; similarly depending on the location, multiple biotic and abiotic factors and interactions with the host, the microbiota secretes essential metabolites such as antimicrobial, anti-inflammatory, and antioxidant substances that not only influence the host health but are a source of multiple phytotherapeutic compounds. According to World Health Organization, a total of 85% of the world's population believes in phytotherapy and uses plants derived medicines to prevent diseases. As phytotherapy is a science of medical practice so rigorous testing, proper legislative measures, multiple clinical trials and certification of the herbal medicines are required to avoid severe side effects that are also associated with some medicinal plant species.

REFERENCES

[1]   Huttenhower C, Gevers D, Knight R. Structure, function and diversity of the healthy human microbiome. Nature 2012; 486(7402): 207-14.
        [http://dx.doi.org/10.1038/nature11234] [PMID: 22699609]

[2]   Mueller UG, Sachs JL. Engineering microbiomes to improve plant and animal health. Trends Microbiol 2015; 23(10): 606-17.
        [http://dx.doi.org/10.1016/j.tim.2015.07.009] [PMID: 26422463]

[3]   Berg G, Rybakova D, Fischer D. Microbiome definition re-visited: Old concepts and new challenges. Microbiome 2020; 8(1): 1-22.
        [PMID: 31901242]

[4]   Simon JC, Marchesi JR, Mougel C, Selosse MA. Host-microbiota interactions: From holobiont theory

to analysis. Microbiome 2019; 7(1): 5.
[http://dx.doi.org/10.1186/s40168-019-0619-4] [PMID: 30635058]

[5]   Zilber-Rosenberg I, Rosenberg E. Role of microorganisms in the evolution of animals and plants: the hologenome theory of evolution. FEMS Microbiol Rev 2008; 32(5): 723-35.
[http://dx.doi.org/10.1111/j.1574-6976.2008.00123.x] [PMID: 18549407]

[6]   Theis KR, Dheilly NM, Klassen JL, *et al.* Getting the hologenome concept right: an eco-evolutionary framework for hosts and their microbiomes. mSystems 2016; 1(2): e00028-16.
[http://dx.doi.org/10.1128/mSystems.00028-16] [PMID: 27822520]

[7]   Marchesi JR, Ravel J. The vocabulary of microbiome research: A proposal. Microbiome 2015; 3(1): 31.
[http://dx.doi.org/10.1186/s40168-015-0094-5] [PMID: 26229597]

[8]   Dupré J, Malley MA. Varieties of living things: life at the intersection of lineage and metabolism.Normandin S, Wolfe C. Vitalism and the Scientific Image in Post-Enlightenment Life Science. Dordrecht: Springer 2013; 2: pp. 311-43.
[http://dx.doi.org/10.1007/978-94-007-2445-7_13]

[9]   Gerardo N, Hurst G Q. BioMed Cen Biol 2013; 15(1): 1-6.

[10]   Kim EK, Lee KA, Hyeon DY, *et al.* Bacterial nucleoside catabolism controls quorum sensing and commensal-to-pathogen transition in the *Drosophila* gut. Cell Host Microbe 2020; 27(3): 345-357.e6.
[http://dx.doi.org/10.1016/j.chom.2020.01.025] [PMID: 32078802]

[11]   Bastías DA, Johnson LJ, Card SD. Symbiotic bacteria of plant-associated fungi: friends or foes? Curr Opin Plant Biol 2020; 56: 1-8.
[http://dx.doi.org/10.1016/j.pbi.2019.10.010] [PMID: 31786411]

[12]   Falkowski PG, Fenchel T, Delong EF. The microbial engines that drive Earth's biogeochemical cycles. Science 2008; 320(5879): 1034-9.
[http://dx.doi.org/10.1126/science.1153213] [PMID: 18497287]

[13]   Apprill A. Marine animal microbiomes: toward understanding host–microbiome interactions in a changing ocean. Front Mar Sci 2017; 4: 222.
[http://dx.doi.org/10.3389/fmars.2017.00222]

[14]   Cho I, Blaser MJ. The human microbiome: At the interface of health and disease. Nat Rev Genet 2012; 13(4): 260-70.
[http://dx.doi.org/10.1038/nrg3182] [PMID: 22411464]

[15]   Groussin M, Mazel F, Sanders JG, *et al.* Unraveling the processes shaping mammalian gut microbiomes over evolutionary time. Nat Commun 2017; 8(1): 14319.
[http://dx.doi.org/10.1038/ncomms14319] [PMID: 28230052]

[16]   Azad MAK, Sarker M, Li T, Yin J, Yin J. Probiotic species in the modulation of gut microbiota: An overview. BioMed Res Int 2018; 2018: 1-8.
[http://dx.doi.org/10.1155/2018/9478630] [PMID: 29854813]

[17]   Gurtler V, Subrahmanyam G. Silkworm gut microbiota: A potential source for biotechnological applications. In: Methods in Microbiol 2021; 49: 1-216.

[18]   Ursell LK, Metcalf JL, Parfrey LW, Knight R. Defining the human microbiome. Nutr Rev 2012; 70(1) (1): S38-44.
[http://dx.doi.org/10.1111/j.1753-4887.2012.00493.x] [PMID: 22861806]

[19]   Proctor LM. The human microbiome project in 2011 and beyond. Cell Host Microbe 2011; 10(4): 287-91.
[http://dx.doi.org/10.1016/j.chom.2011.10.001] [PMID: 22018227]

[20]   Gilbert JA, Blaser MJ, Caporaso JG, Jansson JK, Lynch SV, Knight R. Current understanding of the human microbiome. Nat Med 2018; 24(4): 392-400.

[http://dx.doi.org/10.1038/nm.4517] [PMID: 29634682]

[21]   Lew SQ, Radhakrishnan J. Chronic kidney disease and gastrointestinal disorders. Chronic Renal Dis 2020; 2: 521-39.
[http://dx.doi.org/10.1016/B978-0-12-815876-0.00033-4]

[22]   Purahong W, Orrù L, Donati I, *et al.* Plant microbiome and its link to plant health: host species, organs and *Pseudomonas syringae* pv. actinidiae infection shaping bacterial phyllosphere communities of kiwifruit plants. Front Plant Sci 2018; 9: 1563.
[http://dx.doi.org/10.3389/fpls.2018.01563] [PMID: 30464766]

[23]   Danhorn T, Fuqua C. Biofilm formation by plant-associated bacteria. Annu Rev Microbiol 2007; 61(1): 401-22.
[http://dx.doi.org/10.1146/annurev.micro.61.080706.093316] [PMID: 17506679]

[24]   Dastogeer KM, Tumpa FH, Sultana A, Akter MA, Chakraborty A. Plant microbiome–an account of the factors that shape community composition and diversity.Current Plant Biol. ELSVIER 2020; 23: p. 100161.
[http://dx.doi.org/10.1016/j.cpb.2020.100161]

[25]   Beattie GA, Lindow SE. Bacterial colonization of leaves: A spectrum of strategies. Phytopathology 1999; 89(5): 353-9.
[http://dx.doi.org/10.1094/PHYTO.1999.89.5.353] [PMID: 18944746]

[26]   Mwajita MR, Murage H, Tani A, Kahangi EM. Evaluation of rhizosphere, rhizoplane and phyllosphere bacteria and fungi isolated from rice in Kenya for plant growth promoters. Springerplus 2013; 2(1): 606.
[http://dx.doi.org/10.1186/2193-1801-2-606] [PMID: 24349944]

[27]   El-Ghany A, Masrahi YS, Alawlaqi MM, Al Abboud MA. Rhizosphere and rhizoplane bacteria isolated from subtropical region of Jazan in Saudi Arabia. J Biol Chem Res 2015; 32(2): 934-44.

[28]   Fürnkranz M, Wanek W, Richter A, Abell G, Rasche F, Sessitsch A. Nitrogen fixation by phyllosphere bacteria associated with higher plants and their colonizing epiphytes of a tropical lowland rainforest of Costa Rica. ISME J 2008; 2(5): 561-70.
[http://dx.doi.org/10.1038/ismej.2008.14] [PMID: 18273066]

[29]   Santoyo G, Moreno-Hagelsieb G, del Carmen Orozco-Mosqueda M, Glick BR. Plant growth-promoting bacterial endophytes. Microbiol Res 2016; 183: 92-9.
[http://dx.doi.org/10.1016/j.micres.2015.11.008] [PMID: 26805622]

[30]   Azevedo JL, Maccheroni W Jr, Pereira JO, De Araújo WL. Endophytic microorganisms: A review on insect control and recent advances on tropical plants. Electron J Biotechnol 2000; 3(1): 15-6.
[http://dx.doi.org/10.2225/vol3-issue1-fulltext-4]

[31]   Costa LEO, Queiroz MV, Borges AC, Moraes CA, Araújo EF. Isolation and characterization of endophytic bacteria isolated from the leaves of the common bean (Phaseolus vulgaris). Braz J Microbiol 2012; 43(4): 1562-75.
[http://dx.doi.org/10.1590/S1517-83822012000400041] [PMID: 24031988]

[32]   Araújo WL, Marcon J, Maccheroni W Jr, van Elsas JD, van Vuurde JWL, Azevedo JL. Diversity of endophytic bacterial populations and their interaction with *Xylella fastidiosa* in citrus plants. Appl Environ Microbiol 2002; 68(10): 4906-14.
[http://dx.doi.org/10.1128/AEM.68.10.4906-4914.2002] [PMID: 12324338]

[33]   Carrim AJI, Barbosa EC, Vieira JDG. Enzymatic activity of endophytic bacterial isolates of Jacaranda decurrens Cham. (Carobinha-do-campo). Braz Arch Biol Technol 2006; 49(3): 353-9.
[http://dx.doi.org/10.1590/S1516-89132006000400001]

[34]   Castro RA, Quecine MC, Lacava PT, *et al.* Isolation and enzyme bioprospection of endophytic bacteria associated with plants of Brazilian mangrove ecosystem. Springerplus 2014; 3(1): 382.
[http://dx.doi.org/10.1186/2193-1801-3-382] [PMID: 25110630]

[35]   Martinez-Klimova E, Rodríguez-Peña K, Sánchez S. Endophytes as sources of antibiotics. Biochem
        Pharmacol 2017; 134: 1-17.
        [http://dx.doi.org/10.1016/j.bcp.2016.10.010] [PMID: 27984002]

[36]   Sessitsch A, Hardoim P, Döring J, *et al.* Functional characteristics of an endophyte community
        colonizing rice roots as revealed by metagenomic analysis. Mol Plant Microbe Interact 2012; 25(1):
        28-36.
        [http://dx.doi.org/10.1094/MPMI-08-11-0204] [PMID: 21970692]

[37]   Puri A, Padda KP, Chanway CP. Nitrogen-fixation by endophytic bacteria in agricultural crops: recent
        advances.Nitrogen in Agriculture. intechopen 2018; pp. 1-250.
        [http://dx.doi.org/10.5772/intechopen.71988]

[38]   Khan AL, Hussain J, Al-Harrasi A, Al-Rawahi A, Lee IJ. Endophytic fungi: Resource for gibberellins
        and crop abiotic stress resistance. Crit Rev Biotechnol 2015; 35(1): 62-74.
        [http://dx.doi.org/10.3109/07388551.2013.800018] [PMID: 23984800]

[39]   Ding LJ, Cui HL, Nie SA, Long XE, Duan GL, Zhu YG. Microbiomes inhabiting rice roots and
        rhizosphere. Fed Eur Microbiol Soc Microbiol Ecol 2019; 95(5): 040.

[40]   Bacilio-Jiménez M, Aguilar-Flores S, Ventura-Zapata E, Pérez-Campos E, Bouquelet S, Zenteno E.
        Chemical characterization of root exudates from rice (*Oryza sativa*) and their effects on the
        chemotactic response of endophytic bacteria. Plant Soil 2003; 249(2): 271-7.
        [http://dx.doi.org/10.1023/A:1022888900465]

[41]   Kandel S, Joubert P, Doty S. Bacterial endophyte colonization and distribution within plants.
        Microorganisms 2017; 5(4): 77.
        [http://dx.doi.org/10.3390/microorganisms5040077] [PMID: 29186821]

[42]   Gamalero E, Glick BR. Bacterial ACC deaminase and IAA: Interactions and consequences for plant
        growth in polluted environments.Handbook of Phytoremediation. New York: Nova Science Publishers
        2010; pp. 763-77.

[43]   Leitão AL, Enguita FJ. Gibberellins in *Penicillium* strains: Challenges for endophyte-plant host
        interactions under salinity stress. Microbiol Res 2016; 183: 8-18.
        [http://dx.doi.org/10.1016/j.micres.2015.11.004] [PMID: 26805614]

[44]   Kuo J, Chang CF, Chi WC. Isolation of endophytic fungi with antimicrobial activity from medicinal
        plant *Zanthoxylum simulans* Hance. Folia Microbiol 2021; 66(3): 385-97.
        [http://dx.doi.org/10.1007/s12223-021-00854-4] [PMID: 33544301]

[45]   Pozo MJ, Zabalgogeazcoa I, de Aldan BRV, Martinez-Medina A. Untapping the potential of plant
        mycobiomes for applications in agriculture. Curr Opin Plant Biol 2021; 60: 102034.
        [http://dx.doi.org/10.1016/j.pbi.2021.102034]

[46]   Bruisson S, Zufferey M, L'Haridon F, *et al.* Endophytes and epiphytes from the grapevine leaf
        microbiome as potential biocontrol agents against phytopathogens. Front Microbiol 2019; 10: 2726.
        [http://dx.doi.org/10.3389/fmicb.2019.02726] [PMID: 31849878]

[47]   Ho YN, Mathew DC, Huang CC. Plant-microbe ecology: interactions of plants and symbiotic
        microbial communities.Plant Ecol–Trad App to Recent Trends. intechopen 2017; pp. 93-119.

[48]   Vellend M. Conceptual synthesis in community ecology. Q Rev Biol 2010; 85(2): 183-206.
        [http://dx.doi.org/10.1086/652373] [PMID: 20565040]

[49]   Nemergut DR, Schmidt SK, Fukami T, *et al.* Patterns and processes of microbial community
        assembly. Microbiol Mol Biol Rev 2013; 77(3): 342-56.
        [http://dx.doi.org/10.1128/MMBR.00051-12] [PMID: 24006468]

[50]   Bradáčová K, Weber NF, Morad-Talab N, *et al.* Micronutrients (Zn/Mn), seaweed extracts, and plant
        growth-promoting bacteria as cold-stress protectants in maize. Chem Biol Technol Agric 2016; 3(1):
        19.

[http://dx.doi.org/10.1186/s40538-016-0069-1]

[51]   Saravanakumar D, Kavino M, Raguchander T, Subbian P, Samiyappan R. Plant growth promoting bacteria enhance water stress resistance in green gram plants. Acta Physiol Plant 2011; 33(1): 203-9.
[http://dx.doi.org/10.1007/s11738-010-0539-1]

[52]   Ferreira TS, Moreira CZ, Cária NZ, Victoriano G, Silva WF Jr, Magalhães JC. Phytotherapy: An introduction to its history, use and application. Rev Bras Plantas Med 2014; 16(2): 290-8.
[http://dx.doi.org/10.1590/S1516-05722014000200019]

[53]   Heinrich M. Phytotherapy. Encyclopedia Britannica 2017. Available from: https://www.britannica.com/science/phytotherapy

[54]   Bhat RA, Hakeem K, Dervash MA. Bioactive compounds obtained from plants, their pharmacological applications and encapsulation.Phytomedicine. Elsevier 2021; pp. 181-205.

[55]   Oliveira M, Simoes M, Sassi C. Phytotherapy in the public health system (SUS) in the Sao Paulo State, Brazil. Rev Bras Plantas Med 2006; 8(2): 39-41.

[56]   Idris S, Mishra A, Khushtar M. Phytochemical, ethanomedicinal and pharmacological applications of escin from *Aesculus hippocastanum* L. towards future medicine. J Basic Clin Physiol Pharmacol 2020; 31(5): 20190115.
[http://dx.doi.org/10.1515/jbcpp-2019-0115] [PMID: 32649293]

[57]   Liu XG, Lu X, Gao W, Li P, Yang H. Structure, synthesis, biosynthesis, and activity of the characteristic compounds from *Ginkgo biloba* L. Nat Prod Rep 2022; 39(3): 474-511.
[http://dx.doi.org/10.1039/D1NP00026H] [PMID: 34581387]

[58]   Patil RH, Patil MP, Maheshwari VL. Bioactive secondary metabolites from endophytic fungi: a review of biotechnological production and their potential applications. Studies Nat Prod Chem 2016; 49: 189-205.

[59]   Clomburg JM, Qian S, Tan Z, Cheong S, Gonzalez R. The isoprenoid alcohol pathway, a synthetic route for isoprenoid biosynthesis. Proc Natl Acad Sci 2019; 116(26): 12810-5.
[http://dx.doi.org/10.1073/pnas.1821004116] [PMID: 31186357]

[60]   Palanichamy P, Krishnamoorthy G, Kannan S, Marudhamuthu M. Bioactive potential of secondary metabolites derived from medicinal plant endophytes. Egyp J BasicAppl Sci 2018; 5(4): 303-12.
[http://dx.doi.org/10.1016/j.ejbas.2018.07.002]

[61]   Kharwar RN, Verma VC, Kumar A, *et al.* Javanicin, an antibacterial naphthaquinone from an endophytic fungus of neem, *Chloridium* sp. Curr Microbiol 2009; 58(3): 233-8.
[http://dx.doi.org/10.1007/s00284-008-9313-7] [PMID: 19018591]

[62]   Newman DJ. The influence of microbial endophytes/epiphytes on medicinal plant components. JNat Health Prod Res 2021; 3(1): 1-4.
[http://dx.doi.org/10.33211/jnhpr.18]

[63]   Karthick P, Mohanraju R. Antimicrobial potential of epiphytic bacteria associated with seaweeds of Little Andaman, India. Front Microbiol 2018; 9: 611.
[http://dx.doi.org/10.3389/fmicb.2018.00611] [PMID: 29670590]

[64]   Coburn B, Sekirov I, Finlay BB. Type III secretion systems and disease. Clin Microbiol Rev 2007; 20(4): 535-49.
[http://dx.doi.org/10.1128/CMR.00013-07] [PMID: 17934073]

[65]   Kamber T, Pothier JF, Pelludat C, Rezzonico F, Duffy B, Smits THM. Role of the type VI secretion systems during disease interactions of *Erwinia amylovora* with its plant host. BMC Genomics 2017; 18(1): 628.
[http://dx.doi.org/10.1186/s12864-017-4010-1] [PMID: 28818038]

[66]   Shyntum DY, Theron J, Venter SN, Moleleki LN, Toth IK, Coutinho TA. *Pantoea ananatis* utilizes a type VI secretion system for pathogenesis and bacterial competition. Mol Plant Microbe Interact 2015;

28(4): 420-31.
[http://dx.doi.org/10.1094/MPMI-07-14-0219-R] [PMID: 25411959]

[67]   Bernal P, Llamas MA, Filloux A. Type VI secretion systems in plant-associated bacteria. Environ Microbiol 2018; 20(1): 1-15.
[http://dx.doi.org/10.1111/1462-2920.13956] [PMID: 29027348]

[68]   Pizzorno JE, Murray MT. Pharmacology of natural medicines.In: Textbook of Natural Medicine---Book. Elsevier Health Sciences 2020; 5: pp. (2)411-965.

[69]   Posadzki P, Watson L, Ernst E. Contamination and adulteration of herbal medicinal products (HMPs): An overview of systematic reviews. Eur J Clin Pharmacol 2013; 69(3): 295-307.
[http://dx.doi.org/10.1007/s00228-012-1353-z] [PMID: 22843016]

[70]   Aina O. Herbal medicine use in Ekiti state. Nigeria: epidemiological study and analysis of toxic constituents (Doctoral dissertation, Anglia Ruskin University) 2018; pp. 84-90.

[71]   Neogi SB, Negandhi P, Chopra S, *et al.* Risk factors for stillbirth: Findings from a population based case–control study, Haryana, India. Paediatr Perinat Epidemiol 2016; 30(1): 56-66.
[http://dx.doi.org/10.1111/ppe.12246] [PMID: 26444206]

CHAPTER 3

Soil Bacteria-Medicinal Plants Interaction

Raza Ullah[1,*], Muhammad Rahil Afzal[1], Hafiz Kamran Yousaf[2], Mustansar Aslam[1] and Ayesha Hassan[1]

[1] *Institute of Environmental and Agricultural Science, Faculty of Life Sciences, University of Okara, Okara, 56130, Pakistan*

[2] *Institute of Pure and Applied Zoology, Faculty of Life Sciences, Department of Wildlife and Ecology, University of Okara, Okara, 56130, Pakistan*

Abstract: Regulation of biogeochemical cycles depends on soil micro biota in which numerous and distinct types of bacteria are involved. These bacteria share a common environment in the soil and interact with the plants at three different levels *i.e.* endosphere, phyllosphere, and rhizosphere, resulting in improved soil fertility and plant health. The study of medicinal plants is ignored in Pakistan, though there exists a large number of different avenues for research in this field. Studying the medicinal plant-bacteria relationships in the era of new-generation sequencing paves new ways for understanding their association and facilitates improvement in sustainable production. Answers to new queries like "How bacteria respond to climatic changes" and "How do they interact with one another and with medicinal plants for growth and development" along with the exploration of rhizospheric bacteria in the future for enhancement in the production of secondary metabolites in medicinal plants might be a new vista unlocked for the sustainable agricultural practices. In this chapter, we focused on the role of soil bacteria-medicinal plants interaction in growth, nutrient acquisition, environmental stress alleviation, and quantity and quality of medicinal compounds present in these plants.

Keywords: Alleviation, Bacteria, Interaction, Medicinal plant, Rhizosphere, Soil, Stress and sustainable.

INTRODUCTION

Medicinal plants being a rich source of biologically active chemicals, are considered to be safer for humans as well as the environment than synthetic chemistries for serious diseases' treatments [1] There is a long history of using plant-based remedies throughout Europe and Asia [2]. Around six hundred medicinal plants are listed in Chinese *Materia Medica*, with the first usage of

---

* **Corresponding author Raza Ullah:** Institute of Environmental and Agricultural Science, Faculty of Life Sciences, University of Okara, Okara, 56130, Pakistan; E-mail: raza1838@gmail.com

Zulqurnain Khan, Azra Yasmin & Naila Safdar (Eds.)

medicinal herbs in China in 1100 BC [3, 4].The use of plant-based medications for a wide range of human ailments is expanding because of the increase in population, prices, and hazardous impacts, as well as pathogens of infectious diseases are developing resistance day by day against synthetic medicines. To address the demand for plants with medicinal value, their large-scale cultivations using contemporary farming technologies are being practiced by most of the Asian farmers. The development and quality of pharmaceutically significant plants are being hampered by a variety of plant pests and diseases. Furthermore, pesticide abuse has a negative impact on the quality of therapeutic plant products. All of these factors need the development of novel medicinal plant cultivation technology. Plant growth-promoting rhizobacteria (PGPR) are bacteria that colonize plant rhizosphere and stimulate plant growth by a variety of methods such as nitrogen fixation, quorum sensing, siderophores synthesis, phosphate solubilization, and so on [5]. Because PGPR can be used to replace chemical fertilizers, insecticides, and other chemicals, its popularity has increased rapidly.

Before we get into the interactive strategies of bacteria and medicinal plants, it is important to understand the fundamentals and history of this amazing science. The utilization of plant growth-promoting bacteria can be traced back to Theophrastus (372–287 B.C.) who proposed blending diverse soil samples to repair faults and add vitality to land [6]. The technique behind mixing of various soils can be verified by microscopic examination. Virgil was the first to chronicle the establishment of legumes on cultivable ground [7]. Studies confirmed the existence of rhizosphere bacterial colonization in grasses as well as ascertained that these soil bacteria transformed atmospheric nitrogen and made it available for the plants [8]. The term "rhizobacteria" was coined after study on the radish crop, and it is defined as a population that colonises plant roots competitively, accelerates their development, and decreases plant illnesses by increasing plant immunity [9]. Belligerent colonisation and plant growth promotion properties, as well as biocontrol ability, are just a few of the qualities directly linked to PGPR [10]. PGPR interact with plants in all positive, negative, and neutral ways [11]. These are known as iPGPR and ePGPR. Mostly ePGPR could be found between root cortex cells, rhizoplane, and rhizosphere. ePGPR include bacteria from Agrobacterium, Azotobacter, Caulobacter, and Chromobacterium genera [12]. Endophytes [rhizobium spp.], and Frankia species live in specialized nodular structures of root cells [13, 14]. Plant growth promoting rhizobacteria promote nodule formation and fixing of atmospheric nitrogen in a variety of crops like soyabean (*Glycine max* L. Merr) [15]. Crop plant growth and productivity can be enhanced directly or indirectly by PGPR. Rhizosphere colonization is responsible for the generation of siderophores [16].

This chapter focuses on developing an understanding about medicinal plants, bacterial diversity associated with them and plant-bacterial interactions to get the current state of knowledge in this broad field of study.

BACTERIAL DIVERSITY ASSOCIATED WITH MEDICINAL PLANTS

Research on rhizobacteria from significant medicinal herbs is important because they significantly influence the crop production, synthesis of important plant metabolites, and the quality of medicinal products [17]. Many microbes can produce phyto-therapeutic chemicals [18]. This data is immensely important in the development of biofertilizers' industries for commercially cultivated medicinal plants. The soil microorganisms in the rhizosphere of medicinal plants growing on the desert ecosystem (*Calendula officinalis* L., *Solanum distichum* Schumach. & Thonn. and *Matricaria chamomilla* L.,) contain a high number of G +ve [Gram positive] bacteria that are important for disease suppression. There was a host-specific assortment of microorganisms as well as extremely distinct diazotrophic communities detected in all three plants. The findings revealed that plant species had an essential role in structural and physiological diversity. Additionally, *Bacillus* strains help the plant to develop and increase the synthesis of flavonoids. Significant variations were reported in the endophytic bacterial range in three different species of medicinal plants [19]. The microbial diversity in the rhizosphere of various medicinally significant plants was studied, and 50 strains were categorized into seven genera, with *Corallococcus* and *Myxococcus* being the most common genera [20].

Farming of *Rehmannia glutinosa*, a valuable medicinal plant continuously on the same plot of land reduces its production [21]. A change in the soil microbial population because of *R. glutinosa* cultivation could be one of the major reasons for the limits associated with monoculture cropping [21, 22]. Though, plant, soil and myco-floral interactions are critical for *R. glutinosa* production and quality in a sequential cropping system [23], in the intercropping treatments, addition of Gram-negative bacteria and a decrease in allelochemicals promoted peanut development and enhanced peanut production. There were 120 morphologically different isolates categorised into 21 phylotypes after root endophytes and rhizosphere bacteria were identified [24]. The majority of the isolates were Firmicutes and Gamma-Proteobacteria. The most common species were *Stenotrophomonas* and *Pseudomonas*, which accounted for 27.5 percent of all isolates. *Agrobacterium, Pseudomonas, Enterobacter, Acinetobacter, Pantoea, Sphingobium, Serratia,* and *Stenotrophomonas* were among the 103 bacterial communities isolated and described from the roots and rhizospheric region of Hypericum silenoides Juss [24]. About 123 bacterial species belonging to alpha- and gamma Proteobacteria were isolated from the roots and rhizospheric soil of

Ajuga bracteosa [medicinal plant], with Pseudomonas being the most dominant species, to investigate their plant growth-promoting and biotechnological potentials [25]. The endophytic bacterial community was almost entirely made up of Firmicutes. A pink coloured Gram-negative bacterium isolated from the rhizospheric zone of *Nerium indicum* [*Chuvanna arali*], an Indian medicinal plant, has morphological and chemotaxonomic characteristics with the genus Pontibacter and was identified as a distinct species [26]. Acidobacteria, Proteobacteria, Bacteroidetes, and Actinobacteria were the most common bacterial communities discovered in the rhizosphere of the medicinal plant *Fritillaria thunbergia* [27]. The microbial population of Aloe vera revealed a higher percentage proportion in rhizospheric soil than in non-rhizospheric soil [28]. *Azospirillum, Azotobacter,* and *Pseudomonas* have all been researched as diazotrophic bacteria [29]. Actinobacteria biocontrol strains found in medicinal plants are particularly interesting since they could be a source of antibiotics.

NUTRIENTS ACQUISITION FACILITATION IN MEDICINAL PLANTS

The importance of soil microorganisms in biogeochemical cycles and agricultural development has long been recognized. Plant-bacterial interactions in the rhizosphere reflect both soil fertility and plant health. Plant growth is facilitated by PGPR, a type of free-living soil bacterium that colonizes the roots of plants and promotes growth. These bacterial strains have the ability to help plants grow for a longer period of time. These bacteria perform their functions by [a] producing compounds for plants, [b] boosting the uptake of specific nutrients from the soil, and [c] reducing the plant's susceptibility to various diseases. Plant growth and development are aided and promoted in both direct and indirect ways by bacteria that dwell in the rhizospheric zone. Indirect plant growth promotion requires preventing and shielding plants against phytopathogens' damaging effects, which can be performed by the production of siderophores [*i.e.* small metal-binding molecules]. Several bacteria found in medicinal plants have been found to have biological control properties against soil-borne plant diseases as well as the ability to manufacture antibiotics. Plant hormones such as gibberellins, cytokinins, ethylene, abscisic acid, and auxins are produced by symbiotic and non-symbiotic PGPRs, resulting in direct growth stimulation. Several bacterial genera connected to medicinal plants have been shown to generate indole-3-acetic acid (IAA) or indole-3-ethanol, as well as chemical compounds similar to auxins. PGPRs have also been shown to have a role in the solubilization of mineral phosphates and the uptake of other nutrients, as well as in stress tolerance, soil aggregate stabilization, and the improvement of organic matter content and soil structure. PGPRs can retain more soil organic N and other nutrients in plant-soil systems, reducing the demand for P and N fertilizer amendments while also improving nutrient release by making nutrients available to plants.

Nitrogen Fixation

One of the most critical nutrients for plants to receive is nitrogen [N]. It is essential for plant development, food and feed production, and plays a role in the cellular synthesis of proteins, enzymes, DNA and RNA, and chlorophyll. Microorganism activity in the rooting zone is crucial for nitrogen [N] cycling and plant nitrogen consumption efficiency [30]. Both symbiotic and non-symbiotic rhizospheric bacteria can fix nitrogen in the soil, decreasing the cost of nitrogen fertilizers used in crop cultivation. The symbiotic N-fixing action of Rhizobium bacteria in legume root nodules, for example, is a well-known example [31, 32]. Non-leguminous plants including *Saccharum* spp. hybrids, *Oryza* spp., *Zea maize*, and *Triticeae* spp. have also been reported to fix nitrogen in symbiosis with bacteria like *Bacillus* spp., *Beijerinckia* spp., and *Azotobacter* spp [33, 34]. This BNF [biological nitrogen fixation] mechanism accounts for 65 percent of nitrogen used in agriculture today [35]. Diazotrophs such as *Gluconacetobacter, Azospirillum, Burkholderia, Azoarcus, Cyanobacteria,* and *Pseudomonas* have been discovered to fix nitrogen in the rhizospheric zone of several herbal plants, including *Catharanthus roseus* L. G. Don., coleus [*Coleus forskholii* Briq], *Aloe vera*, and Catharanthus rose [36]. Furthermore, it has been reported that combining five PGPRs [*Bacillus megaterium, Azotobacter chroococcum, Pseudomonas fluorescens, Bacillus polymyxa,* and *Azospirillum lipoferum*] with different concentrations of N fertiliser applied to plots of dill [*Anethum graveolens* L.] resulted in maximum plant growth of dill [37, 38].

Phosphorus Solubilization

Phosphorus [P] is a macro nutrient limiting growth of plants in many soils [39, 40]. Its concentration in the soil ranges from 400 to 1200 mg kg^{-1} of soil. Phosphorus exists in soil both in organic and inorganic forms. Plants need an increased amount of soluble form of P for their optimum growth and development. Soils are relatively abundant in P, however, a major part of this P is not accessible to medicinal and other plants due to an increased precipitation of aluminum and iron in acidic soils, and with calcium in alkaline soils [41]. Plants usually take up phosphorus as HPO_4^{-2} and $H2PO_4^-$, which represent <1% of the total phosphorus in soil. Phosphorus fixation as well as precipitation processes in soil highly depend on the soil types and pH. However, microbes release phosphatases and organic acids that degrade organic P and dissolve inorganic P compounds, respectively, solubilizing P and ultimately enhancing its availability to plants for their better growth [42, 43].

Various research trials have proven a significantly improved growth of medicinal plants when they were introduced with different microorganisms. For example,

Andrographis paniculata [annual herbaceous medicinal plant] when inoculated with *Trichoderma harzianum* and *Glomus mosseae* while grown at two different phosphorus levels (*i.e.* recommended phosphorus contents and 75% of recommended phosphorus contents) under field conditions, it showed significantly improved growth and medicinal attributes including alkaloid [andrographolide] production in comparison to uninoculated plants at both P levels [44]. In another study, the application of AM fungi and phosphate-solubilizing bacteria in a rose-scented geraniums field resulted in yields comparable to the applied P fertilizer. Geranium when co-inoculated with phosphate-solubilizing bacteria and AM fungi showed up with a 33% increase in yield compared to that of the control. Taken together, the application of AM fungi and PGPRs could minimize the amount of phosphate fertilizers needed for the cultivation of aromatic as well as medicinal plants.

Iron Sequestering

Despite its abundance, iron is not readily digested by plants or bacteria in aerobic soils because Fe+3 [ferric ion] is only sparingly soluble in nature, resulting in a very low concentration of iron available for assimilation by living organisms. Iron is required by both plants and microorganisms, and acquiring enough iron in the rhizosphere, where plants, fungi, and bacteria compete for it, is much more difficult [44, 45]. To deal with iron scarcity, bacteria produce low-molecular-mass Siderophores (ranging between 400 to 1500 Da). These molecules have a high affinity for Fe+3 (KKaa values vary from 1023 to 1052), and membrane receptors that may bind the Fe complex and allow bacteria to absorb iron [46, 47]. The organization of bacterial communities in the rhizosphere can be influenced by plant iron feeding. In the rhizosphere, non-transformed tobacco has less accessible iron than transgenic tobacco, which overexpresses ferritin and accumulates more iron. As a result, the bacterial population of converted tobacco lines in the rhizosphere was significantly different from that of non-transformed tobacco lines.

Siderophores Production

Microbes produce iron-chelating siderophores, which provide iron to host plants while limiting pathogenic microbe growth in the rhizosphere, allowing plants to grow and develop. Siderophore-producing bacteria (*Rhodococcus, Bacillus, Pseudomonas,* and *Enterobacter*) transfer iron from the insoluble mineral phase to soluble ferric complexes that are easily absorbed by plants. Phytopathogen growth is inhibited by the removal of iron from the environment by these bacteria. Khamna and colleagues isolated and identified 445 actinomycetes from the rhizospheres of 16 medicinal plants [47]. Streptomyces accounted for almost 89 percent of all bacteria found, with 75 isolates capable of producing siderophores.

REGULATING GROWTH HORMONES IN MEDICINAL PLANTS

During whole life cycles, plants are continuously exposed to an ever-changing environment mainly consisting of unfavorable biotic or biotic factors. Due to these adverse conditions, plants would be unable to fulfil their high potential or sometimes these conditions may lead to plant death. In recent decades, human activities have intensified the frequency of abiotic stress and consequently increased the complexities in ensuring the food supply [48]. In crop production, salinity and drought are considered as the most serious limiting factors. Photosynthesis, cellular metabolism, and nutrient acquisition are adversely affected by drought and consequently, the chlorophyll contents, cell membrane stability and leaf water contents decreased significantly. In response to these adverse conditions, plants respond to drought and try to overcome the stress with growth and productivity which might be the biochemical and physiological changes in plants [49].

Plant hormones or plant growth regulators or PGRs, known as bioinhibitors or biostimulants are the organic compounds that modify the different physiological processes of plants and elevate the biotic and abiotic stress in plants [50]. They stimulate or inhibit the enzyme or enzyme system and assist metabolism in plants [51]. Usually, they are active at extremely low concentrations in plants. To date, nine types of plant hormones have been identified these include cytokinin, abscisic acid, gibberellins, ethylene, salicylates, brassinosteroids, auxin strigolactones and jasmonates. These hormones are produced by plants themselves, but they can also be induced by microorganism and may modulate the immunity, physiology and plant growth. All plant hormones have an important role in plant growth, however, cytokinin has a very important role in plant growth and development such as apical dominance, seed germination, flower and fruit germination, leaf senescence and importantly the plant pathogen interaction.

Cytokinins

Cytokinins are plant hormones that play a critical role in plant growth and development. Since their discovery in maize plants, these have been widely studied for their chemical nature, signal transduction pathways and metabolism [52]. Cytokinins are isoprenoid substituted adenine molecules. Isopentenyltransferases is the 1st enzyme that is involved in catalyzing the isoprenoid to other numerous types of cytokinins *i.e.*, N6-D2-isopentenyl, trans-zeatin, cis-zeatin [Cz] and dihydrozeatin [DZ]. In plants, trans-zeatin mostly occurs in abundance. The types and activity of different cytokinins is remarkably vary according to plants and species under different environmental conditions and developmental stages [53]. Mature tRNAs from plants contain cis zantin as a modified base and breakdown of tRNA was initially suggested as a possible mechanism of for cytokinin synthesis as the released cis zeatin was supposed to be converted to the active form of trans zeatin isomerase, however, further studies suggested that there is small interconversion between cis and trans zeatin. It was initially thought that cytokinins were produced in the roots and later they were transported to the other plant parts including shoots, however, recent studies showed that they are not only synthesized in roots but also in shoots and transported to different plant parts [54]. They are transported to shoots from roots *via* xylem and *vice versa*. Levels of cytokinin in roots rise in reaction to nitrate, and the transfer of this raised cytokinin could represent a long-distance signal to synchronize shoot and root growth [55]. Although trans zeatin riboside is the basic form of cytokine transferred from root to shoot, the active free base trans zeatin is also transferred at very low level. The transported trans zeatin cytokinin regulates the leaf size but does not regulate the meristem in the shoot. In Arabidopsis plant, the cytokinin transport from shoot to root occurs through symplastic connections and regulate the auxin transport and vascular patterning [56]. Although cytokinin has a major role in plant development, however, it also interacts with other hormones especially auxin to regulate the cell division and separation. Cytokinin secreted within the shoots promotes cell proliferation. It also plays an important role in regulating the development of gynoecium, leaf phyllotaxy, female gametophyte development and vascular cambial development. Contrarily, cytokinins may perform important functions in the inhibition of root growth and regulate the root architecture by inhibiting the lateral roots and promoting the primary root elongation.

Gibberellins

Gibberellins, also known as Gas, are phytohormones that are made up of a wider family of diterpenoids with a tetracyclic ent-gibberellane carbon skeleton structure organised in either a four or five-circle system with a variable lactone fifth ring [57]. In general, there are two types of known gibberellins: one group has molecules with 20 carbon atoms, known as C20-Gas, and the other group contains molecules with one lactone and one carbon loss, known as C19-GAs. The second group depicts acidic nature, in which the polarity of each molecule varies depending on the quantity of carboxyl and hydroxyl groups present. All bioactive GAs belong to the same C19-GAs, despite the fact that not all C19-GAs are bioactive. GA1 gibberellins are the most active gibberellins [58]. Gibberellins were first discovered in the pathogenic fungus known as *Gibberella fujikuroi,* which is responsible for foolish seedling [a disease in rice]. These pathogenic fungi produce large quantity of GA then plants grow more in lengthwise and become slender in shape and by these plants are unable to bear their weight and cannot support themselves and become infertile [59]. In plants, GA has significant importance as it plays a crucial role in the germination of seeds, elongation and expansion processes in organs and cell growth, transition from vegetative to reproductive growth, trichome development and seeds, flowers as well as fruits development. Due to cell elongation and expansion properties, GA is commonly used in agriculture and the most remarkable utilization of GA is the introduction of GA into dwarfing alleles in to staple crop like wheat. This introduction in wheat crop has resulted in green revolution and led to mass production of wheat and rice globally. Till date >130 GAs has been identified in plants, bacteria and fungi and only four GAs; GA_1, GA_3, GA_4 and GA_7 are thought to be the active form of GAs. The remaining forms of GAs in plants are the precursors of active forms.

The production of active GAs is a multistep process that makes pinpointing the exact organ or tissue where these phytohormones are generated and delivered to multiple sites difficult. Much research using GUS as a receptor to investigate biosynthetic pathways and expression patterns has yielded a variety of results. Firstly, genes are differently expressed that are responsible for GA biosynthesis among different cell types, and tissues. Secondly, several GA_3ox family members are expressed in elongating and growing the roots and shoot organ. In vegetative organs, GA is synthesized in the growing regions in shoot and roots tip. In certain tissues including cereal scutellar epithelium and anthers, higher rates of GA synthesis took place while in developing seed the exact function of GA is unclear. GA mobility in higher plants was discovered some decades ago, leading to the conclusion that GA is present in the phloem and travels through this medium. GA is thought to flow in acro and basipetal directions within plants, according to research. GA movement is required for a variety of plant developmental processes. There are various examples of GA mobility between organs in which different tissues serve as a source of GA for adjacent GA-nonautonomous organs; for example, the cereal embryo scutellum serves as an aleurone supply. GA9, which is produced in the ovaries, travels to the sepals and petals of female cucumber flowers, where it is converted into GA4, which is responsible for organ enlargement. GA, as a hormone, coordinates the growth of neighbouring organs in this case.

Indoleacetic Acid

Phytohormone auxins were the first plant hormones discovered in plants. Charles Darwin and his son Francis Darwin in 1880 reported that in some plants the growth is regulated by a matter which transmits its impacts from one part to another part in plants. Auxins is taken from the Greek term auxein, which means to grow or increase, and this name was given to that chemical by biochemists in the 1930s [60 - 62]. Indole acetic acid, or IAA, is the most prevalent auxin hormone found in plants and other microorganisms, and it controls many aspects of plant growth and development. IAA is required for plant growth, which includes cell division and elongation, as well as fruit formation and senescence. It can be synthesized not only in plants but also in microorganisms *i.e.*, yeast, filamentous fungi, actinomycetes and bacteria. Although, IAA plays an important role in plant growth and development, however, the information and knowledge of the progression and biosynthesis pathways are still limited studied. IAA can be synthesized by either Ttrp-dependent pathways or Trp-independent pathways by conversion of tryptophan to a variety of arbitrates. Generally, in plants, four Trp-dependent pathways have been suggested in different studies which are indole-3 acetaldoxime [IAOX], tryptamine [TAM], idole-3-pyruvic acid [IPA] and indole-3-acetalamide [IAM].

Ethylene

Illuminating gas that produced from coal was used for lighting in early twentieth century. Leaks from channels including illuminating gas led to premature abscission and senescence in neighboring vegetation. Later Dimitry Neljuboy recognized ethylene as an active constituent in the illuminating gas. Ethylene is the olefin gas and was the first gaseous form of phytohormone [63 - 65]. It is biosynthesized by plants and has effects on various developmental activities such as fruit ripening, seed germination, abscission, and senescence as well as comebacks to various stress such as soil compaction and salinity. Once biosynthesized, ethylene diffuses all the way through the plants and attaches to ethylene receptors to invigorate the ethylene responses. Ethylene involves in numerous activities all over the plant growth and development, however, most of the impacts are burdensome to use for mutant screen [66]. Endogenous hormonal signals and phytohormones regulate leaf growth and improvement, which is influenced by a variety of environmental factors. The relevance of ethylene in leaf growth and development has been confirmed physiologically using ethylene inhibitors and genetically utilising ethylene insensitive mutants or transgenic plants. Experimental evidence has shown that ethylene response factors [ERF5] and ERF6 increase leaf growth in response to various environmental conditions.

ALLEVIATING ENVIRONMENTAL STRESS IN MEDICINAL PLANTS

Heavy metal stress, salt, drought, and high heat are the most crucial limiting variables in agricultural output because they degrade morphological, biochemical, and physiological processes in plants [67]. Plants' physiological, metabolic, nutritional, and hormonal systems, on the other hand, develop survival strategies to adapt to a wide range of environmental conditions [68 - 70]. Plant-beneficial bacteria are also known as PGPBs, BCAs, biofertilizers, and biopesticides [71]. In the face of these obstacles, PGPB can help medicinal plant development [72 - 74]. When compared to a control set of plants developed from un-inoculated seeds, the yield of Madagascar periwinkle [*Catharanthus roseus*] was improved under a water deprivation stress when produced from seeds infected with Pseudomonas fluorescence [75]. When basil [*Ociumum basilicum* L.] was subjected to water stress, researchers noticed that the leaves had more proline and soluble carbohydrate accumulation than basil that had not been treated with PGPB [76]. After simultaneous inoculation with *Piriformospora indica* [77] and *Pseudomonas fluorescens* [*Chlorophytum* spp.,] the survival rate of transplanted musli rises [78]. Metal resistance has been found in several *Pseudomonas* spp. isolated from heavy metal-rich environments. *P. koreensis* AGB, for example, was isolated from *Miscanthus sinensis* roots growing in mine-tailing soil and showed considerable Zn, Cd, As, and Pb resistance [79]. Metal-resistant *Pseudomonads* spp. have also been reported to reduce metal stress in plants in some metal-polluted soils [80 - 84]. In a study of chickpea plant development in Ni-polluted soils, for example [85], discovered that Ni significantly reduced non-inoculated plant growth while PBP inoculation drastically increased plant growth. Certain plant species, on the other hand, adopt a variety of tactics to combat salt stress, including food uptake and/or NaC exclusion [with or without the help of soil microbes]. PBP immunisation did, in fact, provide them with significant protection [86]. A study discovered increased Ca2C, Mg2C, and KC absorption after inoculating plants with P. putida Rs-198 under salt stress. Increased nutrient uptake has been attributed to bacterial-mediated stimulation of the proton pump ATPase [86], changes in the K transporter [HKT1] gene [87], and/or physical barrier modification around the roots [*via* EPS] [88, 89] PBPs excrete organic acids [which alter rhizosphere Ph] and siderophores [which chelate with nutrients], which can solubilize and/or mobilize unavailable soil nutrients, making them more available to plants [90]. Similarly, salinity reduces plant P uptake by lowering P availability in soils, resulting in plant growth retardation [90, 91]. It was discovered that *P. fluorescens* CECT 378 generating siderophores can protect *Zea mays* plants from salt stress and that the ability of *Pseudomonas* spp. to increase plant nutrient [92] discovered that inoculating sunflower plants with *P. fluorescens* CECT 378 generating siderophores improved KC uptake. These findings suggest that PBPs reduce salt stress in plants by increasing nutrient bioavailability (*e.g.*, magnesium, iron, potassium, calcium, and phosphorus) and/or uptake by the plants, and that using nutrient-mobilizing Pseudomonas spp. as a strategy for improving plant development under salinity stress could be a viable option. Inoculating sorghum seedlings with Pseudomonas sp. AKM-P6 boosted cellular metabolite synthesis [proline, carbohydrates, and amino acids], high-molecular-weight proteins, and reduced membrane damage, according to another study [93]. They discovered that inoculating the plants with AKM-P6 protected them against high temperatures by altering certain important events in the plants.

SECONDARY METABOLITE INDUCTION

A rise in phytochemical production in medicinal plants has been connected to the presence of bacteria in the rhizosphere [94, 95]. Although the exact mechanism by which bacteria boost phytochemical production is uncertain, it could be owing to increased plant growth or direct metabolic pathway activation. Bacterial phytohormones and volatile organic compounds [VOCs] can act as induction cues for phytochemical development [96]. The amounts of -terpineol and eugenol in basil were raised by VOCs generated by *Bacillus subtilis* GB03. When inoculated with polysaccharides from Bacillus cereus, the hairy roots of *Salvia miltiorrhiza* accumulate tanshinones [diterpenoid quinines] [97]. A siddha holistic plant enhances the development of secondary metabolites [total phenols, ortho-dihydroxy phenols, flavonoids, alkaloids, and tannins] in the leaves by treating sivakaranthai [*Sphaeranthus amaranthoides*] with *Glomus walkeri* and a variety of other bacteria [98]. Bacteria can help plants grow by producing siderophores, phytohormones, and volatile organic molecules, as well as boosting available P and N. [VOCs]. This can also be accomplished by soil microorganisms. Bacteria can help medicinal plants grow quicker and create more secondary metabolites by acting as environmentally benign bio-fertilizers.

CONCLUSION

The study in literature supports the use of soil bacteria as secondary metabolites booster, which improves the overall quality and worth of medicinal plants while also promoting plant development and stress resistance.

REFERENCES

[1]   Cappelletti S, Daria P, Sani G, Aromatario M. Caffeine: Cognitive and physical performance enhancer or psychoactive drug? Curr Neuropharmacol 2015 Jan; 13(1): 71-88.
[http://dx.doi.org/10.2174/1570159X13666141210215655] [PMID: 26074744]

[2]   Nema R, Khare S, Jain P, Pradhan A, Gupta A, Singh D. Natural products potential and scope for modern cancer research. Am J Plant Sci 2013, 4(6). 1270-7.
[http://dx.doi.org/10.4236/ajps.2013.46157]

[3]   Cragg GM, Newman DJ, Snader KM. Natural products in drug discovery and development. J Nat Prod 1997; 60(1): 52-60.
[http://dx.doi.org/10.1021/np9604893] [PMID: 9014353]

[4]   Toussaint JP, Smith FA, Smith SE. Arbuscular mycorrhizal fungi can induce the production of phytochemicals in sweet basil irrespective of phosphorus nutrition. Mycorrhiza 2007; 17(4): 291-7.
[http://dx.doi.org/10.1007/s00572-006-0104-3] [PMID: 17273856]

[5]   Bhattacharyya PN, Jha DK. Plant growth-promoting rhizobacteria (PGPR): Emergence in agriculture. World J Microbiol Biotechnol 2012; 28(4): 1327-50.
[http://dx.doi.org/10.1007/s11274-011-0979-9] [PMID: 22805914]

[6]   Solaiman ZM, Anawar HM. Plant growth-promoting microbes from herbal vermicompost. In: Egamberdieva D, Shrivastava S, Varma A, Eds. Plant-Growth-Promoting Rhizobacteria (PGPR) and Medicinal Plants Soil Biology. Springer, Cham 2015; 42: pp. 71-88.

[7]   Vessey JK. Plant growth promoting rhizobacteria as biofertilizers. Plant Soil 2003; 255(2): 571-86.
[http://dx.doi.org/10.1023/A:1026037216893]

[8]   Shrivastava S, Egamberdieva D, Varma A. Plant Growth-Promoting Rhizobacteria (PGPR) and Medicinal Plants: The State of the Art. In: Egamberdieva D, Shrivastava S, Varma A, Eds. Plant-Growth-Promoting Rhizobacteria (PGPR) and Medicinal Plants Soil Biology. Springer, Cham 2015; 42: pp. 1-16.
[http://dx.doi.org/10.1007/978-3-319-13401-7_1]

[9]   Kloepper JW, Ryu C-M, Zhang S. The nature and application of biocontrol microbes: Bacillus spp. Induced Systemic Resistance and Promotion of Plant Growth by Bacillus spp 2004; 94: 1259-66.
[http://dx.doi.org/10.1094/PHYTO.2004.94.11.1259]

[10]   Weller DM. Biological control of soilborne plant pathogens in the rhizosphere with bacteria. Annu Rev Phytopathol 1988; 26(1): 379-407.
[http://dx.doi.org/10.1146/annurev.py.26.090188.002115]

[11]   Whipps JM. Microbial interactions and biocontrol in the rhizosphere. J Exp Bot 2001; 52(Spec Issue Suppl. 1): 487-511.
[http://dx.doi.org/10.1093/jxb/52.suppl_1.487] [PMID: 11326055]

[12]   Gray EJ, Smith DL. Intracellular and extracellular PGPR: commonalities and distinctions in the plant–bacterium signaling processes. Soil Biol Biochem 2005; 37(3): 395-412.
[http://dx.doi.org/10.1016/j.soilbio.2004.08.030]

[13]   Verma JP, Yadav J, Tiwari KN, Singh L, Singh V. Impact of PGPR on crop production.pdf. Int J Agric Res 2010; 5: 954-83.
[http://dx.doi.org/10.3923/ijar.2010.954.983]

[14]   Wang ET, Martínez-Romero E. Sesbania herbacea-Rhizobium huautlense nodulation in flooded soils and comparative characterization of S. herbacea-nodulating rhizobia in different environments. Microb Ecol 2000; 40(1): 25-32.
[http://dx.doi.org/10.1007/s002480000010] [PMID: 10977874]

[15]   Zhang X, Xue C, Fang D, *et al.* Manipulating the soil microbiomes during a community recovery process with plant beneficial species for the suppression of Fusarium wilt of watermelon. AMB Exp 2021; 11(1): 87.
[http://dx.doi.org/10.1186/s13568-021-01225-5] [PMID: 34117935]

[16]   Poschenrieder C, Busoms S, Barceló J. How plants handle trivalent (+3) elements. Int J Mol Sci 2019; 20(16): 3984.
[http://dx.doi.org/10.3390/ijms20163984] [PMID: 31426275]

[17]   Bafana A. Diversity and metabolic potential of culturable root-associated bacteria from Origanum vulgare in sub-Himalayan region. World J Microbiol Biotechnol 2013; 29(1): 63-74.
[http://dx.doi.org/10.1007/s11274-012-1158-3] [PMID: 22927014]

[18]   Köberl M, Schmidt R, Ramadan EM, Bauer R, Berg G. The microbiome of medicinal plants: diversity and importance for plant growth, quality and health. Front Microbiol 2013; 4: 400.
[http://dx.doi.org/10.3389/fmicb.2013.00400] [PMID: 24391634]

[19]   Li X, Penttinen P, Zeng Z. LH- PCR 2017.

[20]   Zhao K, Penttinen P, Chen Q, *et al.* The rhizospheres of traditional medicinal plants in Panxi, China,

host a diverse selection of actinobacteria with antimicrobial properties. Appl Microb Cell Physiol 2012; 92: 1321-35.
[http://dx.doi.org/10.1007/s00253-011-3862-6]

[21] Qi J, Yao H, Ma X, Zhou L, Li X. Soil microbial community composition and diversity in the rhizosphere of a chinese medicinal plant. Commun Soil Sci Plant Anal (Internet) 2009; 1462-82. Available from: https://www.tandfonline.com/action/journalInformation?journalCode=lcss20
[http://dx.doi.org/10.1080/00103620902818104]

[22] Wu L, Li Z, Li J, *et al.* Assessment of shifts in microbial community structure and catabolic diversity in response to Rehmannia glutinosa monoculture. Appl Soil Ecol 2013; 67: 1-9.
[http://dx.doi.org/10.1016/j.apsoil.2013.02.008]

[23] Wu L, Wang H, Zhang Z, Lin R, Zhang Z, Lin W. Comparative metaproteomic analysis on consecutively Rehmannia glutinosa-monocultured rhizosphere soil. PLoS One 2011; 6(5): e20611.
[http://dx.doi.org/10.1371/journal.pone.0020611] [PMID: 21655235]

[24] López-Fuentes E. Bacterial community in the roots and rhizosphere of Hypericum silenoides Juss. 1804. Afr J Microbiol Res 2012; 6(11)
[http://dx.doi.org/10.5897/AJMR11.1192]

[25] Kumar G, Kanaujia N, Bafana A. Functional and phylogenetic diversity of root-associated bacteria of Ajuga bracteosa in Kangra valley. Microbiol Res 2012; 167(4): 220-5.
[http://dx.doi.org/10.1016/j.micres.2011.09.001] [PMID: 21968325]

[26] Raichand R, Kaur I, Nitin •, Singh K, Mayilraj S. Pontibacter rhizosphera sp nov, isolated from rhizosphere soil of an Indian medicinal plant Nerium indicum 2010 Jun; 100(1): 129-35.[www.eztaxon.org/]
[http://dx.doi.org/10.1007/s10482-011-9573-2] [PMID: 21409554]

[27] Shi JY, Yuan XF, Lin HR, Yang YQ, Li ZY. Differences in soil properties and bacterial communities between the rhizosphere and bulk soil and among different production areas of the medicinal plant Fritillaria thunbergii. Int J Mol Sci 2011; 12(6): 3770-85.[www.mdpi.com/journal/ijmsArticle]
[http://dx.doi.org/10.3390/ijms12063770] [PMID: 21747705]

[28] Nimnoi P, Lumyong S, Pongsilp N. Impact of rhizobial inoculants on rhizosphere bacterial communities of three medicinal legumes assessed by denaturing gradient gel electrophoresis (DGGE). Ann Microbiol 2011; 61(2): 237-45.
[http://dx.doi.org/10.1007/s13213-010-0128-y]

[29] Karthikeyan B, Jaleel CA, Lakshmanan GMA, Deiveekasundaram M. Studies on rhizosphere microbial diversity of some commercially important medicinal plants. Colloids Surf B Biointerfaces 2008; 62(1): 143-5.
[http://dx.doi.org/10.1016/j.colsurfb.2007.09.004] [PMID: 17951032]

[30] Trabelsi D, Mhamdi R. Microbial inoculants and their impact on soil microbial communities: A review. Biomed Res Int 2013; 2013: 863240.
[http://dx.doi.org/10.1155/2013/863240] [PMID: 23957006]

[31] Gaunt MW, Turner SL, Rigottier-Gois L, Lloyd-Macgilp SA, Young JP. Phylogenies of atpD and recA support the small subunit rRNA-based classification of rhizobia. Int J Syst Evol Microbiol 2001; 51(6): 2037-48.
[http://dx.doi.org/10.1099/00207713-51-6-2037] [PMID: 11760945]

[32] Sessitsch A, Howieson J, Perret X, Antoun H, Martinez-Romero E. Advances in rhizobium research. Crit Rev PlantSci 2002; 21(4): 323-78.
[http://dx.doi.org/10.1080/0735-260291044278]

[33] Bhattacharjee RB, Singh A, Mukhopadhyay S. Advances in rhizobium research. Crit Rev PlantSci 2002; 21(4): 323-78.
[http://dx.doi.org/10.1080/0735-260291044278]

[34]   Thaweenut N, Hachisuka Y, Ando S, Yanagisawa S, Yoneyama T. Two seasons' study on nifH gene expression and nitrogen fixation by diazotrophic endophytes in sugarcane (Saccharum spp. hybrids): Expression of nifH genes similar to those of rhizobia. Plant Soil 2011; 338(1-2): 435-49.
[http://dx.doi.org/10.1007/s11104-010-0557-1]

[35]   Matiru VN, Dakora FD. The rhizosphere signal molecule lumichrome alters seedling development in both legumes and cereals. N phytolog 2005; 166(2): 439-44.
[http://dx.doi.org/10.1111/j.1469-8137.2005.01344.x]

[36]   Orr CH, James A, Leifert C, Cooper JM, Cummings SP. Diversity and activity of free-living nitrogen-fixing bacteria and total bacteria in organic and conventionally managed soils. Appl Environ Microbiol 2011; 77(3): 911-9.
[http://dx.doi.org/10.1128/AEM.01250-10] [PMID: 21131514]

[37]   Hellal F, Mahfouz S, Hassan F. Partial substitution of mineral nitrogen fertilizerbybio-fertilizeron(AnethumgraveolensL.)plant. Agric Biol J 2011; 2(4): 652-60.

[38]   Hameeda B, Harini G, Rupela O, Wani S, Reddy G. Growth promotion of maize by phosphate-solubilizing bacteria isolated from composts and macrofauna. Microbiol Res 2008 Mar 15; 163(2): 234-42.
[http://dx.doi.org/10.1016/j.micres.2006.05.009]

[39]   Richardson AE, Barea JM, McNeill AM, Prigent-Combaret C. Acquisition of phosphorus and nitrogen in the rhizosphere and plant growth promotion by microorganisms. Plant Soil 2009; 321(1-2): 305-39.
[http://dx.doi.org/10.1007/s11104-009-9895-2]

[40]   Sylvia DM, Fuhrmann JJ, Hartel P, Zuberer DAI. In Principles and Applications of Soil Microbiology. 2005.

[41]   Arpana J, Bagyaraj DJ. Response of kalmegh to an arbuscular mycorrhizal fungus and a plant growth promoting rhizomicroorganism at two levels of phosphorus fertilizer. Am-Eurasian J Agric Environ Sci 2007; 2: 33-8.

[42]   Tian F, Ding Y, Zhu H, Yao L, Du B. Genetic diversity of siderophore-producing bacteria of tobacco rhizosphere. Braz J Microbiol 2009; 40(2): 276-84.
[http://dx.doi.org/10.1590/S1517-83822009000200013] [PMID: 24031358]

[43]   Saharan B, Nehra V. Plant growth promoting rhizobacteria: A critical review. LifeSci Med Res 2011; 1;21(1): 30.

[44]   Ma JF. Plant root responses to three abundant soil minerals: silicon, aluminum and iron. Crit Rev Plant Sci 2005; 24(4): 267-81.
[http://dx.doi.org/10.1080/07352680500196017]

[45]   Guerinot ML, Yi Y. Iron: Nutritious, noxious, and not readily available. Plant Physiol 1994; 104(3): 815-20.
[http://dx.doi.org/10.1104/pp.104.3.815] [PMID: 12232127]

[46]   Khamna S, Yokota A, Lumyong S. Actinomycetes isolated from medicinal plant rhizosphere soils: Diversity and screening of antifungal compounds, indole-3-acetic acid and siderophore production. World J Microbiol Biotechnol 2009; 25(4): 649-55.
[http://dx.doi.org/10.1007/s11274-008-9933-x]

[47]   Suzuki N, Rivero RM, Shulaev V, Blumwald E, Mittler R. Abiotic and biotic stress combinations. New Phytol 2014; 203(1): 32-43.
[http://dx.doi.org/10.1111/nph.12797] [PMID: 24720847]

[48]   Kieber JJ, Schaller GE. Cytokinin signaling in plant development. Development 2018; 145(4): dev149344.
[http://dx.doi.org/10.1242/dev.149344] [PMID: 29487105]

[49]   Zhu JK. Abiotic stress signaling and responses in plants. Cell 2016; 167(2): 313-24.

[http://dx.doi.org/10.1016/j.cell.2016.08.029] [PMID: 27716505]

[50] Shao H, Wang H, Tang X. NAC transcription factors in plant multiple abiotic stress responses: Progress and prospects. Front Plant Sci 2015; 6: 902.
[http://dx.doi.org/10.3389/fpls.2015.00902] [PMID: 26579152]

[51] Akhtar SS, Mekureyaw MF, Pandey C, Roitsch T. Role of cytokinins for interactions of plants with microbial pathogens and pest insects. Front Plant Sci 2020; 10: 1777.
[http://dx.doi.org/10.3389/fpls.2019.01777] [PMID: 32140160]

[52] Davière JM, Achard P. Gibberellin signaling in plants. Development 2013; 140(6): 1147-51.
[http://dx.doi.org/10.1242/dev.087650] [PMID: 23444347]

[53] Su Y, Xia S, Wang R, Xiao L. Phytohormonal quantification based on biological principles in Hormone Metabolism and Signaling in Plants. London, UK: Academic Press 2017; pp. 431-70.
[http://dx.doi.org/10.1016/B978-0-12-811562-6.00013-X]

[54] Dowd CD, Chronis D, Radakovic ZS, *et al.* Divergent expression of cytokinin biosynthesis, signaling and catabolism genes underlying differences in feeding sites induced by cyst and root-knot nematodes. Plant J 2017; 92(2): 211-28.
[http://dx.doi.org/10.1111/tpj.13647] [PMID: 28746737]

[55] Großkinsky DK, Tafner R, Moreno MV, *et al.* Cytokinin production by Pseudomonas fluorescens G20-18 determines biocontrol activity against Pseudomonas syringae in Arabidopsis. Sci Rep 2016; 6(1): 23310.
[http://dx.doi.org/10.1038/srep23310] [PMID: 26984671]

[56] Salazar-Cerezo S, Martínez-Montiel N, García-Sánchez J, Pérez-y-Terrón R, Martínez-Contreras RD. Gibberellin biosynthesis and metabolism: A convergent route for plants, fungi and bacteria. Microbiol Res 2018; 208: 85-98.
[http://dx.doi.org/10.1016/j.micres.2018.01.010] [PMID: 29551215]

[57] Sponsel P, Hedden V. Gibberellin biosynthesis and inactivation. plant Horm 2010; 63-94.
[http://dx.doi.org/10.1007/978-1-4020-2686-7_4]

[58] Binenbaum J, Weinstain R, Shani E. Gibberellin localization and transport in plants. Trends Plant Sci 2018; 23(5): 410-21.
[http://dx.doi.org/10.1016/j.tplants.2018.02.005] [PMID: 29530380]

[59] Darwin C, Darwin F. The power of movement in plants. London, England: Murray 2009.
[http://dx.doi.org/10.1017/CBO9780511693670]

[60] Spaepen S, Vanderleyden J. Auxin and plant-microbe interactions. Cold Spring Harb Perspect Biol 2011 Apr 1; 3(4): a001438.
[http://dx.doi.org/10.1101/cshperspect.a001438]

[61] Fu SF, Wei JY, Chen HW, Liu YY, Lu HY, Chou JY. Indole-3-acetic acid: A widespread physiological code in interactions of fungi with other organisms. Plant Signal Behav 2015; 10(8): e1048052.
[http://dx.doi.org/10.1080/15592324.2015.1048052] [PMID: 26179718]

[62] Mattoo AK, Suttle JC, Ori N M, *et al.* The plant hormone ethylene. Development 2019; 141: 4219-30.

[63] Abeles F, Morgan P, Saltveit MJ. Ethylene in Plant Biology. 2nd. San Diego, CA. Academic Press 1992; 111.

[64] Bakshi A, Shemansky JM, Chang C, Binder BM. History of research on the plant hormone ethylene. J Plant Growth Regul 2015; 34(4): 809-27.
[http://dx.doi.org/10.1007/s00344-015-9522-9]

[65] Dubois M, Van den Broeck L, Inzé D. The pivotal role of ethylene in plant growth. Trends Plant Sci 2018; 23(4): 311-23.
[http://dx.doi.org/10.1016/j.tplants.2018.01.003] [PMID: 29428350]

[66]   You J, Chan Z. ROS regulation during abiotic stress responses in crop plants. Front Plant Sci 2015; 6: 01-15.
[http://dx.doi.org/10.3389/fpls.2015.01092]

[67]   Hasanuzzaman M, Nahar K, Alam M, Roychowdhury R, Fujita M. Physiological, biochemical, and molecular mechanisms of heat stress tolerance in plants. Int J Mol Sci 2013; 14(5): 9643-84.
[http://dx.doi.org/10.3390/ijms14059643] [PMID: 23644891]

[68]   Gupta B, Huang B. Mechanism of salinity tolerance in plants: Physiological, biochemical, and molecular characterization. Int J Plant Genomics 2014; 2014: 01-18.
[http://dx.doi.org/10.1155/2014/701596]

[69]   Simontacchi M, Galatro A, Ramos-Artuso F. Plant survival in a changing environment: The role of nitric oxide in plant responses to abiotic stress. Front Plant Sci 2015; 6: 01-19.
[http://dx.doi.org/10.3389/fpls.2015.00977]

[70]   Ma Y, Rajkumar M, Zhang C, Freitas H. Beneficial role of bacterial endophytes in heavy metal phytoremediation. J Environ Manage 2016; 174: 14-25.
[http://dx.doi.org/10.1016/j.jenvman.2016.02.047] [PMID: 26989941]

[71]   Farooq M, Wahid A, Kobayashi N, Fujita D, Basra S. Plant drought stress: Effects, mechanisms and management.Sustainable Agriculture. Netherlands: Springer 2009; pp. 153-88.
[http://dx.doi.org/10.1007/978-90-481-2666-8_12]

[72]   Nadeem SM, Ahmad M, Zahir ZA, Javaid A, Ashraf M. The role of mycorrhizae and plant growth promoting rhizobacteria (PGPR) in improving crop productivity under stressful environments. Biotechnol Adv 2014; 32(2): 429-48.

[73]   Shahzad SM, Arif MS, Ashraf M, Abid M, Ghazanfar MU, Riaz M. Alleviation of abiotic stress in medicinal plants by PGPR.Soil Biology. Switzerland: Springer International Publishing 2015; 42.
[http://dx.doi.org/10.1007/978-3-319-13401-7_7]

[74]   Jaleel CA, Manivannan P, Sankar B, Kishorekumar A, Gopi R, Somasundaram R. Pseudomonas fluorescens enhances biomass yield and ajmalicine production in Catharanthus roseus under water deficit stress. Colloids Surf B 2007 Oct 31; 60(1): 7-11.
[http://dx.doi.org/10.1016/j.colsurfb.2007.05.012]

[75]   Heidari M, Mousavinik SM, Golpayegani A. Plant growth promoting rhizobacteria (PGPR) effect on physiological parameters and mineral uptake in basil (Ociumum basilicm L.) under water stress. ARPN J Agric Biol Sci 2011; 6(5): 6-11.

[76]   Verma S, Varma A, Rexer KH, *et al. Piriformospora indica*, gen. et sp. nov., a new root-colonizing fungus. Mycologia 1998; 90(5): 896-903.
[http://dx.doi.org/10.1080/00275514.1998.12026983]

[77]   Gosal S, Karlupia A, Gosal S, Chhibba I, Varma A. Biotization with Piriformospora indica and Pseudomonas fluorescens improves survival rate, nutrient acquisition, field performance and saponin content of micropropagated Chlorophytum sp. Indian J Biotechnol 2010; 9: 289-97.

[78]   Babu AG, Shea PJ, Sudhakar D, Jung IB, Oh BT. Potential use of Pseudomonas koreensis AGB-1 in association with Miscanthus sinensis to remediate heavy metal(loid)-contaminated mining site soil. J Environ Manage 2015; 151: 160-6.
[http://dx.doi.org/10.1016/j.jenvman.2014.12.045] [PMID: 25575343]

[79]   Islam F, Yasmeen T, Ali Q, Ali S, Arif MS, Hussain S. Influence of Pseudomonas aeruginosa as PGPR on oxidative stress tolerance in wheat under Zn stress. Ecotoxicol Environ Saf 2014 Apr 15;Jun 15; 104: 285-93.
[http://dx.doi.org/10.1016/j.ecoenv.2014.03.008]

[80]   Li K, Pidatala VR, Shaik R, Datta R, Ramakrishna W. Integrated metabolomic and proteomic approaches dissect the effect of metal-resistant bacteria on maize biomass and copper uptake. Environ Sci Technol 2014; 48(2): 1184-93.

[http://dx.doi.org/10.1021/es4047395] [PMID: 24383886]

[81] Shagol CC, Krishnamoorthy R, Kim K, Sundaram S, Sa T. Arsenic-tolerant plant-growth-promoting bacteria isolated from arsenic-polluted soils in South Korea. Environ Sci Pollut Res Int 2014; 21(15): 9356-65.
[http://dx.doi.org/10.1007/s11356-014-2852-5] [PMID: 24737020]

[82] Ma Y, Rajkumar M, Rocha I, Oliveira RS, Freitas H. Serpentine bacteria influence metal translocation and bioconcentration of Brassica juncea and Ricinus communis grown in multi-metal polluted soils. Front Plant Sci 2015 Jan 5; 5: 01-13.
[http://dx.doi.org/10.3389/fpls.2014.00757]

[83] NRJ A, OO B. Effect of bacterial inoculation of strains of Pseudomonas aeruginosa, Alcaligenesfeacalis and Bacillus subtilis on germination, growth and heavy metal (Cd, Cr, and Ni) uptake of Brassica juncea. Int J Phytoremed 2016; 18(2): 200-09.

[84] Tank N, Saraf M. Enhancement of plant growth and decontamination of nickel-spiked soil using PGPR. J Basic Microbiol 2009; 49(2): 195-204.
[http://dx.doi.org/10.1002/jobm.200800090] [PMID: 18798171]

[85] Yao L, Wu Z, Zheng Y, Kaleem I, Li C. Growth promotion and protection against salt stress by Pseudomonas putida Rs-198 on cotton. Eur J Soil Biol 2010; 46(1): 49-54.
[http://dx.doi.org/10.1016/j.ejsobi.2009.11.002]

[86] Zhang H, Kim MS, Sun Y, Dowd SE, Shi H, Paré PW. Soil bacteria confer plant salt tolerance by tissue-specific regulation of the sodium transporter HKT1. Mol Plant Microbe Interact 2008; 21(6): 737-44.
[http://dx.doi.org/10.1094/MPMI-21-6-0737] [PMID: 18624638]

[87] Ashraf M, Hasnain S, Berge O, Mahmood T. Inoculating wheat seedlings with exopolysaccharide-producing bacteria restricts sodium uptake and stimulates plant growth under salt stress. Biol Fertil Soils 2004; 40(3): 157-62.
[http://dx.doi.org/10.1007/s00374-004-0766-y]

[88] Upadhyay SK, Singh JS, Singh DP. Exopolysaccharide producing plant growth promoting rhizobacteria under salinity condition. Pedosphere 2011; 21(2): 214-22.
[http://dx.doi.org/10.1016/S1002-0160(11)60120-3]

[89] Sharpley AN, Meisinger JJ, Power JF, Suarez DL. Root extraction of nutrients associated with long-term soil management. In: JL H, Stewart BA, Eds. Limitations to plant growth Soil Sci. New York: Springer 1992; 19: pp. 151-217.
[http://dx.doi.org/10.1007/978-1-4612-2894-3_6]

[90] Shilev S, Sancho ED, Benlloch-González M. Rhizospheric bacteria alleviate salt-produced stress in sunflower. J Environ Manage 2012; 95: S37-41.
[http://dx.doi.org/10.1016/j.jenvman.2010.07.019]

[91] Bano A, Fatima M. Salt tolerance in Zea mays (L). following inoculation with Rhizobium and Pseudomonas. Biol Fertil Soils 2009; 45(4): 405-13.
[http://dx.doi.org/10.1007/s00374-008-0344-9]

[92] Ali SZ, Sandhya V, Grover M, Kishore N, Rao LV, Venkateswarlu B. Pseudomonas sp. strain AKM-P6 enhances tolerance of sorghum seedlings to elevated temperatures. Biol Fertil Soils 2009; 46(1): 45-55.
[http://dx.doi.org/10.1007/s00374-009-0404-9]

[93] Banchio E, Bogino PC, Zygadlo J, Giordano W. Plant growth promoting rhizobacteria improve growth and essential oil yield in Origanum majorana L. Biochem Syst Ecol 2008; 36(10): 766-71.
[http://dx.doi.org/10.1016/j.bse.2008.08.006]

[94] Vafadar F, Amooaghaie R, Otroshy M. Effects of plant-growth-promoting rhizobacteria and arbuscular mycorrhizal fungus on plant growth, stevioside, NPK, and chlorophyll content of *Stevia*

rebaudiana. J Plant Interact 2014; 9(1): 128-36.
[http://dx.doi.org/10.1080/17429145.2013.779035]

[95] Solano BR, Maicas JB, Mañero JG. Biotechnology of the rhizosphere.Recent Advances in Plant Biotechnology. New York: Springer 2009; pp. 137-62.
[http://dx.doi.org/10.1007/978-1-4419-0194-1_8]

[96] Zhao D, Li X, Zhao L, *et al.* Comparison of zinc and iron uptake among diverse wheat germplasm at two phosphorus levels. Cereal Res Commun 2020; 48(4): 441-8.
[http://dx.doi.org/10.1007/s42976-020-00081-6]

[97] Sumithra P, Selvaraj T. Influence of Glomus walkeri Blaszk and Renker and plant growth promoting rhizomicroorganisms on growth, nutrition and content of secondary metabolites in Sphaeranthes amaranthoides (L.) Burm. Agric Technol Thail 2011; 7: 1685-92.

[98] Adesemoye AO, Kloepper JW. Plant-microbes interactions in enhanced fertilizer-use efficiency. Appl Microbiol Biotechnol 2009; 85: 1-12.
[http://dx.doi.org/10.1007/s00253-009-2196-0]

CHAPTER 4

Soil Fungi-Medicinal Plants Interaction

Samiya Rehman[1,*], Sadia Arshad[2], Saman Zulfiqar[1], Zaib un Nisa[1], Muhammad Saleem[1], Fouzia Tanvir[1] and Tahira Tabassum[1]

[1] *Department of Biochemistry, University of Okara, Punjab, Pakistan*

[2] *The Govt. Sadiq College Women University Bahawalpur, Punjab, Pakistan*

Abstract: Medicinal plants are a natural source of therapeutic compounds and secondary metabolites; therefore, their demand is increasing day by day. Since the last thirty decades, their cultivation as well as preservation with the help of biofertilizers or pesticides is a point of great concern. The rhizosphere is an important area around the roots. It is a habitat for many kinds of microorganisms like fungi. This soil microbial performs a variety of beneficial functions for the growth of plants such as nitrogen fixation, solubilization and removal of toxins. Endophytes are also an important class of microbial flora that helps in the absorption of water and nutrients for the plant. Additionally, they also make plants able to cope with environmental stresses. Fungal endophytes supervise photosynthesis. Certain therapeutically important plants including licorice and white ginger lily can also perform antimicrobial activity depending upon the endophytic composition they have. These types of plants having antimicrobial activity are of great significance as they act as eco-friendly biopesticides.

Keywords: Endophytes, Fungi, Medicinal plants, Rhizosphere, Secondary metabolites.

INTRODUCTION

Medicinal plants have similar parts as other plants but they are characterized by some medicinal values. Herbal medicines relatively use different parts of herbal plants [1]. Medicinal plants used for herbalism have a profound tradition of their application besides conventional medicine. About 30,000 years ago, the use of medicinal plants in curing different types of diseases had been started throughout the world [2]. Medicinal plants contain chemicals in their different parts that can help in the treatment of different diseases in humans [3]. The total number of medicinal plants in the world is quite surprising. The use of herbal medicines is the oldest and still used system of medicines. Plant medicines also known as botanical medicine or Phyto-medicines (WHO) refer to herbal materials which

---

* **Corresponding author Samiya Rehman:** Department of Biochemistry, University of Okara, Punjab, Pakistan; E-mail: samiyarehmans@gmail.com

contain complete plants, or parts of plants like leaves, flowers and roots [4]. It is expected that from lichens to trees, about 70,000 plants are used for medicinal purposes. In developing countries, 80% of people depend on traditional medicines based on plants.

The Rhizosphere, the narrow zone of soil surrounding and influenced by plant roots, is a natural habitat for numerous beneficial microorganisms and represents a biologically complex ecosystem on Earth [5]. This biologically active zone is critical for plant-microbe interactions and, as a consequence, for nutrient cycling, plant growth, and resistance of plants to diseases [6]. During positive plant-microbe interaction, rhizosphere colonization by soil microorganisms is beneficial for both plants and microorganisms. Both partners derive benefits from this intimate association and vitalize each other [7]. The large number of rhizo-deposits released by the plant roots are the key determinants of microbial activity and develop a community structure in the rhizosphere [8]. The rhizosphere microbes utilize the rhizome carbon deposits as a major energy source for their growth and development [9]. Consequently, plant roots can manipulate the rhizosphere microbiome for its benefit by selectively stimulating microorganisms with traits that are beneficial to plant growth and health [10, 11]. Mutual interdependence and the interplay between the rhizosphere microbiome and the plant result in the overall quality of plant productivity.

The rhizospheric microbial forms vary in diversity, which include bacteria, fungi, nematodes, viruses, arthropods, oomycetes, protozoa, algae, and archaea. A beneficial effect of the number of rhizosphere fungi concerning plant growth promotion has long been known [12]. These plant growth-promoting fungi (PGPF) include species of the genera *Aspergillus, Fusarium, Trichoderma, Penicillium, Piriformospora, Phoma*, and *Rhizoctonia,* which have the natural ability to stimulate various growth-related traits of plants [13]. Many studies on dicots and monocots have shown that PGPF mimics the well-studied plant growth-promoting rhizobacteria (PGPR) in their interaction with the host plant [14]. For example, treating seeds with PGPF inoculum can improve the germination and seedling vigor of different plants. They can also induce longer and larger shoots (Fig. **1**). Some may exert an effect on root development and performance. There are PGPFs that may stimulate early and vigorous flowering of plants [15].

Plants have always been used as medicines; from the beginning of life, plants were the first and most important source of medicine. Herbal plants show anticancer activity. Plant-based medications are often low-cost, accessible, and exhibit little reaction after treatment. . Herbal medicines are usually low in cost, abundant and show very little toxicity or side effects in treatment. For a long time, traditional Chinese medicines have been used for the cure of cancer in China and all over the world. In the treatment of cancer, a lot of molecules have been identified that are extracted from plants and their synthetic derivatives. The compounds extracted from plants do not have risk factors or side effects on human health [16].

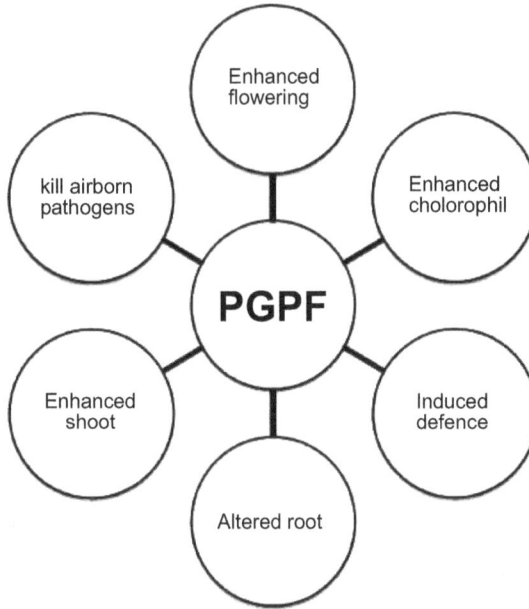

Fig. (1). Role of Plant growth promoting fungi (PGPF) in plants.

Many secondary metabolites were produced by plants which are a further source of many drugs. With the development of these drugs, many therapies are also discovered in this regard. These compounds have epigenetic properties as they have this ability in their genome which helps them to combat cancer. They have u tumour suppressor genes. Substances like mistletoe plant extract, taxanes, podophyllotoxin derivatives, vinca alkaloids, and camptothecin, have inhibitory activity on the growth of all the cancer cell lines and are extracted from marine species. Two of them have activity against mammary epithelial cells, and they also help in the discovery of new compounds and therapies. The healing properties of plants rely on their components for example antibiotic, antipyretic, and antioxidant effects. Natural yield extracts from living organisms (plants and animals) have some active ingredients, an alternative to chemotherapeutic treatment or appropriate with chemotherapeutic treatment [17].

STRUCTURE AND FUNCTION OF SOIL RHIZOSPHERE FUNGAL COMMUNITIES

The microbial population performs a vital role in the appropriate working of plants by inducing beneficial effects on their physiology and development. Although several microbial members of the rhizosphere are helpful for plant growth, furthermore plant deleterious microorganisms occupy the rhizosphere, a defensive system intending to trigger the disease. Certain types of microorganisms that can be found in the rhizosphere are the true human pathogenic bacteria, which can be carried on or in plant tissue. For example, *P. aeruginosa* is a gram-negative bacterium that can affect human lungs as well as elicit soft-rot disease in tomato and onion crops [18].

The plant rhizosphere is the zone where the root system interacts with the soil environment. According to a study conducted by Philipot in 2013, the root zone has a complex population of microbes and invertebrates that interact directly and indirectly with the plant (Fig. **2**). By releasing root exudates into the soil and creating root litter, plants have an impact on the rhizosphere microbiome [19].

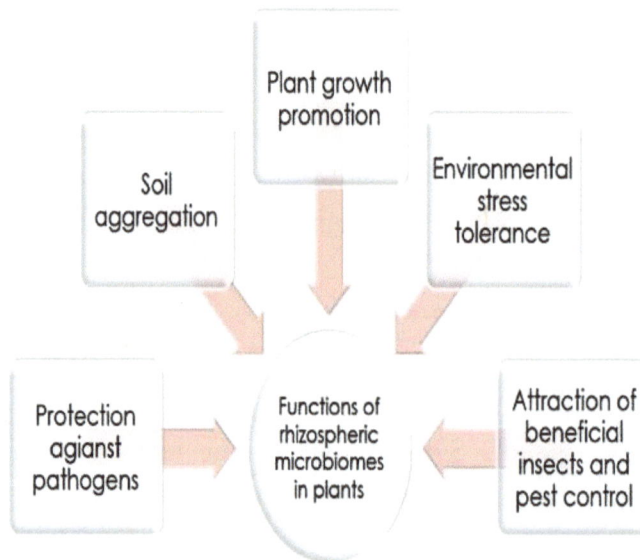

Fig. (2). Functions of the rhizospheric microbiome in the improvement of plants living.

Arbuscular fungi, which play a key role in carbon and nutrient cycling in ecosystems, are one of the most significant and abundant microbial groups in the rhizosphere. Some soil fungi called the ectomycorrhizae may have a symbiotic relationship with plant roots, in which the fungi feed on the organic carbon that the plant transfers to its underground sections. In exchange, the plant obtains nutrients as well as relief from many types of stress [20]. Other fungi populations act as decomposers or saprotrophs like Basidiomycetes, and play a major role in nutrient cycling, thus providing nutrients for plants. Plant pathogenic fungus, on the other hand, feeds on the plant and gives nothing in return, causing the plant to suffer. Fungal taxa and developed tools like FUN Guild are used to examine the function of fungi in a community [21].

Medicinal plants produce secondary plant metabolites in one or more of their parts and affect a wide range of microorganisms. For the past 5000 years, these plants are being used as medicine for humans and they are recognized globally [22]. Many medications created in recent decades are defined as combinations of botanical materials, unaltered natural products, or natural products; many are derived from or are plant components. Surprisingly, the microbiome of plants composes some of the therapeutic chemicals found in plants. Furthermore, medicinal plants can alter the abundance of pathogenic fungi in the soil and their impact on newly rooted plants, as demonstrated in a study by Li *et al.* in 2018, which found that *Atractylodes lancea* suppresses *Fusarium* [23].

Several previous articles documented the fungal rhizosphere population of various medicinal plants for these and other reasons. In 2020, Abdul Latif Khan *et al.* published the rhizosphere fungal communities of *Adenium obesum*, *Aloe dhufarensis*, and *Cleome austroarabica,* all of which had phyla Ascomycota in great abundance [24].

In the rhizospheres of *A. obesum* and *A. dhufarensis*, the genus *Acremonium* was relatively prevalent, while the genus *Corynascus* was abundant in the rhizosphere of *C. austroarabica*. The rhizosphere fungus population of the medicinal plant *Bouvardia ternifolia* was characterized by Villalobos-Flores *et al.* in 2021 [25] and they discovered a high relative abundance of class *Sordariomycetes*. Kushwaha *et al.* compiled a thorough list in 2020 [26].

Medicinal plants can be designated as natural curative resources having advantageous pharmacological effects. Plants essentially have a rhizosphere, developed by the microbial settlers. These - micro-communities are capable of performing diverse beneficial functions for host growth and disease prevention. Root-linked microflora enhances plant obstinacy to forthcoming drought. Rhizosphere is an ecologic domain whereabout friendly microbiota competes with hazardous to do colonization for better cellular growth and differentiation. Regulation of the carbon cycle, nutrition and water supply are the definitive characteristics of a rhizosphere [27].

Nitrogen is present in the chlorophyll, proteins and nucleotides; hence plants and animals cannot survive without it. Nitrogen supply for plants is important, therefore, the maintenance of the nitrogen cycle is prominently done by rhizosphere fungi.

The fertility of plants is increased by microbes in the soil. Plant diversity is a derivative of the root fungal community. Apart from all the benefits of plant soil microbiota, there are many irretrievable damages that can also be caused by harmful or pathogenic microorganisms. These pathogens not only deform the herb and plants but are also instability factors for ecosystems across the globe. Biological control of these harmful microbes can be performed by using a new promising technology of inoculation of antagonistic endophytes [28].

The ratio of plant pathogens is weirdly increasing, causing challenges to the stability of medicinal, herbal and agricultural crops. To develop better strategies for pathogen control, there is a strong need to make reliable diagnostic methods [29]. Throughout life, a plant is protected from pests, parasites, and other hazardous microbes by endophytes. Chemically synthesized pesticides are quite costly; additionally they kill the beneficial microflora. Whereas the antagonistic property of endophytes destroys the disease-causing pathogen only along with the anti-microbial characteristics. Collinge *et al.* in their paper have described the plant endophytes as innately occurring microorganisms, bio competitors present inside the plant tissues with diversified activities [30].

Fungal endophytes have been proven to be a good alternative treatment for the blight disease of cereals and wheat. The use of Microflora related to medicinal herbs and plants in place of chemically prepared pesticides is a new and interesting aspect of plant agronomy as well as a challenge to biotechnologists. Two strains of endophytic fungi named ML37 and ML 38 of *P. olsonii* and *A. alternatum* showed moderate fungicidal activity in wheat crop to prevent Septoria blotch disease [31].

THE ANTIMICROBIAL ACTIVITY OF MEDICINAL PLANTS CORRELATES WITH THE PROPORTION OF ANTAGONISTIC ENDOPHYTES DENSITY

Humans are treating their ailments by using compounds derived from plants, animals and naturally occurring elements. Due to this fact, herbs particularly medicinal plants are in vogue. Since the time of Aristotle and Hippocrates, plants are being used as medicine. It is said that Chinese and Greek people are using these plants 27 centuries before Christ. A Greek physician named Theophrastus in the 1st century AD founded a medicinal plant school and wrote the first encyclopedia about 600 medicinal plants [32].

In Romania over 700 types of medicinal plants are common in their traditional medicinal practice even though they are part of their diet. The medicinal plant's preparations can be syrups, inhalers, tablets, ointments or various other forms based on the needs. Across the globe, 80% population utilizes plants as a primary healthcare resource, additionally above 50,000 flowering plants are considered as medicine [33]. Now plants as medicament are being internationally favoured because they are easily accessible, less expensive with almost no drug resistance and have minimum after-effects [34].

Agronomy of Medicinal Plants

In the present era, medicinal components of organically grown plants were found to be more effective than conventionally cultivated therapeutic plants. *Valeriana Jatamansi* is an antiseptic, pain relieving, digestive and anti-epileptic plant that was found to be cultivated in hilly and mountainous locations of Nepal and China [35].

In the hot weather of Gaomi city, altogether 181 species of medicinal plants were located whereas Faba beans having high medicinal and nutritional values are cultivated in different colder regions of the world. Due to its nitrogen-fixing capability, it has great importance in agronomy as it can add nitrogen-fixing microbe in the rhizosphere. Global warming, drastic climatic changes, and the spread of pathogens are alarming conditions that are decreasing the medicinal plant yield, additionally, increasing population are the main reasons for their illegal cultivation [36].

N. Chinensis has been used traditionally for the cure of several antibacterial, neurological, cardiac and hepatic diseases. It grows at the highest Tibetan plateau at the height of 3500 meters in Monsoon weather [37].

Antimicrobial Activity of Medicinal Plants

Extraction and utilization of antimicrobial compounds from plants are acquiring great attention globally. The cure of infections depends on the antipathogenic activity of novel herbs and the antimicrobial chemicals extracted from plants. At present, the antimicrobial compounds are not much eeffective as microbes have developed resistance against them. These drug-resistant microbes are a cause of increased death rates across the world. Antimicrobial chemical compounds obtained from natural plant products are proven to be safe and effective [38].

Secondary metabolites are the therapeutic substances released by the medicinal plant to control microbial activities. Catechins and epicatechins are examples of these secondary metabolites which can kill the bacteria by disrupting their cell wall [39].

Glycyrrhiza glabra is an important anticarcinogenic, antiviral antimalarial plant that secretes a phenolic compound called Licochalcone A to restrict the oxygen as well as energy supply of microorganisms destroying the pathogen. While medicinal plants belonging to the alkaloid group alter the DNA replication in a microorganism to kill them [40].

Antagonistic Endophytes of a Medicinal Plant and their Antimicrobial Pursuit

Licorice is an established medicinal plant with significant antimicrobial activity due to the presence of more than 100 bacterial endophytes which were isolated from it in a research project held in China in 2018. Among these endophytes, *B. atrophaeus* and *B. mojavensis* showed a positive correlation between their proportion and the antifungal as well as antibacterial activities of Licorice.

Endophytes of *Thymus vulgaris* herb were proved to be good antioxidants beneficial for plant growth. They also helped in controlling the salinity in tomato crops [41].

The antagonistic perspective of endophytes has a mutual type of symbiosis with medicinal plants as endophytes get shelter inside the plant whereas yield and antimicrobial activities are enhanced in plants. White ginger lily is useful in treating diabetes, nasal infections, and anti rheumatism. Its antimicrobial properties are attributed to its antagonistic endophytic fungi which is a resident of its rhizome [42].

SECONDARY METABOLITES PRODUCTION IN MEDICINAL PLANTS BY FUNGI SYMBIOSIS IN SOIL

Secondary metabolites in therapeutic plants are developed as a result of subordinate reactions taking place inside them. They are thought be helpful in relieving disorders ranging from headaches to cancer. Catechins, flavonoids, alkaloids, phenols, and Coumarins are some of the important secondary metabolites found in conventional medicinal plants [43].

Secondary metabolites are small, biologically active chemical compounds that are derivatives of primary metabolites (Fig. **3**). According to a review published in the international journal of pharmaceutical sciences, secondary metabolites are abundant in medicinal plants and have important specialized functions for plants [44].

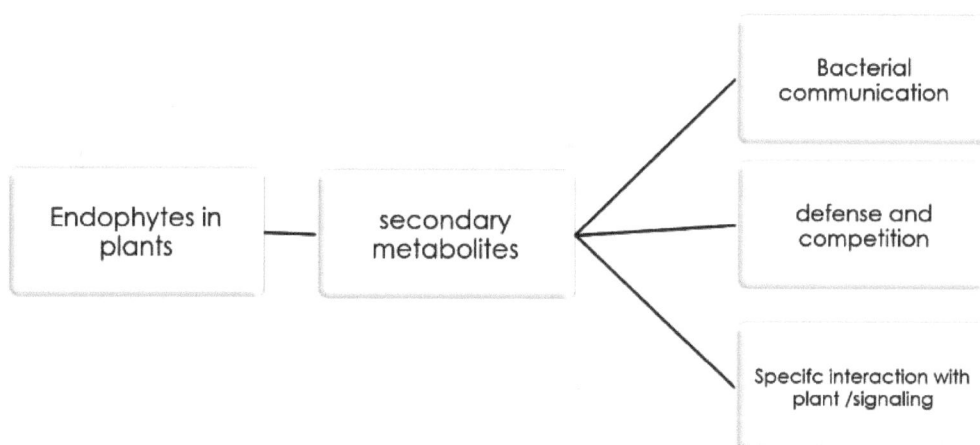

Fig. (3). Role of endophytes in the production and enhancement of secondary metabolites and their benefits in plants.

Plants follow certain subordinated pathways for the production of secondary metabolites including glycolysis and the shikimic acid pathway. Additionally, environmental factors also have an important role in their synthesis. Fungi are found on every plant and on almost every specie of medicinal plant too. They are quite valuable in agriculture. A fungus called Epicoccum is responsible for the production of a secondary metabolite. Epicorazien is known for its antibacterial functionality in therapeutic plants. Trichoderma is another fungus that also produces secondary metabolites necessary for plant growth [45].

Mycorrhizal fungus especially its hyphae are an important constituent of the soil microbiota of plants. Its main function is to distribute carbon evenly in the soil. The excessive supply of carbon may become harmful for the plant so this surplus amount of carbon is used to make sugars and secondary metabolites with the help of mycorrhizal fungi found in soil [46].

The soil of dates tree of temperate climatic regions was also found to be rich in the *Aspergillus* fungus specie which can restrict the growth of non-symbiotic yeast and bacteria and hence preserves the important metabolic components of the plant for its survival [47].

MEDICINAL PLANTS ASSOCIATED MICROFLORA AS AN UNEXPLORED NICHE OF BIOPESTICIDE

A gram of soil has microflora of 40 million prokaryotes, including algae, fungi, bacteria, protozoans and viruses. These microorganisms undoubtedly play a crucial role in plant physiology. Root Microflora of medicinal plants develops an interaction between soil and nutrition reuse also in the putrefaction of organic substances. They also act as signal transmitters for ecological alteration taking place in the soil [48].

The presence of microflora can be affected by climatic changes like temperature, supply of water, nutrition, change in pH or salt stress. On the basis of the pH of a plant, soil microbes can be divided into three classes, namely alkaliphiles, acidophiles, and neutrophils. The high amount of *Herbivorax saccincola*, anaerobic bacteria with an alkaliphilic as well as thermophilic nature was found to be involved in the biomass, especially lignocellulose degradation. *B. marisflavi* is also an alkaliphilic bacterium that helps in the completion of the cyclic process of phosphate solubilization. During phosphate solubilization, the Orthophosphate ions or the non-soluble organic and inorganic phosphate molecules are transformed through hydrolyzation into the soluble phosphate molecules, essentially needed for plant growth [49]. Fungal species also play an important role as a large constituent of microflora in medicinal plants like *C. gigantea* and milkweed for the treatment of asthma, haemorrhoids, and jaundice respectively. In medicinal plants, fungi control drought conditions and allow the plant to cope up with several biotic and abiotic stresses [50].

Unintentionally, the strongly formulated chemical pesticides harm the symbiotic microbial flora, friendly pests, and insects of medicinal plants. One should have a vast knowledge of chemical interactions, and beneficial or hazardous effects of the additional doses or usage of these pesticides. These deleterious impacts can be abolished by the use of biopesticides. In this regard, microalgae are proven to have a significant capacity to kill injurious organisms from plants and soil [51].

Yeast possesses anti-fungal, bugs killing as well as herbicidal properties, hence it can be a good pesticide that can enhance the protective mechanism of plants. *B. thuringiensis* acts as a natural pesticide against pulse beetles. *R. Simulans* is a plant-damaging pest that can be controlled by using *B. subtilis* and more effectively by *B. brevis* bacteria [52]. Biopesticides are more beneficial rather than traditional synthetically prepared chemical fertilizers and pesticides because they are available enormously in nature, less expensive, rarely toxic to other living things, have several modes of action, and can degrade naturally [53].

CONCLUSION

Microorganisms (bacteria and fungi) residing in the soil collectively make the rhizosphere or the soil environment for plants. The health and growth of plants are dependent on the profile and activity of root microbes specially on their fungal species. The microflora especially fungal flora in the roots or rhizosphere of medicinal plants is distinctive as they have an exclusive role in the production and enhancement of biologically significant secondary metabolites. Soil fungi not only act as a decomposer but are also a necessary symbiont in the form of mycorrhizal fungus. Because of the mycorrhizal fungus, a plant can collect more nutrition and water.

Disease, drought, and stress are easily combated by the presence of soil fungi in roots. Moreover, it is found to be the best natural pesticide based on extensive research and experiments. *Metarhizium* increases magnesium availability in the soil which in turn ruptures the insect cuticle, resulting in the death of the insect. Entomopathogenic fungi such as *Ascomycetes* can kill both above and below-the ground pests. Fungal biopesticide is approved to be a better option by many researchers; according to them, it is harmless for field cattle and is eco-friendly. Beauveria kills a range of insects including potato beetle, Pecan and kudzu bug. *Entomophthorales* is a group of pesticidal fungi for aphids. Collectively, root fungi are important for plant health, and plant defence, thus helping to combat abiotic and biotic stresses, and diseases, increasing secondary metabolites' content and enhancing plant hormone profile.

REFERENCES

[1]   Bhardwaj S, Verma R, Gupta J. Challenges and future prospects of herbal medicine. Int Res Med Health Sci 2018; 1(1): 12-5.
[http://dx.doi.org/10.36437/irmhs.2018.1.1.D]

[2]   Kuruppu AI, Paranagama P, Goonasekara CL. Medicinal plants commonly used against cancer in traditional medicine formulae in Sri Lanka. Saudi Pharm J 2019; 27(4): 565-73.
[http://dx.doi.org/10.1016/j.jsps.2019.02.004] [PMID: 31061626]

[3]   Tungmunnithum D, Thongboonyou A, Pholboon A, Yangsabai A. Flavonoids and other phenolic compounds from medicinal plants for pharmaceutical and medical aspects: An overview. Medicines 2018; 5(3): 93.
[http://dx.doi.org/10.3390/medicines5030093] [PMID: 30149600]

[4]   Ozioma E-OJ, Chinwe OAN. Herbal medicines in African traditional medicine. Herb Med 2019; 10: 191-214.

[5]   Kambai C, Joshua VI, Olatidoye OR, Yakubu CK, Adaaja BO, Olaniyi JJ. Comparative study of soil bacteria from the rhizosphere of two selected tree species (anogeissus leiocarpa and pterocarpus erinaceus) in shere hills, plateau state, nigeria. J Appl Sci Environ Manag 2021; 25(7): 1147-53.
[http://dx.doi.org/10.4314/jasem.v25i7.7]

[6]   Shabir T, Baba ZA, Dar SA, Husaini AM. Improving plant growth and quality of plant-products: An interplay of plant- microbe interaction. Int J Curr Microbiol Appl Sci 2020; 9(11): 3759-66.
[http://dx.doi.org/10.20546/ijcmas.2020.911.451]

[7]   Mondal S, Halder SK, Yadav AN, Mondal KC. Microbial consortium with multifunctional plant growth-promoting attributes: Future perspective in agriculture. Adv plant microbiome sustainagricul 2020; 219-58.

[8]   Shi S, Chang J, Tian L, *et al.* Comparative analysis of the rhizomicrobiome of the wild versus cultivated crop: Insights from rice and soybean. Arch Microbiol 2019; 201(7): 879-88.
[http://dx.doi.org/10.1007/s00203-019-01638-8] [PMID: 30963196]

[9]   Dixit VK, Misra S, Mishra SK, Joshi N, Chauhan PS. Rhizobacteria-mediated bioremediation: insights

and future perspectives.Soil Bioremediation: An Approach Towards Sustainable Technology. Wiley Online Library 2021.
[http://dx.doi.org/10.1002/9781119547976.ch9]

[10] Barragán-Fonseca KY, Nurfikari A, van de Zande EM, *et al.* Insect frass and exuviae to promote plant growth and health. Trends Plant Sci 2022; 27(7): 646-54.
[http://dx.doi.org/10.1016/j.tplants.2022.01.007] [PMID: 35248491]

[11] Kumar A, Dubey A. Rhizosphere microbiome: Engineering bacterial competitiveness for enhancing crop production. J Adv Res 2020; 24: 337-52.
[http://dx.doi.org/10.1016/j.jare.2020.04.014] [PMID: 32461810]

[12] Poria V, Singh S, Nain L, Singh B, Saini JK. Rhizospheric microbial communities: Occurrence, distribution, and functions.Microbial Metatranscriptomics Belowground. Singapore: Springer 2021; pp. 239-71.
[http://dx.doi.org/10.1007/978-981-15-9758-9_12]

[13] Chandra P, Singh R. soil salinity and its alleviation using Plant Growth–Promoting Fungi.In Agriculturally Important Fungi for Sustainable Agriculture. Cham: Springer 2020; pp. 101-48.

[14] Vlot AC, Sales JH, Lenk M, *et al.* Systemic propagation of immunity in plants. New Phytol 2021; 229(3): 1234-50.
[http://dx.doi.org/10.1111/nph.16953] [PMID: 32978988]

[15] Basavaraj GL, Murali M, Lavanya SN, Amruthesh KN. Seed priming with biotic agents invokes defense response and enhances plant growth in pearl millet upon infection with Magnaporthe grisea. Biocatal Agric Biotechnol 2019; 21: 101279.
[http://dx.doi.org/10.1016/j.bcab.2019.101279]

[16] Shahid A, Ali S, Zahra T, *et al.* Influence of microbes in progression of cancer and dna damaging effects. haya: The Saudi. J Life Sci 2020.

[17] Momtazi-Borojeni AA, Ghasemi F, Hesari A, Majeed M, Caraglia M, Sahebkar A. Anti-cancer and radio-sensitizing effects of curcumin in nasopharyngeal carcinoma. Curr Pharm Des 2018; 24(19): 2121-8.
[http://dx.doi.org/10.2174/1381612824666180522105202] [PMID: 29788875]

[18] Scherlach K, Hertweck C. Chemical mediators at the bacterial-fungal interface. Annu Rev Microbiol 2020; 74(1): 267-90.
[http://dx.doi.org/10.1146/annurev-micro-012420-081224] [PMID: 32660387]

[19] Philippot L, Raaijmakers JM, Lemanceau P, van der Putten WH. Going back to the roots: The microbial ecology of the rhizosphere. Nat Rev Microbiol 2013; 11(11): 789-99.
[http://dx.doi.org/10.1038/nrmicro3109] [PMID: 24056930]

[20] Song Z, Bi Y, Zhang J, Gong Y, Yang H. Author Correction: Arbuscular mycorrhizal fungi promote the growth of plants in the mining associated clay. Sci Rep 2020; 10(1): 18373.
[http://dx.doi.org/10.1038/s41598-020-75521-8] [PMID: 31913322]

[21] Nilsson RH, Anslan S, Bahram M, Wurzbacher C, Baldrian P, Tedersoo L. Mycobiome diversity: High-throughput sequencing and identification of fungi. Nat Rev Microbiol 2019; 17(2): 95-109.
[http://dx.doi.org/10.1038/s41579-018-0116-y] [PMID: 30442909]

[22] Applebaum I, Jeyaraman M, Sherman C, Doniger T, Steinberger Y. Structure and function of the soil rhizosphere fungal communities in medicinal plants : A preliminary study. Agri 2022; 12(2): 152.

[23] Li X, de Boer W, Zhang Y, Ding C, Zhang T, Wang X. Suppression of soil-borne Fusarium pathogens of peanut by intercropping with the medicinal herb Atractylodes lancea. Soil Biol Biochem 2018; 116: 120-30.
[http://dx.doi.org/10.1016/j.soilbio.2017.09.029]

[24] Khan AL, Asaf S, Abed RMM, *et al.* Rhizosphere microbiome of arid land medicinal plants and extra cellular enzymes contribute to their abundance. Microorganisms 2020; 8(2): 213.

[http://dx.doi.org/10.3390/microorganisms8020213] [PMID: 32033333]

[25]   Villalobos-Flores LE, Espinosa-Torres SD, Hernández-Quiroz F, *et al.* The bacterial and fungal microbiota of the mexican rubiaceae family medicinal plant bouvardia ternifolia. Microb Ecol 2021; 84(2): 510-26.
[PMID: 34553243]

[26]   Kushwaha RK, Rodrigues V, Kumar V, Patel H, Raina M, Kumar D. Soil microbes-medical plants interactions: Ecological diversity and future prospects.Plant Microbiome Symbiosis. Cham, Switzerland: Springer 2020; pp. 263-86.
[http://dx.doi.org/10.1007/978-3-030-36248-5_14]

[27]   Kuzyakov Y, Razavi BS. Rhizosphere size and shape: Temporal dynamics and spatial stationarity. Soil Biol Biochem 2019; 135: 343-60.
[http://dx.doi.org/10.1016/j.soilbio.2019.05.011]

[28]   Wang XQ, Zhao DL, Shen LL, Jing CL, Zhang CS. Application and mechanisms of Bacillus subtilis in biological control of plant disease.Role of rhizospheric microbes in soil. Singapore: Springer 2018; pp. 225-50.
[http://dx.doi.org/10.1007/978-981-10-8402-7_9]

[29]   Luchi N, Ioos R, Santini A. Fast and reliable molecular methods to detect fungal pathogens in woody plants. Appl Microbiol Biotechnol 2020; 104(6): 2453-68.
[http://dx.doi.org/10.1007/s00253-020-10395-4] [PMID: 32006049]

[30]   Collinge HJL, Jørgensen M, Latz EC, Rojas A, Manzotti B, Jensen N. Fani Searching for novel fungal biological control agents for plant disease control among endophytes-T.Hodkinson F, Doohan M, Saunders B. Endophytes a Growing World. Cambridge Univ 2019.

[31]   Latz MAC, Jensen B, Collinge DB, Lyngs Jørgensen HJ. Identification of two endophytic fungi that control Septoria tritici blotch in the field, using a structured screening approach. Biol Control 2020; 141: 104128.
[http://dx.doi.org/10.1016/j.biocontrol.2019.104128]

[32]   Sharma S, Sharma A, Sagar MK, Kumar A. Herbal plants: Source of medicinal value (A Review). Int J Pharmacol Clin Res 2019; 1(1): 9-11.

[33]   Yeşil Y, İnal İ. Ethnomedicinal plants of hasankeyf (batman-turkey). Front Pharmacol 2021; 11: 624710.
[http://dx.doi.org/10.3389/fphar.2020.624710] [PMID: 33776756]

[34]   Abubakar A, Haque M. Preparation of medicinal plants: Basic extraction and fractionation procedures for experimental purposes. J Pharm Bioallied Sci 2020; 12(1): 1-10.
[http://dx.doi.org/10.4103/jpbs.JPBS_175_19] [PMID: 32801594]

[35]   Charmakar S, Kunwar RM, Sharma HP, *et al.* Production, distribution, use and trade of Valeriana jatamansi Jones in Nepal. Glob Ecol Conserv 2021; 30: e01792.
[http://dx.doi.org/10.1016/j.gecco.2021.e01792]

[36]   Applequist WL, Brinckmann JA, Cunningham AB, *et al.* Scientists' warning on climate change and medicinal plants. Planta Med 2020; 86(1): 10-8.
[http://dx.doi.org/10.1055/a-1041-3406] [PMID: 31731314]

[37]   Li J, Wu J, Peng K, *et al.* Simulating the effects of climate change across the geographical distribution of two medicinal plants in the genus *Nardostachys*. PeerJ 2019; 7: e6730.
[http://dx.doi.org/10.7717/peerj.6730] [PMID: 31024763]

[38]   Casciaro B, Calcaterra A, Cappiello F, *et al.* Nigritanine as a new potential antimicrobial alkaloid for the treatment of *staphylococcus aureus*-induced infections. Toxins 2019; 11(9): 511.
[http://dx.doi.org/10.3390/toxins11090511] [PMID: 31480508]

[39]   Alonso-Esteban JI, Pinela J, Barros L, *et al.* Phenolic composition and antioxidant, antimicrobial and cytotoxic properties of hop (Humulus lupulus L.) Seeds. Ind Crops Prod 2019; 134: 154-9.

[http://dx.doi.org/10.1016/j.indcrop.2019.04.001]

[40]   Khameneh B, Iranshahy M, Soheili V, Fazly Bazzaz BS. Review on plant antimicrobials: A mechanistic viewpoint. Antimicrob Resist Infect Control 2019; 8(1): 118.
[http://dx.doi.org/10.1186/s13756-019-0559-6] [PMID: 31346459]

[41]   Abdelshafy Mohamad OA, Ma JB, Liu YH, *et al.* Beneficial endophytic bacterial populations associated with medicinal plant Thymus vulgaris alleviate salt stress and confer resistance to Fusarium oxysporum. Front Plant Sci 2020; 11: 47.
[http://dx.doi.org/10.3389/fpls.2020.00047] [PMID: 32117385]

[42]   Lutfia A, Munir E, Yurnaliza . Antagonistic endophytic fungi of hedychium coronarium j. koenig from hutan sibayak and taman hutan raya, north sumatra against staphylococcus aureus atcc® 29213 ™. IOP Conf Ser Earth Environ Sci 2019; 305(1): 012002.
[http://dx.doi.org/10.1088/1755-1315/305/1/012002]

[43]   Szewczyk A, Mirek A, Mateja K, Bętkowska S, Kwiecień A, Ekiert H. The accumulation of secondary metabolites in *in vitro* cultures of some species of rue. Acta Biol Cracoviensia 2019; 31.

[44]   Jain C, Khatana S, Vijayvergia R. Bioactivity of secondary metabolites of various plants: A review. Int J Pharm Sci Res 2019; 10(2): 494-504.

[45]   Elkhateeb WA, Daba GM. Where to Find? A report for some terrestrial fungal isolates, and selected applications using fungal secondary metabolites. Biomed J 2018; 2: 4.

[46]   Prescott CE, Grayston SJ, Helmisaari HS, *et al.* Surplus carbon drives allocation and plant–soil interactions. Trends Ecol Evol 2020; 35(12): 1110-8.
[http://dx.doi.org/10.1016/j.tree.2020.08.007] [PMID: 32928565]

[47]   Orfali R, Perveen S. Secondary metabolites from the Aspergillus sp. in the rhizosphere soil of Phoenix dactylifera (Palm tree). BMC Chem 2019; 13(1): 103.
[http://dx.doi.org/10.1186/s13065-019-0624-5] [PMID: 31410414]

[48]   Toor MD, Adnan M. Role of soil microbes in agriculture : A Review. Open access J BiogenericRes 2020; 10.

[49]   Prabhu N, Borkar S, Garg S. Phosphate solubilization mechanisms in alkaliphilic bacterium Bacillus marisflavi FA7. Curr Sci 2018; 114(4): 845-53.
[http://dx.doi.org/10.18520/cs/v114/i04/845-853]

[50]   Yadav G, Meena M. Bioprospecting of endophytes in medicinal plants of thar desert: An attractive resource for biopharmaceuticals. Biotechnol Rep 2021; 30: e00629.
[http://dx.doi.org/10.1016/j.btre.2021.e00629] [PMID: 34136363]

[51]   Costa JAV, Freitas BCB, Cruz CG, Silveira J, Morais MG. Potential of microalgae as biopesticides to contribute to sustainable agriculture and environmental development. J Environ Sci Health B 2019; 54(5): 366-75.
[http://dx.doi.org/10.1080/03601234.2019.1571366] [PMID: 30729858]

[52]   Gokturk T, Tozlu E, Kotan R. Prospects of entomopathogenic bacteria and fungi for biological control of Ricania simulans (Walker 1851) (Hemiptera: Ricaniidae). Pak J Zool 2018; 50.

[53]   Thakur N, Kaur S, Tomar P, Thakur S, Yadav AN. Microbial biopesticides: Current status and advancement for sustainable agriculture and environment.New and Future Developments In Microbial Biotechnology And Bioengineering. Elsevier 2020; pp. 243-82.
[http://dx.doi.org/10.1016/B978-0-12-820526-6.00016-6]

Endophytic Bacteria: Value Addition in Medicinal Plants

Ahmad Mahmood[1,*], Hafiz Shahzad Ahmad[1], Waleed Asghar[2], Tanveer ul Haq[1], Ali Hamid[3], Zulqurnain Khan[5], Oğuz Can Turgay[4] and Ryota Kataoka[2]

[1] *Department of Soil and Environmental Sciences, Muhammad Nawaz Shareef University of Agriculture, Multan, Pakistan*

[2] *Department of Environmental Sciences, Faculty of Life & Environmental Sciences, University of Yamanashi, Takeda, Kofu, Yamanashi, Japan*

[3] *Institute of Plant Breeding and Biotechnology, Muhammad Nawaz Shareef University of Agriculture, Multan, Pakistan*

[4] *Department of Soil Science and Plant Nutrition, Ankara University, Ankara, Turkey*

[5] *Department of Biotechnology, Institute of Plant Breeding and Biotechnology (IPBB), Muhammad Nawaz Shareef University of Agriculture, Multan-60000, Pakistan*

Abstract: Plants and microorganisms have long evolved together and our most recent discoveries using advanced techniques have allowed us to understand the chemical interface within the plant further explaining the relationship between them. As we discover the interaction between the plants and the associated microorganisms, it has been clearer to us that there has been a synergy more evident than that of antagonism among them. A lot of chemicals or metabolites are being released favouring both the host as well as the microbes during this contact. Such an interaction which leads to the release of certain metabolites can be managed and manipulated in bringing about positive effects for the biosphere and environment. One instance of this positive effect can be the use of medicinal plants and their microbe-facilitated associated metabolites which may be regulated through the application of different endophytic microorganisms. If we can control the release of different metabolites from plants particularly from those of medicinal plants, we can harvest significant benefits for human and animal health as we are utilizing endophytes for their role as biofertilizers. The food for medicine concept has been emerging and requires quick and efficient identification of metabolites as well as chemicals that may be used in addressing multiple diseases in human beings and other animals.

Keywords: Chemical interface, Endophytic bacteria, Metabolites, Medicinal plants.

---
[*] **Corresponding author Ahmad Mahmood:** Department of Soil and Environmental Sciences, Muhammad Nawaz Shareef University of Agriculture, Multan, Pakistan; E-mail: ahmad.mahmood@mnsuam.edu.pk

Zulqurnain Khan, Azra Yasmin & Naila Safdar (Eds.)
All rights reserved-© 2023 Bentham Science Publishers

INTRODUCTION

Human beings have long been depending on different natural resources for their survival, among which plants have played a significant role. Plants alongside being the food have also served as fibre as well as medicine needed for many types of diseases. It is a blessing to have these plants from which a multitude of diseases are cured through the metabolites released by these plants [1-4]. The field of medicine has already progressed in identifying many of these chemicals, and harvested them successfully from a wide variety of plants. Several classes, and purposes have already been defined for the said metabolites [5, 6] and more recently when COVID-19 struck, the focus was again shifted towards plants and their metabolites in finding a quick cure and remedies [7, 8]

In a more recent history, it has been observed that plants as well as microorganisms work in synergy in producing different metabolites [9 - 11]. For instance, some of the microorganisms when associated with different plants lead to the production of certain chemicals; which sometimes are released by the plants while on other instances, these metabolites are produced by the microorganisms. The advancements in technology have deciphered the chemical interface between host plants and microorganisms, and we can trace back to the original source of these metabolites. For example, microbes release certain types of metabolites for their metabolism, in their defence, in their host's defence, in facilitating their hosts, and in accomplishing their niche. Similarly, plants produce different primary and secondary metabolites serving various purposes: plant metabolism, recruiting beneficial microbes, facilitating nutrient uptake, and restricting different molecules' entry among others. The chemical interface for both entities is complex, with the combined responses, to which we are recently exposed.

The microorganisms produce different primary compounds; alcohols, organic acids, amino acids, nucleotides, and polysaccharides, and secondary compounds; bacteriocins, peptide antibiotics, polyketides, and cyclic lipopeptides among others. On the other hand, plants produce carbohydrates, lipids, proteins, and nucleic acids among the primary chemicals, while alkaloids, phenolics, sterols, steroids, essential oils and lignin in the secondary chemicals. All these secretions within the plant or in the environment possess multiple functions all the way from plant health to animal, microbial and environmental health.

A class of plants used for their therapeutic functions termed usually medicinal plants have a long history and have been serving animals. Either these medicinal plants or the microbes associated with them lead to the production of the chemicals which are directly or indirectly utilized for the cure of certain diseases. The role of microbes particularly those of endophytes has been receiving an increased focus due to the fact that they are in a closer interaction with the host plants and may allow us to regulate such chemicals for our use. This chapter focuses on the identification of these endophytes, their niche and roles, alongside putting forward their role in medicinal plants' metabolite regulation.

WHAT ARE ENDOPHYTES

Different niches of microorganisms have been defined and discovered as we go along further discoveries in this realm (Fig. **1**). Particularly in the plant sphere, the microbes can reside in the bulk soil, rhizosphere, phyllosphere, and endosphere. The phyllosphere can further be divided into different plant parts, and recent discoveries have put forward the microbiome incident in the flowering parts. Microorganisms are present in all parts of the plant especially those of roots, stem and leaves; where the majority of them are incident in the roots [12, 13]. The presence of endophytic microorganisms in the roots or the rootzone ensures that they are in close interaction with the plants, thus helping the plants in overcoming different kinds of stresses which may include biotic and abiotic issues. It has been widely established that there have been vertical as well as horizontal transfer of microbes *i.e.*, from one generation of the plants to the next, or from one plant to another plant [14 - 16]. Although the seeds which are dehydrated may have the presence of endophytic microbes [17, 18] which may reside in a dormant phase before they become functional when the plants need them and provide them with particular carbohydrates or exudates.

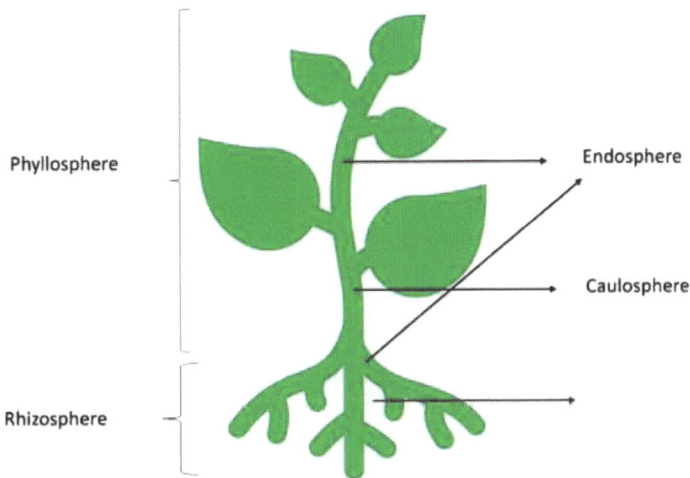

Phyllosphere

Endosphere

Caulosphere

Rhizosphere

Fig. (1). Incidence of plant microbiome in different spheres of the plant.

Endophytic bacteria are further classified into culturable and non-culturable where the culturable can be cultured in laboratory media, however, the non-culturable endophytic bacteria are unable to be cultured. It has been believed that culturable endophytic bacteria constitute only a small part of the large community which is mainly non-culturable. However, both of them may have been contributing significantly towards plant health. The next-generation sequencing and other lateral techniques have led to further discoveries of endophytic microbial classes which have never been noticed before and they have been playing significant supports towards plant growth [19 - 22]. However, the comprehension of their mechanisms or role in plant growth and health is still in the primitive stage. As we go along and discover more through the utilization of OMIC approaches such as transcriptomics, proteomics, metabolomics and metagenomics, we are exploring more and more of this black box.

Endophytic bacteria have been supporting the plants through various mechanisms which include direct and indirect mechanisms. The direct mechanisms of the rhizosphere microorganisms include nitrogen fixation, phosphate and potassium solubilization, and production of different hormones and antibiotics, among others, while the indirect mechanisms include iron chelation, 1-aminocyclopropane-1-carboxylate deaminase (ACC) deaminase activity and biocontrol of pathogens among others [23]. The endophytic bacteria are in closer interaction with host plants as compared to their counterparts residing in bulk soil as well as the rhizosphere. This closer interaction ensures that they have priority in being recruited as well as supporting plant growth and development. It is further elaborated that endophytic bacteria are commonly a subset of bulk soil or rhizosphere bacteria, which with the establishment of plants lead their way into the endosphere. This property of endophytic bacteria also paves the way towards their application into the soil, which ultimately leads to their entry into the plants. Previously, it has been discussed that the endophytic bacteria can be applied directly to the seeds, however, soil application may also be effective considering the previously discussed recruiting mechanisms. In case of seedlings, dipping the seedlings into the endophytic bacterial inoculums may also have the answer towards the quick entry of endophytes into the plants. All these capabilities make them a priority for the plants and may lead the future biofertilizer development initiatives.

Role of Endophytic Bacteria in Metabolite Production

It has widely been discussed that the endophytic bacteria may lead to the production of different metabolites within the plants [24 - 26]. They can also decrease the concentration of different chemicals through mechanisms unknown to us at the moment. In the concept of quorum sensing as well as gene switching on and off by the incident of different endophytic bacteria, it can be estimated that bacteria may lead to the production of different chemicals within the plant and help us to control the release of different metabolites within the plant (Table **1**).

Table 1. Role of endophytic bacteria in metabolite regulation.

Bacteria	Isolation Source	Metabolite	References
Lysinibacillus sp. and *Bacillus cereus*	*Miqueliadentata Bedd*	Camptothecine	[27]
Streptomyces sp.	Mangroves	Indolocarbazoles	[28]
Streptomyces sp.	*Isodon eriocalyx*	Anthracyclin	[29]
Streptomyces sp.	*Aplysina fistularis*	Saadamycin	[30]
Streptomyces sp.	Mangroves	Xiamycin	[31]

As discussed above, some bacteria may lead to the secretion and excretion of different metabolites aimed at different functions. The production and release of the metabolites within as well as outside of the plants may be affected by different factors such as prevalent bacterial community, niche of the incident endophytic bacteria, host plant's requirements, secretion of exudates by the plants, and environmental factors. Plants being superior in this interaction put their requirements ahead and opt to release certain chemicals which are sensed by the present endophytic bacteria and in response, the bacteria produce or regulate the required chemicals. The environment also plays a significant role in the functionality of the endophytic bacteria. For example, if the endosphere uptakes the pollutants, that may lead to lower or no functionality of the endophytic bacteria. Therefore, in efforts of harnessing the discussed benefits we may need to look beyond the conventional isolation and characterization techniques and investigate the whole picture covering the plant, endophytic bacterial community as well as the niche.

Endophytic Bacteria and Medicinal Plants

The focus has recently been shifted towards 'food for medicine' and in these circumstances, endophytes residing within the plants may help us in regulating different chemicals inside the plants. In recent studies, a lot of medicinal plants have been the isolation source of endophytic bacteria where significant production *in vitro* or *in planta* has been observed through the support from endophytic bacteria (Table **2**).

Table 2. Role of endophytic bacteria in metabolite regulation in medicinal plants.

Bacteria	Host Plant	Metabolites Regulated	References
• *Streptomyces* sp. • *Aeromicrobium ponti* • *Streptomyces* sp. • *Microbacterium* sp. • *Kytococcus schroeteri* • *Streptomyces* sp. • *Pseudomonas fluorescens* • *Microbacterium* sp. • *Burkholderia* sp.	• *Allium tuberosum* • *Vochysia divergens* • *Glycine max* • *Catharanthus roseus* • *Ephedra foliata* • *Lychnophora ericoides* • *Atractylodes lancea* • *Coptis teeta*	Alkaloids	[32-39]
• *Streptomyces* sp. • *Guignardia mangiferae*	• *Bruguiera gymnorrhiza* • *Gelsemium elegans*	Sesquiterpenes	[39]
• *Guignardia mangiferae*	• *Gelsemium elegans*	Polyketones	[40-41]
• *Streptomyces* sp. • *Paenibacillus polymyxa*	• *Aucuba japonica* • *Ephedra foliata* • *Maytenus hookeri*	Lactones	[42]
• *Nocardiopsis* sp. • *Bacillus atrophaeus* • *Bacillus mojavensis*	• *Zingiber officinale* • *Glycyrrhiza uralensis*	Organic acids	[42- 44]
• *Streptomyces* sp.	• *Inula cappa* • *Bruguiera gymnorrhiza*	Cyclopeptides	[45]
• *Streptomyces* sp.	• *Boesenbergia rotunda*	Flavonoids	[46]
• *Burkholderia* sp. • *Agrobacterium* sp.	• *Panax ginseng*	Saponins	[47]

Considering the diverse metabolite groups, endophytic bacteria and isolation sources, it can be said that this synergy of endophytic bacteria with those of their host plants possesses an excellent potential which needs to be exploited. It needs to be more about the diversity, than the quantity of the metabolites which must be taken into account as we come across unprecedented diseases and health issues in animals including human beings.

PROSPECTS AND CONCLUSION

The major focus has been recently shifted towards using foods as medicine, therefore in the coming future, we may need to look beyond just the use of plants, but the metabolites they possess which play a major role in addressing the diseases as well as other health issues. Sometimes, the quantity of our desired metabolites within the plants is not that much high and we need to produce more and more plants to meet such high requirements. Therefore, if we can try to regulate the metabolites by the use of different microorganisms particularly those of endophytes, we may have an edge towards realization of the food for medicine concept. The endophytic bacteria-assisted regulation of different metabolites within the medicinal plants may lead to significant progress in medication of different diseases. For this, we need to look toward the metabolite profiling of the said plants which will not only lead to identification of new chemicals but also help us identify their regulation within the plants. Such an intervention in the medicinal plants will be a great addition to the science as well as to the society. However, research into diverse plants, integration of techniques in understanding the complete picture, and efforts to uncover the hidden capabilities of culturable as well as non-culturable endophytic bacteria may provide answers to the questions we have been looking and striving for.

REFERENCES

[1]   González-Lamothe R, Mitchell G, Gattuso M, Diarra M, Malouin F, Bouarab K. Plant antimicrobial agents and their effects on plant and human pathogens. Int J Mol Sci 2009; 10(8): 3400-19.
[http://dx.doi.org/10.3390/ijms10083400] [PMID: 20111686]

[2]   Raghuveer I, Anurag K, Anumalik Y, *et al*. Metabolites in plants and its classification. World J Pharm Pharm Sci 2015; 4: 287-305.

[3]   Bhuiyan FR, Howlader S, Raihan T, Hasan M. Plants metabolites: Possibility of natural therapeutics against the COVID-19 pandemic. Front Med 2020; 7: 444.
[http://dx.doi.org/10.3389/fmed.2020.00444] [PMID: 32850918]

[4]   Mahmood A, Kataoka R. Metabolite profiling reveals a complex response of plants to application of plant growth-promoting endophytic bacteria. Microbiol Res 2020; 234: 126421.
[http://dx.doi.org/10.1016/j.micres.2020.126421] [PMID: 32006789]

[5]   Kabera JN, Semana E, Mussa AR, He X. Plant secondary metabolites: Biosynthesis, classification, function and pharmacological properties. J Pharm Pharmacol 2014; 2: 377-92.

[6]   Shin SA, Moon SY, Kim WY, Paek SM, Park HH, Lee CS. Structure-based classification and anti-cancer effects of plant metabolites. Int J Mol Sci 2018; 19(9): 2651.
[http://dx.doi.org/10.3390/ijms19092651] [PMID: 30200668]

[7]   Alam S, Sarker MMR, Afrin S, *et al*. Traditional herbal medicines, bioactive metabolites, and plant products against COVID-19: Update on clinical trials and mechanism of actions. Front Pharmacol 2021; 12: 671498.
[http://dx.doi.org/10.3389/fphar.2021.671498] [PMID: 34122096]

[8]   Khan T, Khan MA, Karam K, *et al*. Plant *in vitro* culture technologies; a promise into factories of secondary metabolites against COVID-19. Front Plant Sci 2021; 12: 610194.
[http://dx.doi.org/10.3389/fpls.2021.610194]

[9]   Wurtzel ET, Kutchan TM. Plant metabolism, the diverse chemistry set of the future. Science 2016; 353(6305): 1232-6.
[http://dx.doi.org/10.1126/science.aad2062] [PMID: 27634523]

[10]   Singh R, Kumar M, Mittal A, Mehta PK. Microbial metabolites in nutrition, healthcare and agriculture. Biotech 2017; 7: 1-14.

[11]   Mahmood A, Kataoka R. Application of endophytes through seed priming.Priming and Pretreatment of Seeds and Seedlings: Implication in Plant Stress Tolerance and Enhancing Productivity in Crop Plants. Springer Singapore, Singapore 2019; pp. 509-21.
[http://dx.doi.org/10.1007/978-981-13-8625-1_25]

[12]   Simmons T, Caddell DF, Deng S, Coleman-Derr D. Exploring the root microbiome: Extracting bacterial community data from the soil, rhizosphere, and root endosphere. J of Visualized Exp 2018; 135: 57561.

[13]   Mahmood A, Takagi K, Ito K, Kataoka R. Changes in endophytic bacterial communities during different growth stages of cucumber *Cucumis sativus L.*. World J Microbiol Biotechnol 2019; 35(7): 104.
[http://dx.doi.org/10.1007/s11274-019-2676-z] [PMID: 31236765]

[14]   Truyens S, Weyens N, Cuypers A, Vangronsveld J. Bacterial seed endophytes: genera, vertical transmission and interaction with plants. Environ Microbiol Rep 2015; 7(1): 40-50.
[http://dx.doi.org/10.1111/1758-2229.12181]

[15]   Frank A, Saldierna Guzmán J, Shay J. Transmission of bacterial endophytes. Microorganisms 2017; 5(4): 70.
[http://dx.doi.org/10.3390/microorganisms5040070] [PMID: 29125552]

[16]   Samreen T, Naveed M, Nazir MZ, *et al.* Seed associated bacterial and fungal endophytes: Diversity, life cycle, transmission, and application potential. Appl Soil Ecol 2021; 168: 104191.
[http://dx.doi.org/10.1016/j.apsoil.2021.104191]

[17]   Rosenblueth M, López-López A, Martínez J, Rogel M, Toledo I, Martínez-Romero E. Seed bacterial endophytes: Common genera, seed-to-seed variability and their possible role in plants. XXVIII International Horticultural Congress on Science and Horticulture for People (IHC2010): International Symposium on. 938: 39-48.

[18]   Verma SK, White J. The role of seed-vectored endophytes in seedling development and establishment. Seed endophytes Springer 2019; 10: 973-8.
[http://dx.doi.org/10.1007/978-3-030-10504-4]

[19]   Muresu R, Polone E, Sulas L, *et al.* Coexistence of predominantly nonculturable rhizobia with diverse, endophytic bacterial taxa within nodules of wild legumes. FEMS Microbiol Ecol 2008; 63(3): 383-400.
[http://dx.doi.org/10.1111/j.1574-6941.2007.00424.x] [PMID: 18194345]

[20]   Thomas P, Swarna GK, Patil P, Rawal RD. Ubiquitous presence of normally non-culturable endophytic bacteria in field shoot-tips of banana and their gradual activation to quiescent cultivable form in tissue cultures. plant cell tiss org 2008; 93(1): 39-54.
[http://dx.doi.org/10.1007/s11240-008-9340-x]

[21]   Thomas P, Swarna GK, Roy PK, Patil P. Identification of culturable and originally non-culturable endophytic bacteria isolated from shoot tip cultures of banana cv. grand naine. plant cell tiss org 2008; 93(1): 55-63.
[http://dx.doi.org/10.1007/s11240-008-9341-9]

[22]   Podolich O, Ardanov P, Zaets I, Pirttilä AM, Kozyrovska N. Reviving of the endophytic bacterial community as a putative mechanism of plant resistance. Plant Soil 2015; 388(1-2): 367-77.
[http://dx.doi.org/10.1007/s11104-014-2235-1]

[23]   Mahmood A, Turgay OC, Farooq M, Hayat R. Seed biopriming with plant growth promoting rhizobacteria: A review. FEMS Microbiol Ecol 2016; 92(8): fiw112.
[http://dx.doi.org/10.1093/femsec/fiw112] [PMID: 27222220]

[24]   Brader G, Compant S, Mitter B, Trognitz F, Sessitsch A. Metabolic potential of endophytic bacteria. Curr Opin Biotechnol 2014; 27(100): 30-7.
[http://dx.doi.org/10.1016/j.copbio.2013.09.012] [PMID: 24863894]

[25]   Krishnamoorthy A, Agarwal T, Kotamreddy JNR, *et al.* Impact of seed-transmitted endophytic bacteria on intra- and inter-cultivar plant growth promotion modulated by certain sets of metabolites in rice crop. Microbiol Res 2020; 241: 126582.
[http://dx.doi.org/10.1016/j.micres.2020.126582] [PMID: 32882536]

[26]   Wu W, Chen W, Liu S, *et al.* Beneficial relationships between endophytic bacteria and medicinal plants. Front Plant Sci 2021; 12: 646146.
[http://dx.doi.org/10.3389/fpls.2021.646146] [PMID: 33968103]

[27]   Shweta S, Bindu JH, Raghu J, *et al.* Isolation of endophytic bacteria producing the anti-cancer alkaloid camptothecine from Miquelia dentata Bedd. *Icacinaceae.* Phytomedicine 2013; 20(10): 913-7.
[http://dx.doi.org/10.1016/j.phymed.2013.04.004] [PMID: 23694750]

[28]   Xu DB, Ye WW, Han Y, Deng ZX, Hong K. Natural products from mangrove actinomycetes. Mar Drugs 2014; 12(5): 2590-613.
[http://dx.doi.org/10.3390/md12052590] [PMID: 24798926]

[29]   Li W, Yang X, Yang Y, Zhao L, Xu L, Ding Z. A new anthracycline from endophytic Streptomyces sp. YIM66403. J Antibiot 2015; 68(3): 216-9.
[http://dx.doi.org/10.1038/ja.2014.128] [PMID: 25248729]

[30]   El-Gendy MMA, EL-Bondkly AMA. Production and genetic improvement of a novel antimycotic agent, saadamycin, against dermatophytes and other clinical fungi from endophytic streptomyces sp. hedaya48. J Ind Microbiol Biotechnol 2010; 37(8): 831-41.
[http://dx.doi.org/10.1007/s10295-010-0729-2] [PMID: 20458610]

[31]   Ding L, Münch J, Goerls H, *et al.* Xiamycin, a pentacyclic indolosesquiterpene with selective anti-HIV activity from a bacterial mangrove endophyte. Bioorg Med Chem Lett 2010; 20(22): 6685-7.
[http://dx.doi.org/10.1016/j.bmcl.2010.09.010] [PMID: 20880706]

[32]   Yan Y, Zhang S, Zhang J, Ma P, Duan J, Liang Z. Effect and mechanism of endophytic bacteria on growth and secondary metabolite synthesis in *salvia miltiorrhiza* hairy roots. Acta Physiol Plant 2014; 36(5): 1095-105.
[http://dx.doi.org/10.1007/s11738-014-1484-1]

[33]   Conti R, Chagas FO, Caraballo-Rodriguez AM, *et al.* Endophytic actinobacteria from the brazilian medicinal plant lychnophora ericoides mart. and the biological potential of their secondary metabolites. Chem Biodivers 2016; 13(6): 727-36.
[http://dx.doi.org/10.1002/cbdv.201500225]

[34]   Gos FMWR, Savi DC, Shaaban KA, *et al.* Antibacterial activity of endophytic actinomycetes isolated from the medicinal plant Vochysia divergens *Pantanal, Brazil.* Front Microbiol 2017; 8: 1642.
[http://dx.doi.org/10.3389/fmicb.2017.01642] [PMID: 28932210]

[35]   Singh R, Dubey AK. Diversity and applications of endophytic actinobacteria of plants in special and other ecological niches. Front Microbiol 2018; 9: 1767.
[http://dx.doi.org/10.3389/fmicb.2018.01767] [PMID: 30135681]

[36]   Zhou JY, Sun K, Chen F, Yuan J, Li X, Dai CC. Endophytic pseudomonas induces metabolic flux changes that enhance medicinal sesquiterpenoid accumulation in atractylodes lancea. Plant Physiol Biochem 2018; 130: 473-81.
[http://dx.doi.org/10.1016/j.plaphy.2018.07.016] [PMID: 30081324]

[37]   Anjum N, Chandra R. Endophytic bacteria of catharanthus roseus as an alternative source of vindoline and application of response surface methodology to enhance its production. Arch Biol Sci 2019; 71(1): 27-38.
[http://dx.doi.org/10.2298/ABS180802044A]

[38]   Ghiasvand M, Makhdoumi A, Matin MM, Vaezi J. Exploring the bioactive compounds from endophytic bacteria of a medicinal plant: Ephedra foliata *ephedrales: Ephedraceae* Advan in Trad Med 2020; 20(1): 61-70.
[http://dx.doi.org/10.1007/s13596-019-00410-z]

[39]   Liu TH, Zhang XM, Tian SZ, Chen LG, Yuan JI. Bioinformatics analysis of endophytic bacteria related to berberine in the Chinese medicinal plant. Coptis teeta Wall 3 Biotech 2020; 10(3): 96.

[40]   Liu Y, Liu W, Liang Z. Endophytic bacteria from *Pinellia ternata*, a new source of purine alkaloids

and bacterial manure. Pharm Biol 2015; 53(10): 1545-8.
[http://dx.doi.org/10.3109/13880209.2015.1016580] [PMID: 25868618]

[41]   Yang YH, Yang DS, Li GH, *et al.* New secondary metabolites from an engineering mutant of endophytic streptomyces sp. cs. Fitoterapia 2018; 130: 17-25.
[http://dx.doi.org/10.1016/j.fitote.2018.07.019] [PMID: 30076887]

[42]   Sasaki T, Igarashi Y, Saito N, Furumai T. Cedarmycins a and b, new antimicrobial antibiotics from streptomyces sp. tp-A0456. J Antibiot 2001; 54(7): 567-72.
[http://dx.doi.org/10.7164/antibiotics.54.567] [PMID: 11560375]

[43]   Sabu R, Soumya KR, Radhakrishnan EK. Endophytic nocardiopsis sp. from zingiber officinale with both antiphytopathogenic mechanisms and antibiofilm activity against clinical isolates. 3 biotech. 2017; 7(2): 1-3.

[44]   Sabu R, Soumya KR, Radhakrishnan EK. Endophytic nocardiopsis sp. from zingiber officinale with both antiphytopathogenic mechanisms and antibiofilm activity against clinical isolates. 3 biotech. 2017; 7(2): 1-3.

[45]   Zhou H, Yang Y, Yang X, *et al.* A new cyclic tetrapeptide from an endophytic *streptomyces* sp. YIM67005. Nat Prod Res 2014; 28(5): 318-23.
[http://dx.doi.org/10.1080/14786419.2013.863198] [PMID: 24304298]

[46]   Fu Y, Yin ZH, Yin CY. Biotransformation of ginsenoside rb1 to ginsenoside rg3 by endophytic bacterium *burkholderia* sp. ge 17-7 isolated from *panax ginseng*. J Appl Microbiol 2017; 122(6): 1579-85.
[http://dx.doi.org/10.1111/jam.13435] [PMID: 28256039]

[47]   Yan H, Jin H, Fu Y, Yin Z, Yin C. Production of rare ginsenosides rg3 and rh2 by endophytic bacteria from panax ginseng. J Agric Food Chem 2019; 67(31): 8493-9.
[http://dx.doi.org/10.1021/acs.jafc.9b03159] [PMID: 31310523]

Probiotic Bacteria and Plants

Muhammad Shahbaz[1,*], **Jawad Ali**[1], **Hammad Naeem**[1], **Shamas Murtaza**[1], **Nighat Raza**[1] and **Umar Farooq**[1]

[1] Department of Food Science and Technology, Muhammad Nawaz Sharif University of Agriculture Multan, Pakistan

Abstract: Probiotics are microorganisms, when consumed, give health benefits due to improvement in the activity of gut microflora. Various health claims are associated with probiotics *e.g.* modulation of the immune system, mitigation of lactose intolerance, protection from infections and maintenance of healthy gut microflora. They have also been demonstrated to be helpful in treating a wide range of illnesses, including cancer, inflammatory bowel syndrome, diarrhea brought on by antibiotics, and infantile diarrhea. *Streptococcus, Bacillus, Enterococcus, Escherichia coli, Bifidobacterium, Lactobacillus*, and several strains of the fungus *Saccharomyces* are significant probiotic bacterial genera. In fibrous parts of plants and probiotic bacteria, the bacteriocins play a major synergistic antimicrobial role. Prebiotics are non-digestible plant materials *i.e.*, complex carbohydrates, fermented in the colon, thus yield short chain fatty acids and energy, and enhance the growth of probiotics. Inulin and fructans are important plant prebiotics. The indirect health benefits of prebiotics are immunomodulatory characteristics, mineral absorption, cancer prevention, and modulation of the metabolism of gut flora, and the prevention of constipation and diarrhea. Many fruits, tuber crops, root crops as well as vegetables contain a huge reservoir of prebiotic carbohydrates. The function of probiotic microbes in foods includes modulation of the immune system, normalization of gastrointestinal activity, and the inhibition of the growth of pathogenic microbes and harmful metabolites. The function of prebiotic food material is to promote the growth of healthy bacteria in the intestinal tract. This chapter highlights the potential need of probiotics and prebiotics in our diet, and it also discusses their health benefits, mode of action, sources, food applications, distinct types, and future perspectives.

Keywords: Bacteriocins, Fructans, Inulin, Metabolites, Probiotics, Prebiotics, Synergistic effect.

INTRODUCTION

The definition of probiotics is "constituents of microbial cells or microbial cultures which when consumed exert beneficial impact on human health and

* **Corresponding author Muhammad Shahbaz**: Department of Food Science and Technology, Muhammad Nawaz Sharif University of Agriculture Multan, Pakistan; E-mail: shahbaz.ft@mnsuam.edu.pk

well-being" [1]. Regular use of probiotics in the diet has shown numerous health benefits to human health; they play a major role in maintaining regular digestive processes and nourishing the health of animals as well. Nevertheless, numerous uncertainties with respect to regulatory, microbiological, and technological aspects of probiotics exist [2]. After birth, gastrointestinal flora is gained quickly, it is necessary for maintaining normal human homeostasis and remains quite stable during the entire life. During the development of gastrointestinal flora, the association between the host and this microflora gives rise to the development of a distinct and unique intestinal immune system. Major challenge challenge in this host mucosal immune system is to differentiate between benign microorganisms and pathogens by triggering protective immunity without much inflammatory response, which can disturb the stability of the gastrointestinal mucosa [3]. The use of irradiation, immunosuppressive therapy, antibiotics as well as other means of treatment, can alter the composition of gut microflora and has adverse effects on their growth. Hence, to prevent diseases and to reestablish the equilibrium of gut microflora, an attractive option is to introduce beneficial microbial strains into the gastrointestinal tract [4].

Prebiotics include fiber, resistant starches, poly and oligosaccharides, as well as sugar polyols are non-digestible carbohydrates that exert a beneficial impact on the growth and maintenance of microflora present in the gut. Prebiotics are recognized for their potential to nutrify microbes harboring in the gastro-intestinal tract while significantly boosting metabolic activity, thus strengthening the immune system and improves nutrient absorption as well as digestion in the host while retarding the proliferation of pathogenic microorganisms [5]. Prebiotics can survive in the acidic medium/environment, and they are very resistant to various enzymes of the digestive system present in small-intestines. This ability makes prebiotics a very efficient tool to enhance the multiplication and growth of helpful gut microbes that ferment the prebiotics, which in turn leads to the production of vitamins, short-chain FAs (Fatty acids) and various fragmented substances [6]. Anyhow, different microorganisms utilize prebiotics differently because each microbe has its own nutritional requirement to survive in the gastrointestinal tract. In general, gut microflora utilizes prebiotics as nutritional supplements for their metabolic activities and proliferation, so in food industries they have been used widely as supplements of functional foods in various formulations [7]. In this sense, this chapter discusses the need and potential for probiotics and prebiotics in our daily diet, and it also discusses their health benefits, mode of action, sources, food applications, distinct types, and future perspective.

OVERVIEW OF PROBIOTICS

Probiotics are live microbes which can be used in the formulation of various products like dietary supplements, drugs, and foods. Probiotic is somewhat a new term which is specifically used for the bacteria related to beneficial effects on human health as well as animals. Meaning of the term probiotic is "for life" and an Expert Committee defined this term as "live microbes which when ingested in certain quantity exert positive impact on health beyond providing general nutrition" [8]. Probiotics which are commonly used in feed and food are *Lactobacillus* and *Bifidobacterium*. Among other microorganisms which are used as probiotics are yeast (*Saccharomyces cerevisiae*) and few species of *Bacillus* and *Escherichia coli*. For food fermentation Lactic-acid producing bacteria are being used since primeval time, they can perform dual functions by providing health benefits and acting as food fermenting agent [9].

HEALTH BENEFITS OF PROBIOTICS

In addition to meeting basic dietary needs, probiotics provide a number of positive health effects. Health benefits provided by probiotics are strain-specific, so there is no specific strain that would exhibit all projected health benefits and within the same specie, not all strains are effective against specified health conditions. In dairy products, the most used cultures of probiotics which have proven human health benefits are *L. casei*, *Bifidobacterium* spp. and *L. acidophilus* while *Saccharomyces cerevisiae* and some species of *Bacillus* and *E. coli* are also used [9]. Furthermore, probiotics are aimed to aid the naturally occurring gut microflora of the body. Antibiotic-related diarrhea can be prevented by using different probiotic preparations or they can be used as part of the treatment for such ailments. Recently, it has been reported that the functional properties of probiotics of potential importance are anti-carcinogenicity, antimutagenicity, and antigenotoxicity. Experimental data indicated that consuming fermented dairy items is linked with minor chances of colon cancers, which implied that probiotics are capable to inhibit carcinogens and procarcinogens and probiotics also inhibit the growth of those bacteria which can convert procarcinogens to carcinogens thus decreasing the risks of cancer [10].

MECHANISM OF ACTION OF PROBIOTICS

Probiotics have been recognized for providing numerous potential health-beneficial effects. The list of claimed effects of probiotics ranges from mitigation of constipation to prevention of major diseases like cardiovascular ailments, cancer, and inflammatory bowel disease. Among these claims, some claims are considered well recognized like relief from lactose intolerance or reducing intestinal transit time, whereas some claims require further scientific research like the effect of probiotics on blood cholesterol level or cancer prevention. Each probiotic strain possesses a specific mechanism of action and mostly it is a combination of different activities which makes it a complicated and difficult task to investigate the responsible mechanisms. Generally, three levels of action of probiotics can be differentiated which influence human health; maintaining the immune system of the host, strengthening the mucosal barriers and by the interaction of probiotics with other microbes present in the host the figure illustrated the mechanisms of action of probiotics by which they exert action in the gut of the host.

Probiotics exert a beneficial impact on epithelial cells of the intestine in various ways as explained in this study. (A); some strains can provide a physical barrier thus blocking the entry of pathogens in the epithelial cells, it is referred to as colonization resistance. (B); other probiotics increase the intercellular strength of apical tight junctions which maintain the permeability of intestines, for example, prevention of protein redistribution on tight junction, or by expression upregulation of zona-occludens 1 can block the entry of molecules in lamina propria. (C); Some strains can provide signals to dendritic cells which trigger the natural immune system, these cells induce TREG cells and produce anti-inflammatory cytokines, like TGF-β and IL-10 by entering mesenteric lymph nodes [11]. This is illustrated in Fig. (**1**) for better understanding.

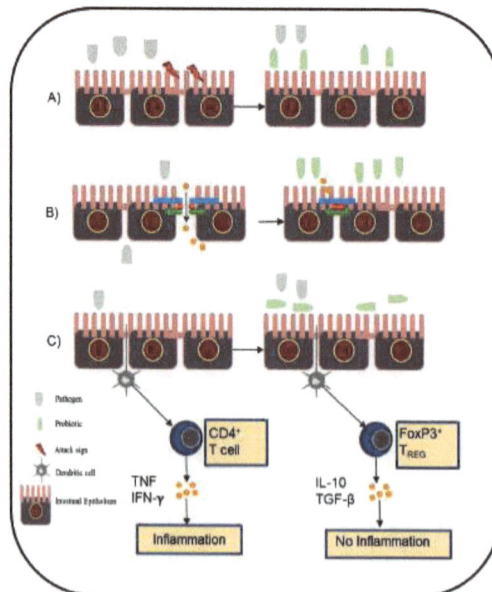

Fig. (1). Beneficial effects of probiotics on intestinal epithelial cells.

FOOD APPLICATIONS OF PROBIOTICS

There are many applications of probiotics in dairy and non-dairy products, some of them are described below:

Yogurt

Yogurt is a popular probiotic product and is considered the original source of probiotics. Yogurt is well recognized owing to its numerous health benefits and nutritional value. Bacterial cultures *Streptococcus salivarius* subsp. *thermophilus* and *Lactobacillus delbrueckii* subsp. *bulgaricus* are used in the production of yogurt. Moreover, other *Bifidobacteria* and *Lactobacilli* are also sometimes used after or during inoculation. The proteolytic activities as well as the viability of probiotics must be considered in the production of yogurt. Survival of *Bifidobacterium* and *Lactobacillus* spp. in the yogurt depends on various factors such as storage temperature, the buffering capacity of media, the concentration of metabolites like acetic acid and lactic acid, the presence of dissolved oxygen and hydrogen peroxide, pH as well as strains of probiotic bacteria. Even though yogurt is recognized as a probiotic vehicle but at the time of consumption, most commercial products of yogurt have considerably low viable cells [12]. Anyhow, encapsulation in a whey protein matrix, alginate-pectin, alginate-prebiotic, alginate-starch, chitosan-coated alginate, plain alginate beads, or the addition of cysteine or prebiotics could possibly enhance the stability and viability of probiotics in yogurt [13].

Cheese

Among dairy products, milk and yogurt are considered the most common vehicles for probiotics. Nevertheless, cheese is also well suited as an alternative carrier of probiotics. Comparison of cheese with fermented milk (Cow, sheep, *etc.*) and Yogurt shows that cheeses are more advantageous because they have longer Shelf life, higher fat content, more solid consistency, high energy are highly nutritious, and have a higher buffering capacity as well as pH [14]. Nonetheless, the comparison of the serving size of cheese with that of yogurt shows that both provide the same health benefits, while cheese needs to have greater viability of probiotic cultures as well as greater density of probiotic cells. In the probiotic industry, cheese was introduced in the year 2006 when Danisco tested the survival and growth of various strains of probiotics in cheese [15]. Only a few probiotic cheeses were available in the market at that time. Presently, 200 different varieties of commercial probiotic cheeses are available in the market in numerous forms like hard, semi-hard and fresh cheese. Hard and semi-hard cheese requires a somewhat high level of probiotics inoculation (nearly 4 to 5 times) as compared to yogurt and it has a somewhat low recommended daily intake. Cottage cheese is a fresh cheese that has a shorter life span even at refrigerated storage temperature and it has a relatively high recommended daily intake. So, it can serve as a vehicle of probiotics having a high potential to carry probiotics.

Vegetable based probiotic products

Since ancient times vegetable fermentation has been well recognized. Fermented vegetables can be used as a reliable medium for delivering probiotics. Anyhow, vegetable fermentation requires low incubation temperature which is a major problem to introduce traditional *Bifibacterium* and *L. acidophilus* probiotic bacteria. In vegetable fermentation, the better adapted probiotics are *L. plantarum*, *L. casei,* and *L. rhamnosus*. However, in plant-based substrates bacteria proliferate quite rapidly when the temperature is adjusted to around 37°C [16]. *Bifidobacterium* strains were investigated in the production of probiotic food using carrot juice as raw material. It was revealed that in carrot juice *Bifidobacteria* cause several biochemical activities without any nutrient supplementation [17]. The suitability of tomato juice was checked in several research studies to produce a product by using *L. delbrueckii, L. casei, L. plantarum,* and *L. acidophilus*. Studies revealed that these lactic acid bacteria rapidly utilize tomato juice to produce lactic acid and cell synthesis without adjustment of pH and nutrient supplementation. Anyhow, conventional approaches to production may cause the inactivation of probiotic cells/cultures and such fermented probiotic vegetables should be stored at low temperatures.

CHALLENGES IN THE APPLICATION OF PROBIOTICS

The most crucial challenges of probiotic products are to ensure health benefits and commercial success. Viability and sensory acceptance of probiotic products has been discussed in this section.

Viability and Survival

Probiotics are well recognized to deliver several health benefits, but these claimed benefits cannot be gained without a huge number of viable cells in probiotic-products. Many strains of probiotic bacteria die in probiotic food products after exposure to the acidic environment in the human stomach, oxygen during refrigerated storage of products, and/or low pH after fermentation. The probiotic bacteria's concentration has an impact on how effective they are in the product. They should be viable for the duration of storage and be able to withstand the harsh conditions in the gut [18]. Numerous studies investigated the effect of adding specific compounds in enhancing the viability of probiotics. Results revealed that fructooligosaccharides (FOS), oligosaccharides, and inulin exert a positive impact on the viability of probiotics. However, these compounds impact the specific strains of probiotic bacteria. In fermented milk, the impact of raffinose on the survival of probiotic bacteria such as *L. acidophilus* and *Bifidobacteria* was investigated. Results revealed that the addition of raffinose in fermented milk enhances the viability of *L. acidophilus* and *Bifidobacteria*. Additionally, the supplementation of probiotic products with pectin, maltodextrin, mannitol, and FOS were found to have a positive impact on the survivability of probiotic bacteria [19].

The survival and viability of probiotic bacteria are strain specific. Encapsulation is the best-suited option to retain the viability of very sensitive bacterial strains, especially microencapsulation is preferred as it does not cause any adverse effect on sensory attributes of the food product. Techniques of microencapsulation have been applied successfully using different food matrices to protect the probiotic bacterial cells from the destruction caused by the severe external environment [20]. In general, microencapsulation enhances the survival rate of probiotic bacteria when exposed to mild heat treatment, bile salts, and acidic environments. Probiotics get immobilised by microencapsulation, which increases the survival of these probiotic bacteria in food items throughout storage, processing, and digesting[21].

Sensory Acceptance

Probiotic food products should perform exactly as conventional foods in terms of sensory characteristics. Consequently, it is crucial to develop tests for checking the sensory quality of probiotic food products which can be followed by definite sensory analysis. Sensory analysis of probiotic food products must cover all attributes regarding changes over time during the storage period. Studies revealed that there is a possibility to obtain similar or even better performance of probiotic foods as compared to conventional foods like milk fermented with *L. acidophilus* La-5 and *B. animalis* supplemented with inulin [22], curdled milk with *L. acidophilus* and inulin and functional yogurt supplemented with *L. rhamnosus* GR-1 and *L. reuteri* RC-14 [23]. Sensory analyses are carried to obtain important data regarding the development of probiotic foods at a commercial scale. Mostly, the developed probiotic products are required to compete with the commercial products in parallel. Generally, the metabolism of probiotic bacteria results in the development of such components that negatively affect the taste and aroma of the probiotic product, it is termed a probiotic off-flavor. As for example, *Bifidobacterium* spp. produce acetic acid in the product which can cause vinegary off-flavor thus affecting the sensory quality of the product [9].

FUTURE PERSPECTIVE OF PROBIOTICS

Dairy-based products with live bacteria are the main sources of probiotics that are now accessible. The next food category would be non-dairy beverages where probiotic bacteria will prosper. Technologies of microencapsulation have delivered the required protection to probiotics which ultimately make them food ingredients apart from supplemental and pharmaceutical use.

Nanotechnology, Encapsulation, and Probiotics

The term "nano" is derived from ancient Greek which means "dwarf". One nanometer is equal to a billionth of a meter (10^{-9}). This advanced technology of probiotics is attracting the interest of researchers and scientists and it provided new and advanced prospects for applications of probiotics. Probiotics have been used in pharmaceuticals and drug delivery but recently they also find their applications in the food and agriculture sector. Until now, little is known about the impact of nanotechnology on human health and the environment so, the application of nanotechnology in the production of organic food requires precaution. Due to the lack of information about the risks of nanotechnology, there are no regulations currently that limit or control the production of nanosized particles [24]. Nanoencapsulation is 'a technique in which substances are packed in miniature by using technologies like nanoestructuration, nanoemulsification, and nanocomposite which provide better functionality of the final product as well as controls the release of the core. Encapsulating the food components may extend the shelf life of the probiotic food product. Delivering probiotic bacterial strains to certain regions of the stomach where they may interact with specific receptors is a desirable strategy. As these nanoencapsulated probiotic bacterial strains possess the ability to modulate immune responses so, they can also act as de novo vaccines [25]. Alginate can be effectively used in microencapsulation of various probiotic strains and results revealed that these probiotic strains show better survival rates as compared to free cells at mild heat treatment of about 65°C, high concentration of bile salts, and at low pH of 2. In acidic food items, the technique of microencapsulation proves to be crucial in improving the stability and viability of probiotic bacterial strains and it also helps to deliver viable bacterial cells to the gastrointestinal tract of the host. Moreover, this technique effectively protects the bacterial cells from moderate heat treatment thus this technique can be applied in the production of functional food products which require a mild heat treatment. In this technique, a protective coating is used to separate probiotic bacterial cells from their surrounding environment. Numerous studies indicated that vegetable gum or gelatin is used in the microencapsulation technique which protects acid-sensitive *Lactobacillus* and *Bifidobacterium* [21].

Biotechnology and Probiotics

With the advancement and expansion in the field of biotechnology and sequencing techniques, it is well-timed and reasonable to launch projects of genome sequencing for characteristic probiotic microbes. If information regarding genome sequencing of a specific probiotic strain of interest is available, then it can be used to examine the transcription profile of that specific strain during fermenter growth. This information will enable optimization and better control of the growth of probiotic bacteria which is currently impossible. Transcriptional profiling of probiotic strains will enable us to understand the behavior of genes that are important for the survival of probiotics during processing steps (*e.g.*, acid and stress-tolerant genes) and it also recognize novel genes which play an important role in technological functionality of probiotics [26]. Growing knowledge of genes play a crucial role in technological functionality and advancement in the development of tools for the manipulation of genetic properties of Bifidobacterium and Lactobacillus species will enable us to tailor the technological characteristics of probiotic strains in the future. Anyhow, safety issues of these novel strains are of utmost importance in probiotic food products, and for each novel strain, these issues need to be considered seriously [27].

OVERVIEW OF PREBIOTICS

The term "prebiotics" has evolved significantly over the past few decades. The concept originated as "specifically fermentable ingredients which enhance the composition and activity of gut microorganisms and provide health benefits to host", this definition describes the conditions that are required to show a beneficial impact on the subject's host. As per this definition, prebiotics should be fermented by the intestinal microorganisms and they should be capable to withstand the host digestion process [28]. In 2010, information regarding the diversity and density of bacterial species and advancement in molecular approaches persuaded the ISAPP (International Scientific Association for Prebiotics and Probiotics) to release a statement that revised the definition of dietary prebiotics as "specific fermented ingredients which cause specific alterations in the activity and composition of gastro-intestinal microorganisms, thus confer beneficial health effects on the host" [29]. This definition includes the non-specific species of bacteria which expands from the colon and throughout the length of the gastrointestinal tract. In 2015, a new definition of prebiotics was proposed as "non-digestible carbohydrates that are metabolized in the gut by microorganisms and modulate the activity and composition of the gut microbiota, which confer beneficial physiological impact on the host". According to this the specificity of microorganisms and selective fermentation processes has been eliminated as an essential requirement, it also limited the interaction of prebiotics with only gut microbes without including extra-intestinal microbes living on the skin, vagina, and respiratory tract [30]. In 2017, progression in scientific and clinical development tend ISAPP to again update the concept of prebiotic as "a substrate which is used by microorganisms of host selectively and confer beneficial effects on the hosts' health while retaining the health benefits mediated by microflora". As per this definition, the prebiotics are not only limited to foods and carbohydrates but they are also no longer bound to the gastrointestinal tract; rather, they also include extra-intestinal tissues and non-food elements [31].

TYPES OF PREBIOTICS

Various forms of prebiotics possess discrete health benefits *e.g.*, fructooligosaccharides, galactooligosaccharides, and inulin have long been recognized as major prebiotics. Anyhow, dietary fibers and several other compounds are also considered essential prebiotics which provide multiple health benefits to the host.

Synthetic

Different types of synthetic prebiotics along with their production method, chemical structure and health benefits are shown in Table **1**.

Table 1. Type of prebiotics and their effects in human body.

Type of Prebiotics	Effects	References
Galacto-oligosaccharides	Faecal excretion of bifidobacteria and lactobacilli increased	[37]
Oligofructans	Positively affect the selective stimulation of probiotic bacteria in the colon	[38]
Fructans	Modulate gut physiology to provide protection from pathogens, improve the level of glucose	[39]
Pectino-oligosaccharides	Anti-inflammatory effects	[40]
Xylooligosaccharides	Improvement of intestinal bacteria & Beneficial effects on intestinal functions	[41]

GOS: Galacto-oligosaccharides, FOS: Fructo-oligosaccharides.

Fructooligosaccharides (FOS)

Fructooligosaccharides possess prebiotic potential as they contain low-calorie dietary fibers, they are also recognized by other names like oligofructose or oligifructan. At present, owing to their numerous health benefits for humans and animals, FOS is regarded as crucial natural food ingredients. The linear fructose chains in the FOS structure are joined by (2-1) bonds, and the fructose units in these chains can range from 2 to 60, terminating in glucose fructooligosaccharides. In addition, FOS are synthesized by a process called enzyme-catalyzed transglycosylation from sucrose, and sucrose-containing reduced ends are present at the termination of the individual molecule [32]. During the process of synthesis of FOS from sucrose, the fucosyltransferase enzyme plays a crucial role, this enzyme expresses hydrolytic activity when the concentration of sucrose is low. They are produced commercially, and in numerous foods, they are added as supplements as well as used as nutraceuticals because in the gastrointestinal tract, they remain undigested and in the large intestine, they are fermented by intestinal bacteria into lactate and short-chain fatty acids [33].

Galactooligosaccharides (GOS)

Galactooligosaccharides contain oligo galactosyl lactose, oligogalactose, and oligolactose. Lactulose (cow milk) is converted into GOS *via* isomerization and transglycosylation. GOS possess bifidogenic potential because they are fermented by *Bifidobacteria* (probiotics) but remain enzymatically undigested in the gut. At the industrial level, GOS were synthesized *via* nucleophilic and electrophilic displacement previously, but these processes are uneconomical [34]. Operative enzymes which are involved in the formation of GOS are galactosyltransferase and galactosidase. The enzyme which is used for large-scale production of GOS is galactosyltransferase. Although, the catalytic reaction which is involved in the synthesis of GOS by the enzyme galactosyltransferase is expensive approach because it needs nucleotide sugars to act as donors. Consequently, globotriose production and oligosaccharides from human milk are commonly used to reduce costs [35].

Fructans

Fructans are extensively used prebiotics to improve the health of humans, they are natural polymers, present in various functional foods like onion, leek, garlic, chicory roots, asparagus, and artichoke. The linear fructose chains seen in the structure of fructans are connected by (2-1) bonds. Fructans enhance the growth and proliferation of *Lactobacilli* and *Bifidobacteria* which improves the physiology of the gut thus provide protection from pathogenic microorganisms. Furthermore, regular consumption of prebiotics such as fructans regulates lipid metabolism and improves glucose levels considerably as well as reduces the levels of diacylglycerol and lipopolysaccharides in the plasma membrane [36].

Natural

Various plants contain non-digestible carbohydrates naturally [42].

Garlic

The scientific name of garlic is *Allium sativum* and it is used in the treatment of numerous ailments such as gastrointestinal disorders and flu. The high amount of fructooligosaccharides is present in garlic which leads to the prevention of numerous ailments as well as protects the gastrointestinal tract. One of the essential components of garlic is garlic fructan (GF) which accounts for about 75% of the dry weight of garlic, and it has been described to exert positive impact on gut microbiota and exhibit prebiotic potential. An investigative study on the effect of garlic fructan on gut microbiota revealed that they enhance proliferation of *Bifidobacteria* whereas, it restrains the growth and proliferation of pathogens like *Clostridia* species [43].

Flaxseeds

Flaxseed is also known by the name of linseed and its scientific name is *Linum usitatissimum.* Flaxseeds are recognized as functional foods as these seeds provide health beneficial effects, and are rich in nutrients. Numerous functional components are present in these seeds like α-linoleic acid, phenolic compounds, high-quality proteins, soluble fibers, and minerals [44]. A research study revealed that the consumption of flaxseed can exert a positive impact on alcoholic liver condition, and it can reduce the growth of *Proteobacteria* and *Porphyromonadaceae* in the gut [45].

Onion

Onion belongs to the family *Liliaceae* and its scientific name is *Allium cepa.* Apart from its high nutritional value it also possesses medicinal attributes, and its consumption provides numerous functional components such as minerals, vitamins, dietary fibers, and carbohydrates. Onion has a significant amount of soluble carbohydrates, primarily fructooligosaccharides and monosaccharides (sucrose, fructose, and glucose). These compounds have a strong prebiotic potential, which helps to enhance the host's gut flora and overall health [46].

Oats

Oats are a huge reservoir of non-starch polysaccharides, and its scientific name is *Avena sativa.* Oats are regarded as healthy cereal grains owing to the enormous quantity of proteins, vitamins, minerals, and fiber they contain. Among the soluble non-starch polysaccharides which are present in oats, the chief component is β-glucan which is of utmost importance. β-glucan possess the ability to form highly viscous solutions which exert health beneficial effect in the gut. Furthermore, oats play a crucial part in insulin resistance, hypertension, obesity, and dyslipidemia [47].

MODE OF ACTION OF PREBIOTICS

Prebiotics improve human health by positively influencing the growth of beneficial gut microbes as well as their metabolic activities. In general, prebiotics are readily available for fermentation by gut microflora but remain undigested by host enzymes. They increase calcium ions absorption by improving lipid metabolism which ultimately exerts a positive impact on immunological and bowel activities [48]. The exact mechanism of prebiotic action has not been demonstrated yet. Although, it is assumed that gut microflora uses prebiotics as carbon sources and energy depending on their compositional and structural characteristics. To assess the action of prebiotics on various body organs, different models are being used. These models revealed that, prebiotics enhance the production of short-chain fatty acids like propionic and butyric acids owing to fermentation by regulating the lipogenic enzymes of the liver. These fermented products enhance the growth and multiplication of beneficial gut microbes by increasing the expression of transcriptional genes [49].

Moreover, prebiotics such as fructooligosaccharides enhance lymphocyte and leukocyte count in gut-associated lymphoid tissues (GALTs) and peripheral blood. Immunoglobulin A (IgA) is synthesized by GALTs thus triggering the phagocytic action of intra-inflammatory macrophages directly. Beneficial gut microflora uses prebiotics as nutrients and synthesizes certain antimicrobial compounds which ultimately increase their abundance in the gut as compared to pathogenic microorganisms [50]. Various studies illustrated the positive impact of prebiotics in the modulation of cytokine expression. A study was conducted to evaluate the impact of prebiotics on mice afflicted with obesity. Results of subject study revealed enhanced production of pro-inflammatory cytokines (TNF-α, IL-6, IL-1b, IL-1a, and INF-γ), low profile of plasma LPS and reduced expression of inflammatory markers and oxidative stress [51].

HEALTH BENEFITS OF PREBIOTICS

There are numerous health beneficial effects of prebiotics, some of them are described as follows:

Effect of Prebiotics on Gut Microbes

The prime target of dietary supplements is to provide healthy gut microbe which remarkably improve the health of an individual. A remarkable gut colonizer is *Lactobacilli* which has been recognized to prevent irritable bowel syndrome, traveler's diarrhea, relieve constipation, reduce lactose intolerance, and decrease inflammation of gut mucosa. Moreover, in the gastrointestinal tract of healthy humans, the most common probiotics are *Bifidobacteria* which play a crucial role in the fermentation of selective oligosaccharides, this characteristic makes them usual markers for prebiotic potential [52]. Substantial colonizers of the gut also include commensal *Clostridia* which belongs to the phylum *Firmicutes*, they are well-recognized for their role in the modulation of metabolic, physiologic, and immune processes. In the gut, prebiotics enhance the proliferation of these beneficial microbes which ultimately help to treat various digestive disorders and boost the activity of the immune system [53].

Effect of Prebiotics on Metabolite Production

Generation of primary and secondary metabolites is achieved by direct and indirect fermentation of certain compounds, which have several beneficial effects on human health. In the gut, amino acids, carbohydrates and few of the nutrients are not absorbed in small intestine, they are fermented by microbes of the gut to synthesize short-chain fatty acids. Enteric microorganisms of the gut synthesize short-chain fatty acids such as propionate, butyrate, and acetate by primary anaerobic fermentation of prebiotics. These short-chain fatty acids play a crucial part as a substrate for lipid, cholesterol, and glucose metabolism. For peripheral tissues, propionate, and acetate act as substrates [54]. Furthermore, propionate, butyrate, and acetate also elevate G-protein-coupled-receptors (GPRs). The major role in the modulation of vital metabolic hormones such as peptide YY (PYY) and GLP 1 [55].

Effect of Prebiotics on Mineral Absorption

The major target of consuming prebiotics is to enhance the bioavailability and absorption of calcium in the elderly population and infants to make their bones healthy. Regular ingestion of prebiotics has lowered the risks of osteoporosis and fractures (bone) worldwide. Primary absorption of calcium takes place in the distal intestine and this phenomenon is triggered by several chemical changes caused by various microbes as well as acidic fermentation of prebiotic dietary fibers, however, mixed results have been provided by clinical evaluations regarding mineral absorption in relation to prebiotics [56]. Studies illustrated that there are no considerable changes in calcium absorption following the consumption of fructooligosaccharides, inulin, galactooligosaccharides, and oligofructose, while some studies suggested that calcium absorption increases significantly by consuming oligosaccharide components with lactulose [57].

Effect of Prebiotics on Necrotizing Enterocolitis

Necrotizing enterocolitis specifically affects preterm infants, it is a gastrointestinal disorder that is described by necrosis, localized infection and inflammation of the bowel in affected individuals, and it is a leading cause of mortality. The growth of beneficial gut microorganisms such as *Bifidobacteria* can be triggered by consuming prebiotics like fructooligosaccharides and galactooligosaccharides which ultimately restrain the growth of pathogenic microorganisms in premature neonates thus preventing them from necrotizing enterocolitis [58]. Moreover, short-chain fatty acids improve the emptying of gastric elements and bowel motility which increases the feeding tolerance in infants [59].

CONCLUSION

Regular consumption of probiotics can prevent allergic, neoplastic, inflammatory, and infectious conditions. There are many challenges in the application of probiotics, namely sensory acceptance and viability but most importantly, the processor should keep in mind the strain of bacteria being used, inoculation of starter cultures, and processing parameters. An addition of a particular quantity of probiotics in the food does not guarantee beneficial effects on health. In this regard, the viability of probiotics through the entire storage period and recovery levels in the GI tract are crucial factors. Proper evaluation of probiotic products is required before bringing them into routine usage and various standards of reliability and quality must be met. Prebiotics are thought of as nutritious food components for gut microorganisms. Short-chain oligosaccharides that are part of their chemical makeup are utilised by gut microorganisms for proliferation uponupon fermentation. Various plants contain prebiotics and their high demand leads to the industrial production of prebiotics. Additionally, these prebiotics can be used in the formulation of various food products which help to gain better technological and sensorial attributes. Mechanism of action of prebiotics is still debatable because it is dependent on different gut microbes which are involved in the fermentation of these indigestible carbohydrates that ultimately exert health-beneficial effects. Associated health benefits of prebiotics make them an important ingredient in the formulation of numerous value-added food products thus enabling industries to develop novel functional foods with distinctive ingredients. This approach enables food processors to make products with various combinations of probiotics and prebiotics to vary the level of therapeutic effect. However, subject formulations are required to be studied at the nutrigenomics level to provide more information about the response of various nutrients.

REFERENCES

[1]   Marteau P, Cuillerier E, Meance S, *et al. Bifidobacterium animalis* strain DN-173 010 shortens the colonic transit time in healthy women: A double-blind, randomized, controlled study. Aliment Pharmacol Ther 2002; 16(3): 587-93.
        [http://dx.doi.org/10.1046/j.1365-2036.2002.01188.x] [PMID: 11876714]

[2]   Krockel L. Use of probiotic bacteria in meat products. meat industry 2006; 86: 109-13.

[3]   Delia P, Sansotta G, Donato V, *et al.* Use of probiotics for prevention of radiation-induced diarrhea. World J Gastroenterol 2007; 13(6): 912-5.
        [http://dx.doi.org/10.3748/wjg.v13.i6.912] [PMID: 17352022]

[4]   Gupta V, Garg R. Probiotics. Indian J Med Microbiol 2009; 27(3): 202-9.
        [http://dx.doi.org/10.4103/0255-0857.53201] [PMID: 19584499]

[5]   Lockyer S, Stanner S. Prebiotics: An added benefit of some fibre types. Nutr Bull 2019; 44(1): 74-91.
        [http://dx.doi.org/10.1111/nbu.12366]

[6]   Yadav S, Jha R. Strategies to modulate the intestinal microbiota and their effects on nutrient utilization, performance, and health of poultry. J Anim Sci Biotechnol 2019; 10(1): 2.
        [http://dx.doi.org/10.1186/s40104-018-0310-9] [PMID: 30651986]

[7]   Rolim PM. Development of prebiotic food products and health benefits. Food Sci Technol 2015; 35(1): 3-10.
        [http://dx.doi.org/10.1590/1678-457X.6546]

[8]   FAO/WHO. Health and nutritional properties of probiotics in food including powder milk with live lactic acid bacteria. Cordoba, Argentina: Food and Agriculture Organization of the United Nations and World Health Organization Expert Consultation Report. 2001.

[9]   Song D, Ibrahim S, Hayek S. Recent application of probiotics in food and agricultural science. Probiotics 2012; 10: 11-34.
        [http://dx.doi.org/10.5772/50121]

[10]   Vasiljevic T, Shah NP. Probiotics-from metchnikoff to bioactives. Int Dairy J 2008; 18(7): 714-28.
        [http://dx.doi.org/10.1016/j.idairyj.2008.03.004]

[11]   Gareau MG, Sherman PM, Walker WA. Probiotics and the gut microbiota in intestinal health and disease. Nat Rev Gastroenterol Hepatol 2010; 7(9): 503-14.
        [http://dx.doi.org/10.1038/nrgastro.2010.117] [PMID: 20664519]

[12]   Donkor ON, Nilmini SLI, Stolic P, Vasiljevic T, Shah NP. Survival and activity of selected probiotic organisms in set-type yoghurt during cold storage. Int Dairy J 2007; 17(6): 657-65.
        [http://dx.doi.org/10.1016/j.idairyj.2006.08.006]

[13]   Sandoval-Castilla O, Lobato-Calleros C, García-Galindo HS, Alvarez-Ramírez J, Vernon-Carter EJ. Textural properties of alginate–pectin beads and survivability of entrapped lb. casei in simulated gastrointestinal conditions and in yoghurt. Food Res Int 2010; 43(1): 111-7.
        [http://dx.doi.org/10.1016/j.foodres.2009.09.010]

[14]   Ong L, Henriksson A, Shah NP. Chemical analysis and sensory evaluation of cheddar cheese produced with *lactobacillus acidophilus, lb. casei, lb. paracasei* or *bifidobacterium sp.* Int Dairy J 2007; 17(8): 937-45.
        [http://dx.doi.org/10.1016/j.idairyj.2007.01.002]

[15]   Mäkeläinen H, Ibrahim F, Forssten S, Jorgensen P, Ouwehand AC. Probiotic cheese. Nutrafoods 2010; 9(3): 15-9.
        [http://dx.doi.org/10.1007/BF03223337]

[16]   Savard T, Gardner N, Champagne C. Growth of *lactobacillus* and *bifidobacterium* cultures in a vegetable juice medium, and their stability during storage in a fermented vegetable juice. Sci Aliments 2003; 23: 273-83.

[http://dx.doi.org/10.3166/sda.23.273-283]

[17]   Kun S, Rezessy-Szabó JM, Nguyen QD, Hoschke Á. Changes of microbial population and some components in carrot juice during fermentation with selected bifidobacterium strains. Process Biochem 2008; 43(8): 816-21.
[http://dx.doi.org/10.1016/j.procbio.2008.03.008]

[18]   Shah NP. Functional cultures and health benefits. Int Dairy J 2007; 17(11): 1262-77.
[http://dx.doi.org/10.1016/j.idairyj.2007.01.014]

[19]   Yeo SK, Liong MT. Effect of prebiotics on viability and growth characteristics of probiotics in soymilk. J Sci Food Agric 2010; 90(2): 267-75.
[http://dx.doi.org/10.1002/jsfa.3808] [PMID: 20355041]

[20]   Del Piano M, Morelli L, Strozzi GP, et al. Probiotics: from research to consumer. Dig Liver Dis 2006; 38 (Suppl. 2): S248-55.
[http://dx.doi.org/10.1016/S1590-8658(07)60004-8] [PMID: 17259085]

[21]   Ding WK, Shah NP. Acid, bile, and heat tolerance of free and microencapsulated probiotic bacteria. J Food Sci 2007; 72(9): M446-50.
[http://dx.doi.org/10.1111/j.1750-3841.2007.00565.x] [PMID: 18034741]

[22]   Oliveira LB, Jurkiewicz CH. Influence of inulin and acacia gum on the viability of probiotic bacteria in synbiotic fermented milk. Braz J Food Technol 2009; 12: 138-44.
[http://dx.doi.org/10.4260/BJFT20095808]

[23]   Hekmat S, Reid G. Sensory properties of probiotic yogurt is comparable to standard yogurt. Nutr Res 2006; 26(4): 163-6.
[http://dx.doi.org/10.1016/j.nutres.2006.04.004]

[24]   Sozer N, Kokini JL. Nanotechnology and its applications in the food sector. Trends Biotechnol 2009; 27(2): 82-9.
[http://dx.doi.org/10.1016/j.tibtech.2008.10.010] [PMID: 19135747]

[25]   Sekhon BS. Food nanotechnology: An overview. Nanotechnol Sci Appl 2010; 3: 1-15.
[PMID: 24198465]

[26]   Klaenhammer TR, Barrangou R, Buck BL, Azcarate-Peril MA, Altermann E. Genomic features of lactic acid bacteria effecting bioprocessing and health. FEMS Microbiol Rev 2005; 29(3): 393-409.
[http://dx.doi.org/10.1016/j.fmrre.2005.04.007] [PMID: 15964092]

[27]   Ahmed FE. Genetically modified probiotics in foods. Trends Biotechnol 2003; 21(11): 491-7.
[http://dx.doi.org/10.1016/j.tibtech.2003.09.006] [PMID: 14573362]

[28]   Gibson GR, Probert HM, Loo JV, Rastall RA, Roberfroid MB. Dietary modulation of the human colonic microbiota: Updating the concept of prebiotics. Nutr Res Rev 2004; 17(2): 259-75.
[http://dx.doi.org/10.1079/NRR200479] [PMID: 19079930]

[29]   Gibson GR, Scott KP, Rastall RA, et al. Dietary prebiotics: Current status and new definition. Food Sci Technol Bull 2010; 7(1): 1-19.
[http://dx.doi.org/10.1616/1476-2137.15880]

[30]   Bindels LB, Delzenne NM, Cani PD, Walter J. Towards a more comprehensive concept for prebiotics. Nat Rev Gastroenterol Hepatol 2015; 12(5): 303-10.
[http://dx.doi.org/10.1038/nrgastro.2015.47] [PMID: 25824997]

[31]   Gibson GR, Hutkins R, Sanders ME, et al. Expert consensus document: The international scientific association for probiotics and prebiotics isapp consensus statement on the definition and scope of prebiotics. Nat Rev Gastroenterol Hepatol 2017; 14(8): 491-502.
[http://dx.doi.org/10.1038/nrgastro.2017.75] [PMID: 28611480]

[32]   Kumar CG, Sripada S, Poornachandra Y. Status and future prospects of fructooligosaccharides as nutraceuticals.Role of materials science in food bioengineering. Academic Press 2018; pp. 451-503.

[http://dx.doi.org/10.1016/B978-0-12-811448-3.00014-0]

[33]   Singh R, Kumar M, Mittal A, Mehta PK. Microbial metabolites in nutrition, healthcare and agriculture. 3 Biotech 2017; 7(1): 15.
[http://dx.doi.org/10.1007/s13205-016-0586-4] [PMID: 28391479]

[34]   Slavin J. Fiber and prebiotics: Mechanisms and health benefits. Nutrients 2013; 5(4): 1417-35.
[http://dx.doi.org/10.3390/nu5041417] [PMID: 23609775]

[35]   Oliveira DL, Wilbey RA, Grandison AS, Roseiro LB. Milk oligosaccharides: A review. Int J Dairy Technol 2015; 68(3): 305-21.
[http://dx.doi.org/10.1111/1471-0307.12209]

[36]   Delzenne NM, Kok N. Effects of fructans-type prebiotics on lipid metabolism. Am J Clin Nutr 2001; 73(2) (Suppl.): 456s-8s.
[http://dx.doi.org/10.1093/ajcn/73.2.456s] [PMID: 11157357]

[37]   Macfarlane GT, Steed H, Macfarlane S. Bacterial metabolism and health-related effects of galacto-oligosaccharides and other prebiotics. J Appl Microbiol 2008; 104(2): 305-44.
[PMID: 18215222]

[38]   Patel S, Goyal A. The current trends and future perspectives of prebiotics research: A review. Biotech 2012; 2(2): 115-25.

[39]   Franco-Robles E, López MG. Implication of fructans in health: immunomodulatory and antioxidant mechanisms. ScientificWorldJournal 2015; 2015: 1-15.
[http://dx.doi.org/10.1155/2015/289267] [PMID: 25961072]

[40]   Chung WSF, Meijerink M, Zeuner B, *et al*. Prebiotic potential of pectin and pectic oligosaccharides to promote anti-inflammatory commensal bacteria in the human colon. FEMS Microbiol Ecol 2017; 93(11).
[http://dx.doi.org/10.1093/femsec/fix127] [PMID: 29029078]

[41]   Lin SH, Chou LM, Chien YW, Chang JS, Lin CI. Prebiotic effects of xylooligosaccharides on the improvement of microbiota balance in human subjects. Gastroenterol Res Pract 2016; 2016: 1-6.
[http://dx.doi.org/10.1155/2016/5789232] [PMID: 27651791]

[42]   Prado SBR, Castro-Alves VC, Ferreira GF, Fabi JP. Ingestion of non-digestible carbohydrates from plant-source foods and decreased risk of colorectal cancer: A review on the biological effects and the mechanisms of action. Front Nutr 2019; 6: 72-85.
[http://dx.doi.org/10.3389/fnut.2019.00072] [PMID: 31157230]

[43]   Zhang N, Huang X, Zeng Y, Wu X, Peng X. Study on prebiotic effectiveness of neutral garlic fructan *in vitro*. Food Sci Hum Wellness 2013; 2(3-4): 119-23.
[http://dx.doi.org/10.1016/j.fshw.2013.07.001]

[44]   Dzuvor C, Taylor J, Acquah C, Pan S, Agyei D. Bioprocessing of functional ingredients from flaxseed. Molecules 2018; 23(10): 2444.
[http://dx.doi.org/10.3390/molecules23102444] [PMID: 30250012]

[45]   Zhang X, Wang H, Yin P, Fan H, Sun L, Liu Y. Flaxseed oil ameliorates alcoholic liver disease *via* anti-inflammation and modulating gut microbiota in mice. Lipids Health Dis 2017; 16(1): 44.
[http://dx.doi.org/10.1186/s12944-017-0431-8] [PMID: 28228158]

[46]   Vinke PC, El Aidy S, van Dijk G. The role of supplemental complex dietary carbohydrates and gut microbiota in promoting cardiometabolic and immunological health in obesity: lessons from healthy non-obese individuals. Front Nutr 2017; 4: 34-42.
[http://dx.doi.org/10.3389/fnut.2017.00034] [PMID: 28791292]

[47]   Kaur R, Sharma M, Ji D, Xu M, Agyei D. Structural features, modification, and functionalities of beta-glucan. Fibers 2019; 8(1): 1.
[http://dx.doi.org/10.3390/fib8010001]

[48]   Nath A, Molnár MA, Csighy A, *et al.* Biological activities of lac-tosebased prebiotics and symbiosis with probiotics on controlling osteoporosis, blood-lipid and glucose levels. Medicina 2018; 54(6): 98-103.
[http://dx.doi.org/10.3390/medicina54060098] [PMID: 30513975]

[49]   Zhang T, Yang Y, Liang Y, Jiao X, Zhao C. Beneficial effect of intestinal fermentation of natural polysaccharides. Nutrients 2018; 10(8): 1055.
[http://dx.doi.org/10.3390/nu10081055] [PMID: 30096921]

[50]   Peng M, Tabashsum Z, Anderson M, *et al.* Effectiveness of probiotics, prebiotics, and prebiotic-like components in common functional foods. Compr Rev Food Sci Food Saf 2020; 19(4): 1908-33.
[http://dx.doi.org/10.1111/1541-4337.12565] [PMID: 33337097]

[51]   Cani PD, Possemiers S, Van de Wiele T, *et al.* Changes in gut microbiota control inflammation in obese mice through a mechanism involving GLP-2-driven improvement of gut permeability. Gut 2009; 58(8): 1091-103.
[http://dx.doi.org/10.1136/gut.2008.165886] [PMID: 19240062]

[52]   Oak SJ, Jha R. The effects of probiotics in lactose intolerance: A systematic review. Crit Rev Food Sci Nutr 2019; 59(11): 1675-83.
[http://dx.doi.org/10.1080/10408398.2018.1425977] [PMID: 29425071]

[53]   Markowiak P, Śliżewska K. Effects of probiotics, prebiotics, and synbiotics on human health. Nutrients 2017; 9(9): 1021.
[http://dx.doi.org/10.3390/nu9091021] [PMID: 28914794]

[54]   Dalile B, Van Oudenhove L, Vervliet B, Verbeke K. The role of short-chain fatty acids in microbiota–gut–brain communication. Nat Rev Gastroenterol Hepatol 2019; 16(8): 461-78.
[http://dx.doi.org/10.1038/s41575-019-0157-3] [PMID: 31123355]

[55]   Christiansen CB, Gabe MBN, Svendsen B, Dragsted LO, Rosenkilde MM, Holst JJ. The impact of short-chain fatty acids on glp-1 and pyy secretion from the isolated perfused rat colon. Am J Physiol Gastrointest Liver Physiol 2018; 315(1): G53-65.
[http://dx.doi.org/10.1152/ajpgi.00346.2017] [PMID: 29494208]

[56]   Dubey MR, Patel VP. Probiotics: A promising tool for calcium absorption. Open Nutr J 2018; 12(1): 59-69.
[http://dx.doi.org/10.2174/1874288201812010059]

[57]   Scholz-Ahrens KE, Ade P, Marten B, *et al.* Prebiotics, probiotics, and synbiotics affect mineral absorption, bone mineral content, and bone structure. J Nutr 2007; 137(3) (Suppl. 2): 838S-46S.
[http://dx.doi.org/10.1093/jn/137.3.838S] [PMID: 17311984]

[58]   Niño DF, Sodhi CP, Hackam DJ. Necrotizing enterocolitis: New insights into pathogenesis and mechanisms. Nat Rev Gastroenterol Hepatol 2016; 13(10): 590-600.
[http://dx.doi.org/10.1038/nrgastro.2016.119] [PMID: 27534694]

[59]   Indrio F, Riezzo G, Raimondi F, Bisceglia M, Cavallo L, Francavilla R. The effects of probiotics on feeding tolerance, bowel habits, and gastrointestinal motility in preterm newborns. J Pediatr 2008; 152(6): 801-6.
[http://dx.doi.org/10.1016/j.jpeds.2007.11.005] [PMID: 18492520]

SECTION 2: GENOME EDITING OF MEDICINAL PLANTS

CHAPTER 7

Medicinal Plants and Molecular Techniques

Fatima Javeria[1], Saira Karimi[2], Shoaib ur Rehman[1], Furqan Ahmad[1], Akash Fatima[1], Muhammad Ashfaq[3], Babar Farid[1] and Zulqurnain Khan[4,*]

[1] *Institute of Plant Breeding and Biotechnology, MNS University of Agriculture, Multan, Pakistan*

[2] *Department of Biosciences, COMSATS University, Islamabad, Pakistan*

[3] *Institute of Plant Protection, MNS University of Agriculture, Multan, Pakistan*

[4] *Department of Biotechnology, Institute of Plant Breeding and Biotechnology (IPBB), Muhammad Nawaz Shareef University of Agriculture, Multan-60000, Pakistan*

Abstract: Medicinal plants provide a substantial source of bioactive compounds which serve greatly in the pharmaceutical industry. Before revolutionary advancements in medicines, traditional biotechnology approaches have been used in the breeding of significant therapeutic plants. The challenge is to incorporate effective, efficient, and resilient breeding techniques to enhance the production of phytochemicals by medicinal plants. Genetics and biotechnology can aid in the rapid advancement of therapeutic plants by assessing genetic diversity, conservation, proliferation, and overproduction. Hence, the use of advanced technologies is crucial for selecting, multiplying, and preserving medicinal plants.

Keywords: Conservation, Genetic diversity, Medicinal plants, Molecular techniques, Proliferation, Pharmaceutical industry.

INTRODUCTION

Medicinal plants are a significant group of plants that have been overlooked by humans and plant breeders for many years. They are an important source of novel medications all around the world [1]. In Europe, more than 1300 therapeutic plants are used, 90% of which are accumulated from wild assets; in the USA (United States of America), regular sources represent around 118 of the main 150 professionally prescribed drugs [2]. Besides, up to 80% of individuals in underdeveloped countries depend just on herbal medicines for their essential treatment, while more than 25% of suggested medications in developed nations come from wild herbal plants [3]. The usage of medical plants is quickly expan-

* **Corresponding author Zulqurnain Khan:** Department of Biotechnology, Institute of Plant Breeding and Biotechnology (IPBB), Muhammad Nawaz Shareef University of Agriculture, Multan-60000, Pakistan; E-mail: zulqurnain.khan@mnsuam.edu.pk

ding globally, due to the strong demand for herbal medications, healthcare products, and bioactive components of medicinal plants.

In recent decades, the adverse side effects of chemical medications have prompted a greater focus on medicinal plants and their utilization in diet, pharmaceuticals, and other fields [4]. Considering the importance and need for these natural chemicals, genetic modification of this crop to increase secondary metabolite yield is critical. Plant biotechnology and genetic engineering approaches can help to compensate for the time it takes to develop therapeutic plants. Recently, "green" extraction techniques, have developed interest as they cause the elimination of toxic substances and volatile organic solvents.

Multiple molecular practices like DNA fingerprinting techniques have been established for the identification of traditional medicinal plants, these techniques are RFLP, AP-PCR, RAPD, DALP, AFLP, PCR-RFLP, ISSR, SCAR, and isothermal amplification) [5], DNA microarray [6], DNA barcoding, and Forensic informative nucleotide sequencing (FINS). Formerly, morphological traits were used to identify medicinal compounds, as described in Shengnong Bencaojing (200 AD) [7]. Morphological traits were graphically illustrated in Bencao Gangmu (1,593 AD) [8]. With the advancement of technology, a variety of identification methods have been developed. The development of molecular tools in the 1990s was a turning point in the history of traditional medicine identification. DNA barcodes have recently been proposed for identifying living species, including therapeutic compounds [9]. Nowadays, morphological and microscopic characteristics are coupled to identify therapeutic materials initially. To improve the accuracy of identification, these procedures are combined with chemical profiles acquired from thin-layer chromatography (TLC), high-performance liquid chromatography (HPLC), or liquid chromatography/mass spectrometry (LC/MS) [10]. The subject of the possible application of herbal fingerprinting techniques has been brought by the expanding market for herbal remedies, the growth of Latvia's foreign commerce, and the absence of acceptable analytical procedures. Strategies for collecting the chromatographic fingerprints of four medicinal plants, for example (*Hibiscus sabdariffa L., Calendula officinalis L., Matricaria recutita L., Achillea millefolium L.*) that are taxonomically and evolutionarily distinct were established using high-performance liquid chromatography (HPLC) and thin layer chromatography (TLC) [11].

DNA Fingerprinting

The DNA polymorphism is investigated using DNA fingerprinting. Capillary electrophoresis [12] agarose gel electrophoresis [13], or polyacrylamide gels [14] are commonly used to detect polymorphism patterns. Due to post-harvest processing and storage, the quality and quantity of the DNA in medicinal materials may be compromised as compared to fresh materials [5]. DNA fingerprints that do not require DNA amplification, such as restriction fragment length polymorphism (RFLP), are ineffectual due to the lower yields and poor integrity of extracted DNA. Hence PCR-based fingerprinting is preferred by scientists. Some of the available fingerprints are arbitrary primed PCR (AP-PCR), random amplified polymorphic DNA (RAPD), amplified fragment length polymorphism (AFLP), direct amplification of length polymorphism (DALP), inter-simple sequence repeat (ISSR), PCR restriction fragment length polymorphism (PCR, RFLP), and sequence-characterized amplification region (SCAR). Modern techniques have been used for population structure and genetic diversity in medicinal plants, such as in Iran, SCoT markers was used for the analysis of medicinal plant *Melissa officinalis* [15].

Polymerase Chain Reaction

PCR is an *in vitro* technique used for the amplification of DNA fragments using digestive enzymes [16]. Two oligonucleotide primers *i.e.* forward and reverse primers are used to initiate a DNA polymerase which is a thermostable enzyme that copies each strand of DNA throughout the PCR process [17]. A three-step reaction occurs when the DNA double helix is denatured, oligonucleotide primers are annealed to each complementary strand, and DNA polymerase synthesizes new strands at their best temperature. PCR is conducted in thermocyclers that are completely programmable to alter the reaction temperature at each step.

Arbitrarily Primed PCR (AP-PCR)

A whole-genome fingerprinting technique "arbitrarily primed PCR" (AP-PCR) was initially reported in 1990. It is identical to RAPD, except it uses two sets of primers greater than eighteen nucleotides of arbitrary sequence. The dry roots of Oriental ginseng (*Panax ginseng*) were successfully distinguished from those of American ginseng (*Panax quinquefolius*) using 3 primers (M13 forward and reverse, and Gal-K primer) [18]. A similar technique was used to categorize additional medicinal species, such as Kudidan (*Elephantopi Herba*), Pugongying (*Taraxaci Herba*), and Dangshen (*Codonopsis Radix*) [5].

Random Amplified Polymorphic DNA (RAPD)

Another whole-genome fingerprint is random amplified polymorphic DNA (RAPD). The approach of RAPD is analogous to AP-PCR, except for the usage of a single 10-nucleotide primer with random sequence resulting in amplification at different sites of the genome that are evaluated by gel electrophoresis. RAPD can be used to evaluate several loci at the same time and does not require preliminary information of the genome. In AP-PCR and RAPD, the DNA quality and integrity stays a momentous issue. Furthermore, dominant markers as AP-PCR and RAPD are not capable of differentiating between the homozygous and heterozygous loci.

Restriction Fragment Length Polymorphism

In RFLP, sequence-specific DNA restriction endonucleases are used to cleave genomic D NA, resulting in a multitude of small fragments of varying lengths that are separated by gel electrophoresis based on their molecular size. A DNA fingerprint of a source is a band pattern created using a specific DNA from a specific source using a specific restriction enzyme. *Dasamula*, an herbal medication formulation made up of the roots of ten plants, is extensively used to treat a variety of ailments. However, there is no reliable way to check the existence of various needed plant materials in the drug. Six of the ten Dasamula plants, namely *Desmodium giganicum, Aegle marmelos, Solanum xanthocarpum, Solanum indicum, Tribulus terresteris,* and *Oroxylum indicum*, were studied using RFLP technique [19].

Inter Simple Sequence Repeat-Anchored PCR (ISSR-PCR)

Simple sequence repeats (SSRs), commonly called microsatellites, are short tandem repeats of 2-6 bp found across the genome. ISSR fingerprinting is a whole-genome scanning fingerprint that uses PCR primers based on other species. ISSR-PCR amplifies the DNA regions between the flanking SSR sequences using primers attached to SSR sequences (*e.g.*, CACACACA) [20]. *Oroxylum indicum* is an important traditional medicinal plant native to the Indian subcontinent that is in danger of extinction. The genetic diversity and population structure of 39 *O. indicum* obtained from South and Northeast India were assessed using Inter Simple Sequence Repeats (ISSR) markers. (UPGMA) grouping also revealed that the South and North East Indian populations are genetically distinct [21].

Amplification Refractory Mutation System (ARMS) and Multiplex Amplification Refractory Mutation System

Primers bind to their targeted sequence in ARMS only when their 3′ ends are complimentary. The "typical" target sequence will not bind to mismatched or mutated 3′ end and no amplification will occur in ARMS [22] whereas ARMS uses a multiplex PCR utilizing an ordinary primer and numerous mutant specific primers [23]. Dendrobium is a significant genus in traditional Chinese herbal medicine, and the exact identification of Dendrobium species is necessary for therapy and pharmacological research. The six medicinal *Dendrobium* species were successfully identified using multiplex PCR and the (ARMS) was utilized to separate *D. tosaense* from the other 19 *Dendrobium* species [24]. The existing PCR-based (multiplex and ARMS) studies can be employed to verify the authenticity of raw materials from various species of medicinal Dendrobium.

Start Codon Targeted Markers (SCoT)

SCoT marker system has proven to be better, when compared to RAPD. SCoT can provide great repeatability even if it uses a random marker system like RAPD. SCoT also has a greater annealing temperature than RAPD. This is explained by the fact that SCoT's 18-mer primers are longer than RAPD's 10-mer primers. When used to access genetic linkages in certain cultivars, SCoT has proven to be more effective than ISSR. Start Codon Targeted Polymorphism produces useful information. As a gene focused marker system, it generates extremely accurate and trustworthy data [25]. *Melissa officinalis* is a valuable medicinal herb that is widely utilized by locals of Iran. This species is spreading throughout Iran and has formed various local populations. Based on SCoT markers, there was no information available on the population genetics or genetic diversity of this species from Iran. So, a study was performed using ten geographical populations and the discriminating capability of start codon targeted (SCoT) marker analysis discovered certain SCoT loci with adaptive potential. According to this study, populations of *M. officinalis* are being affected by a variety of population genetic divergence, genetic drift, restricted gene flow, and adaptability. Such information may be utilized to protect this significant facility for the nation [15].

DNA Microarray

DNA microarray, also known as biochip or gene chip is a technique for analyzing multiple genes at the same time. It is a hybridization-based approach that hybridizes single or several loci in a target genome using labeled nucleotide probes [6]. A DNA microarray is a solid surface with numerous tiny DNA patches bonded to it. Short nucleotide fragments derived *via* restriction digestion or synthesized oligonucleotides are used as probes. The word "probe" refers to small gene sequences. The size of these spots is picomolar (1012 moles). These probes are permitted to hybridize with the target sequence of the biological samples. They are attached to a supporting matrix, which is where probes and tested DNA samples are hybridized [26].

DNA Barcoding

DNA barcoding is a systematic approach that differs a brief genetic marker from a specific organism and then associates it with a specific species. The technique's main purpose is to detect the unknown species and link them to a recognized systematic classification system [6, 27].

1st Generation Sequencing

Techniques for sequencing have progressed speedily. Fig. (**1**) shows different generation of sequencing. The 1st generation sequencing methods essentially improved Sanger's sequencing approaches with fluorophores employed for fluorescence-based detection.

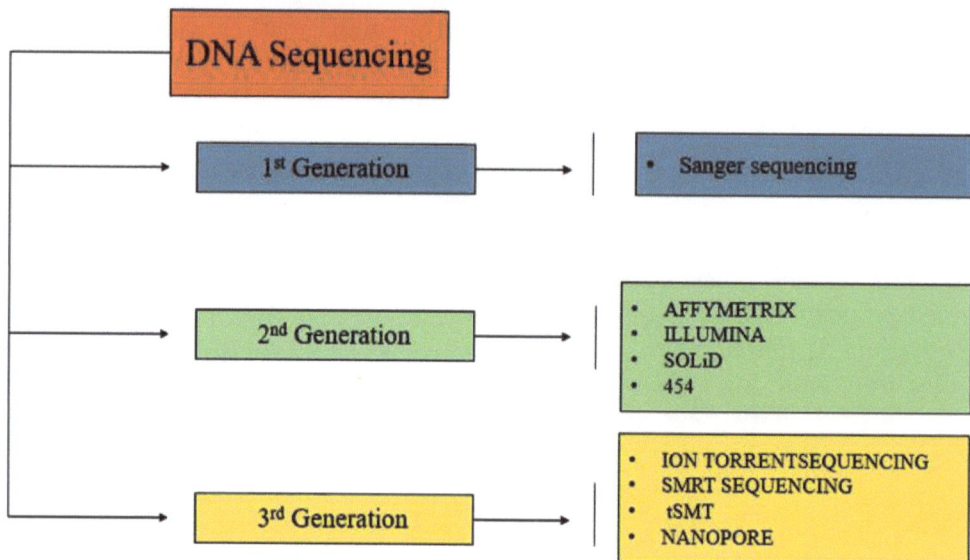

Fig. (1). Generations of DNA sequencing.

Sanger Sequencing

The most prominent DNA sequencing procedure is the dideoxy chain termination technology (Sanger technique). This method involves copying the DNA template to be sequenced and then inserting dideoxy nucleotide bases to prevent replication and the results are visualized on agarose gel electrophoresis. Sanger and Next-Generation Sequencing (NGS) were used to confirm the taxonomic identity of fifteen herbal supplements derived from five therapeutic plants: *Echinacea purpurea, Valeriana officinalis, Ginkgo biloba*, and *Hypericum perforatum*. To remove background contamination, three distinct DNA extraction procedures, two lysate dilutions, an internal amplification control, and several negative controls were employed. HPLC-MS was also used to detect the presence of active medicinal components in ginkgo supplements [28].

2nd Generation Sequencing

Reversible Terminator Sequencing (Illumina)

In 2006, Solexa pioneered reversible terminator sequencing, often called sequencing by synthesis. Fluorescently tagged modified nucleotides are employed as reversible terminators. 3'-O-blocked reversible terminators and 3'-unblocked reversible terminators are two different forms of reversible terminators. Each DNA base out of four has a distinct fluorophore connected to a nitrogenous base in addition to a 3'-O-azidomethyl group. This fluorescent tag is split using tris [2-carboxyethyl) phosphine (TCEP), which eliminates the 3'-O-azidomethyl group while simultaneously regenerating 3'-OH, and the cycle can be repeated. The nucleotide, as well as the fluorescent group, are attached to 3' reversible terminators, which act as both a termination group and a reporter. The fluorophore is the same for all four bases, and each altered base is flooded sequentially rather than all at once since the 3'-position is not crowded (*i.e.*, the base has a free 3'-OH).

Sequencing by Ligation (SOLiD)

SOLiD was introduced in 2007 and entails a genomic library, ligating it, and then sequencing it. The target sample is amplified using EmPCR, then agarose beads are anchored to a glass surface. The position is then inundated with fluorescently tagged oligonucleotides once it has been connected. Because the template and oligonucleotides are complementary to each other hence, annealing is followed by DNA ligase-mediated ligation. The fluorescent dye is extracted from the fragment using silver ions (Ag+) and the phosphonothioate connection between bases. This enables the detection of four distinct fluorescence peaks, each of which corresponds to a different nucleotide. A 5'- becomes vulnerable to further ligation when the fluorophore is removed [29].

3rd Generation Sequencing

Certain unique approaches have been developed using single real-time sequencing and single-molecule sequencing (SMS) with the least cost and minimizing possible biases. Among the firms that use SMS are Pacific Biosciences, Oxford Nanopore Technologies, and Helicos Biosciences.

Ion Torrent Semiconductor Sequencing

Ion torrent sequencing is the primary practical approach that does not rely on camera scanning or fluorescence, instead directly translating nucleotide composition into digital information (0, 1) on the device. As a result, it is both quicker and less costly than earlier sequencing platforms. When nucleotides are incorporated into expanding DNA, a hydronium ion is released, which is recognized by the sequencer's pH sensor. The semiconductor chip immediately converts this to a voltage [30].

NGS methods may also be used to assess metabolisms in plants that are not model plants or haven't been studied before such as medicinal plants [31]. NGS technology may be used to profile transcriptomes or sequence cDNA (RNA-seq). The full transcriptomes of the target populations will be sequenced using RNA-seq. NGS-based transcriptome analysis outperforms microarray approaches in terms of identifying undiscovered genes as well as expression differences between paralogous and homoeologous copies of the gene. This approach has been efficaciously used to discover the uni-genes involved in the production of monoterpenoids in the medicinal plant ajowan (*Trachyspermum ammi L.*) [32]. The findings of such studies can be utilized to improve the functional breeding of medicinal plants.

Plant Tissue Culture (PTC) and Medicinal Plants

(PTC) is the backbone of therapeutic plant biotechnology and a major platform for the production of secondary metabolites [33]. This strategy is the most encouraging rescuer of therapeutic plants with restricted yields and sensitive to biotic pressure [34]. PTC prompts the progress *in situ* and *ex situ* preservation, polyploidy induction, micropropagation, gene transformation, and bioreactor applications in the unexploited genome of therapeutic plants [35]. PTC can provide a secure environment in which to examine the influence of multiple experimental circumstances and materials on the synthesis of secondary products in therapeutic plants [36, 37]. PTC also aids in the study of endogenous hormone, transport, and metabolism signaling [38, 39].

Plant Tissue Culture-based *ex-situ* Conservation

The undiscriminating use and overharvesting of the therapeutic plants have put them at threat even in their natural habitats [40]. Other human activities that put wild populations of various medicinal plants in danger include human-driven climate change, ecological damage by overexploitation of natural resources, deforestation, wild fires, industrialization, and land development for agriculture. [41]. *In situ* conservation alone will not suffice to conserve significant therapeutic plants. *Ex-situ* conservation is therefore critical in medicinal plants [42]. *Ex-situ* conservation is a backup of unique plant genetics that might be lost in their natural habitat.

Cryopreservation

Cryopreservation is the long-term storage of biotic materials at extremely low temperatures (below 196 ∘C) [43]. Plant germplasm is protected through seeds, bulbs, corms, buds, cuttings, tubers, roots, rhizomes, and other organs [44]. Cryopreservation is being used for *ex-situ* preservation and eternal storage of therapeutic plants with resistant seeds. The three major aims of cryopreservation in medicinal plants are the preservation of biosynthetic capacity, the conservation of germplasm, and the genetic constancy of somaclones [45]. *Dendrobium candidum* [46], *Hypericum perforatum* [47] and *Eruca sativa* [48] are some important medicinal plants that can be stored *via* cryopreservation.

Synthetic Seeds Production in Medicinal Plants

Plant *ex-situ* conservation can take several forms, including seed gene banks, field gene banks, cryopreserved tissues, and *in-vitro* collections, which conserve germplasm in fields or botanical gardens [49]. Genetic stability, simplicity of management and transportation, and efficacy concerning the space, time, labor, and cost are all significant characteristics of synthetic seeds [50]. As a result, synthetic seeds, which relate to the encapsulation of explant in alginate, are the most effective for medicinal plant conservation [51]. *Clitoria ternatea* Linn [52], Black-oil tree (*Celastrus paniculatus* Wild) are highly valued medicinal plant species which are conserved by synthetic seed technology [53].

Omic Approaches

Genomics

The term "genome" comprises two terms "gene" and "chromosome." The genome is each living entity's full set of hereditary information, and it is required for its growth and function. The study of genomics allows researchers to delve further into the difficulties posed by the genetic foundation. The full genome of a free-living microorganism, *Haemophilus influenzae*, was sequenced in the 1990s, and this is considered an important contribution to genomics [29]. Genomic research characterizes and measures expression as a whole, as well as its regulatory network. In short, this study makes it simpler to assess transcriptomic, proteomic, and epigenomic data in prokaryotes and eukaryotes.

Transcriptomics

Transcriptomics is an example of Integromics, in which data from several omic branches is combined to provide precise information. Transcriptomics provides the most informative foundation for beginning any research project, and with the introduction of new high-throughput methods, generating a lot of data and information has become very simple and quick. The transcriptome contains all the transcripts existing in a cell which include mRNA, miRNA, short RNAs and noncoding RNAs, Transcriptomics measures the quantity of RNA and the transcriptional structure to examine the differential expressions of all the transcripts at different stages of development at different physiological circumstances. Diversification, noncoding RNAs, and transcriptional unit structure in coding regions are all covered.

The transcriptomic study starts with a basic technology termed as EST, which stands for expressed sequence tags, and was followed by SAGE, which stands for Serial Analysis of Gene Expression which depends upon Sanger sequencing. Both were time-consuming and only resolute a limited number of transcripts, delivering just half of the transcriptome's information. The development of a technical innovation known as microarray in the 1990s marked the beginning of a transformative decade in transcriptomics.

A collaboration of US research groups began the project Transcriptome Characterization, Sequencing, and Assembly of Medicinal Plants Relevant to Human Health in 2011 with funding from the National Institutes of Health (NIH) [54]. The database currently contains transcriptome data for 31 medicinal species, including *Cinchona pubescens* Vahl common name the quinine tree with the family name Rubiaceae, *Datura stramonium* L. with the family name "Solanaceae", *Colchicum autumnale* L. with the family name Colchicaceae, the source of colchicine, and *Podophyllum peltatum* L. Berberidaceae family (mayapple; the roots of Podophyllum) [55].

A database was created in Phytometasyn project (www.phytometasyn.ca). It includes *de novo* transcript assemblies of about twenty therapeutic plants, including *Eschscholzia californica* (Cham) [56]. California poppy, a Papaveraceae species accumulates many active BIAs, primarily of the pavine-type for example eschscholtzidine [57].

Proteomics

Proteomics is the study of whole proteins from a source, as well as the methodologies used to examine these proteins and their interactions. The relevance and function of proteins in an organism are defined by a 3-D map of their connections [58]. These investigations begin at the protein level and may occasionally be traced down to their genes. In eukaryotes, the alternative splicing event boosts the proteome diversity selectively.' The 3-D structure of proteins is considered in structural proteomics, which aids in the structure-based rational drug design process.

Functional proteomics, on the other hand, is primarily concerned with cellular protein expression, protein changes, protein interactions, signaling, and disease processes. With the advancement of technology, procedures such as NMR, X-rays, HPLC, mass spectroscopy, and 2-D PAGE have led to the collection of massive amounts of data [59]. The large amount of new experimental data generated by diverse protein detection technologies makes it challenging for scientists. Bioinformatics processes comprising algorithms, databases, and pipelines for computational analysis allow for a quicker and more accurate study in only a few days. Databases and resource portals have made data administration, storage, and sharing easier for academics, resulting in a rise in proteomics research.

Metabolomics

Metabolomics is a high-throughput analysis of metabolites that play a role in metabolism. Metabolomics aids in the improvement of genetically modified crops as well as the identification of the risks associated with their use by providing a snapshot of plant growth across time. Plant metabolite identification is particularly essential since it aids in the identification of primary and secondary metabolite functions [60].

Genome Editing

Sequence-specific nucleases include ZFNs, TALENs, and CRISPR/Cas, that may generate DSBs at specific locations in a DNA sequence. These fusion proteins feature a generic DNA-cleavage domain as well as a programmable and sequence-specific DNA-binding domain. The FokI restriction enzyme is used to create a DNA-cleavage domain in both ZFNs and TALENs. DNA binding in ZFNs is mediated by a group of Cys2His2 zinc finger proteins, whereas DNA binding in TALENs is mediated by Xanthamonas sp. A single guide RNA (sgRNA) and a Cas9 endonuclease collectively create the CRISPR/Cas system [61]. The 20-nucleotide sgRNA is accountable for directing Cas9 cleavage and serving as a DNA-binding domain. All three ZFNs, TALENs, and CRISPR/Cas have a DNA-binding domain that can be designed to identify and link to any sequence of interest. DNA constructs encoding sequence-specific nucleases and desired foreign DNA are co-transformed into plant tissues, resulting in precise site-specific foreign DNA integration. In comparison to transgenesis, gene editing has resulted in a rise in some crop species and a decrease in others among crops that have been authorized, not regulated, or commercialized. Breeding characteristics show a similar divergent tendency. New private businesses have benefited from gene editing as well. The patenting landscape is primarily controlled by China, particularly by its governmental sector, as opposed to the approved/marketed landscape, which is dominated by the USA. The statistics suggest that regulatory frameworks will encourage or restrict the innovation [62].

CONCLUSION

DNA technologies are trustworthy and effective methods for identifying taxa at different taxonomic levels (*e.g.*, species, subspecies, variety, strain), since they produce consistent findings regardless of age, tissue origin, physiological circumstances, environmental factors, harvest, storage, and processing of materials. The necessity for DNA authentication will increase along with the need for high-quality herbs in order to guarantee medicinal effectiveness, fair drug trade, and increased customer confidence. Table **1** shows some success stories in which medicinal plants are identified using different molecular techniques. This identification will also help in the maintenance of genetic diversity which is necessary for the survival of the plant. With the help of molecular techniques, scientists are now able to modify the genome with value-added traits which will help in the production of desired high-quality drugs from the medicinal plant. These drugs can be used for the treatment of different diseases effectively with minimum side effects. So, molecular methods have great importance in the maintenance of medicinal plants.

Table 1. Success Reports.

Sr. No.	Medicinal Plant	Sample	Technique Used	Purpose/ Identification/Gene	References
1	*Aconitum carmichaeli*	Fresh leaves	PCR, sequencing; microarray (silicon)	5S gene spacer	[63]
2	*Alisma nanum*	Dried Rhizome	PCR, sequencing; RFLP; ARMS	ITS (Internal transcribed spacer)	[64]
3	*Angelica acutiloba var. acutiloba*	Fresh leaves	PCR, sequencing	Spacer between atpF-atpA	[65]
4	*Actaea podocarpa*	Leaves	AFLP	Identification of medicinal plant	[66]
5	*Angelica sinensis*	Dried Root	RAPD; RFLP	*In vitro* propagation and DNA analysis of Angelica plants	[67]
6	*Astragalus membranaceus var. mongholicus*	Fresh roots	3′ untranslated region sequence-based amplified polymorphism (UAP)	Authentication of *Radix astragali*	[68]

REFERENCES

[1]   Chacko SM, Thambi PT, Kuttan R, Nishigaki I. Beneficial effects of green tea: A literature review. Chin Med 2010; 5(1): 13.
[http://dx.doi.org/10.1186/1749-8546-5-13] [PMID: 20370896]

[2]   Balunas MJ, Kinghorn AD. Drug discovery from medicinal plants. Life Sci 2005; 78(5): 431-41.
[http://dx.doi.org/10.1016/j.lfs.2005.09.012] [PMID: 16198377]

[3]   Hamilton AC. Medicinal plants, conservation and livelihoods. Biodivers Conserv 2004; 13(8): 1477-517.
[http://dx.doi.org/10.1023/B:BIOC.0000021333.23413.42]

[4]   Nabavi S, Di Lorenzo A, Izadi M, Sobarzo-Sánchez E, Daglia M, Nabavi S. Antibacterial effects of cinnamon: From farm to food, cosmetic and pharmaceutical industries. Nutrients 2015; 7(9): 7729-48.
[http://dx.doi.org/10.3390/nu7095359] [PMID: 26378575]

[5]   Li M, Jiang C, Pui-Hay P, Shaw P-C, Yuan Y. Molecular identification of traditional medicinal materials. Molecular Pharmacognosy: Springer 2019; pp. 13-39.

[6]   Sarwat M, Yamdagni MM. DNA barcoding, microarrays and next generation sequencing: recent tools for genetic diversity estimation and authentication of medicinal plants. Crit Rev Biotechnol 2016; 36(2): 191-203.
[http://dx.doi.org/10.3109/07388551.2014.947563] [PMID: 25264574]

[7]   Tang W, Eisenbrand G. Chinese drugs of plant origin: chemistry, pharmacology, and use in traditional and modern medicine. Springer Science & Business Media 2013.

[8]   Jaiswal Y, Liang Z, Ho A, *et al.* Distribution of toxic alkaloids in tissues from three herbal medicine Aconitum species using laser micro-dissection, UHPLC–QTOF MS and LC–MS/MS techniques. Phytochemistry 2014; 107: 155-74.
[http://dx.doi.org/10.1016/j.phytochem.2014.07.026] [PMID: 25172517]

[9]   Hollingsworth PM, Forrest LL, Spouge JL, *et al.* A DNA barcode for land plants. Proc Natl Acad Sci 2009; 106(31): 12794-7.
[http://dx.doi.org/10.1073/pnas.0905845106] [PMID: 19666622]

[10]   Nikam PH, Kareparamban J, Jadhav A, Kadam V. Future trends in standardization of herbal drugs. J Appl Pharm Sci 2012; 2(6): 38-44.

[11]   Bārzdiņa A, Paulausks A, Bandere D, Brangule A. The potential use of herbal fingerprints by means of HPLC and TLC for characterization and identification of herbal extracts and the distinction of latvian native medicinal plants. Molecules 2022; 27(8): 2555.
[http://dx.doi.org/10.3390/molecules27082555] [PMID: 35458753]

[12]   Heller C. Principles of DNA separation with capillary electrophoresis. Electrophoresis 2001; 22(4): 629-43.
[http://dx.doi.org/10.1002/1522-2683(200102)22:4<629::AID-ELPS629>3.0.CO;2-S]   [PMID: 11296917]

[13]   Lee PY, Costumbrado J, Hsu C-Y, Kim YH. Agarose gel electrophoresis for the separation of DNA fragments. J Vis Exp 2012; (62): e3923.
[PMID: 22546956]

[14]   Green MR, Sambrook J. Polyacrylamide gel electrophoresis. Cold Spring Harbor Protocols 2020; 2020(12): pdb. prot100412.
[http://dx.doi.org/10.1101/pdb.prot100412]

[15]   Koohdar F, Sheidai M. Genetic diversity and population structure in medicinal plant Melissa officinalis L. (Lamiaceae). Genet Resour Crop Evol 2022; 69(5): 1753-8.
[http://dx.doi.org/10.1007/s10722-021-01338-7]

[16]   Bej AK, Mahbubani MH, Atlas RM. Amplification of nucleic acids by polymerase chain reaction

(PCR) and other methods and their applications. Crit Rev Biochem Mol Biol 1991; 26(3-4): 301-34.
[http://dx.doi.org/10.3109/10409239109114071] [PMID: 1718663]

[17]   Dieffenbach CW, Dveksler GS. PCR primer: A laboratory manual. Cold Spring Harbor Laboratory Press 2003.

[18]   Ka-Shing C, Hoi-Shan K, Pang-Chui S, Shaw P-C, *et al.* Pharmacognostical identification of american and oriental ginseng roots by genomic fingerprinting using arbitrarily primed polymerase chain reaction (AP-PCR). J Ethnopharmacol 1994; 42(1): 67-9.
[http://dx.doi.org/10.1016/0378-8741(94)90025-6] [PMID: 8046946]

[19]   Biswas K, Biswas R. Identification of medicinal plants using PCR-RFLP in Dasamula—an Ayurvedic drug. J PharmacBioSci 2013; 3: 93-8.

[20]   Kumar A, Mishra P, Singh SC, Sundaresan V. Efficiency of ISSR and RAPD markers in genetic divergence analysis and conservation management of Justicia adhatoda L., a medicinal plant. Plant Syst Evol 2014; 300(6): 1409-20.
[http://dx.doi.org/10.1007/s00606-013-0970-z]

[21]   Rajasekharan P, Kareem VA, Ravish B, Mini S. Genetic diversity in Oroxylum indicum (L.) Vent., a threatened medicinal plants from India by ISSR analysis. 2017.

[22]   Sucher N, Carles M. Genome-based approaches to the authentication of medicinal plants. Planta Med 2008; 74(6): 603-23.
[http://dx.doi.org/10.1055/s-2008-1074517] [PMID: 18449847]

[23]   Heubl G. New aspects of DNA-based authentication of Chinese medicinal plants by molecular biological techniques. Planta Med 2010; 76(17): 1963-74.
[http://dx.doi.org/10.1055/s-0030-1250519] [PMID: 21058240]

[24]   Chiang CH, Yu TA, Lo SF, Kuo CL, Peng WH, Tsay HS. Molecular authentication of Dendrobium species by multiplex polymerase chain reaction and amplification refractory mutation system analysis. J Am Soc Hortic Sci 2012; 137(6): 438-44.
[http://dx.doi.org/10.21273/JASHS.137.6.438]

[25]   Sankhla AK, Malik C, Parashar M. A review on start codon targeted (SCoT) marker. J Plant Sci Res 2015; 31(2).

[26]   Li CC, Lo HY, Hsiang CY, Ho TY. DNA microarray analysis as a tool to investigate the therapeutic mechanisms and drug development of Chinese medicinal herbs. Biomedicine 2012; 2(1): 10-6.
[http://dx.doi.org/10.1016/j.biomed.2012.02.002]

[27]   Sarwat M, Yamdagni M. DNA barcoding: Novel tool for genetic fingerprinting of medicinal plants. DBT sponsored National Seminar on Biotechnology and Pharma Industry.

[28]   Ivanova NV, Kuzmina ML, Braukmann TWA, Borisenko AV, Zakharov EV. Authentication of herbal supplements using next-generation sequencing. PLoS One 2016; 11(5): e0156426.
[http://dx.doi.org/10.1371/journal.pone.0156426] [PMID: 27227830]

[29]   Arivaradarajan P, Misra G. Omics ApproachesTechnologies And Applications. Springer 2018.
[http://dx.doi.org/10.1007/978-981-13-2925-8]

[30]   Merriman B, Rothberg JM. Progress in ion torrent semiconductor chip based sequencing. Electrophoresis 2012; 33(23): 3397-417.
[http://dx.doi.org/10.1002/elps.201200424] [PMID: 23208921]

[31]   Liu Y, Sun G, Zhong Z, *et al.* Overexpression of AtEDT1 promotes root elongation and affects medicinal secondary metabolite biosynthesis in roots of transgenic Salvia miltiorrhiza. Protoplasma 2017; 254(4): 1617-25.
[http://dx.doi.org/10.1007/s00709-016-1045-0] [PMID: 27915455]

[32]   Soltani Howyzeh M, Sadat Noori SA, Shariati J V, Amiripour M. Comparative transcriptome analysis to identify putative genes involved in thymol biosynthesis pathway in medicinal plant Trachyspermum

ammi L. Sci Rep 2018; 8(1): 13405.
[http://dx.doi.org/10.1038/s41598-018-31618-9] [PMID: 30194320]

[33]   Chandran H, Meena M, Barupal T, Sharma K. Plant tissue culture as a perpetual source for production of industrially important bioactive compounds. Biotechnol Rep 2020; 26: e00450.
[http://dx.doi.org/10.1016/j.btre.2020.e00450] [PMID: 32373483]

[34]   Suman S. Plant tissue culture: A promising tool of quality material production with special reference to micropropagation of banana. Biochem Cell Arch 2017; 17(1): 1-26.

[35]   Grzegorczyk-Karolak I, Kuźma Ł, Skała E, Kiss AK. Hairy root cultures of Salvia viridis L. for production of polyphenolic compounds. Ind Crops Prod 2018; 117: 235-44.
[http://dx.doi.org/10.1016/j.indcrop.2018.03.014]

[36]   Verma N, Shukla S. Impact of various factors responsible for fluctuation in plant secondary metabolites. J Appl Res Med Aromat Plants 2015; 2(4): 105-13.
[http://dx.doi.org/10.1016/j.jarmap.2015.09.002]

[37]   Espinosa-Leal CA, Puente-Garza CA, García-Lara S. *In vitro* plant tissue culture: means for production of biological active compounds. Planta 2018; 248(1): 1-18.
[http://dx.doi.org/10.1007/s00425-018-2910-1] [PMID: 29736623]

[38]   Kumari A, Baskaran P, Plačková L, *et al.* Plant growth regulator interactions in physiological processes for controlling plant regeneration and *in vitro* development of Tulbaghia simmleri. J Plant Physiol 2018; 223: 65-71.
[http://dx.doi.org/10.1016/j.jplph.2018.01.005] [PMID: 29505949]

[39]   Muhammad B, Muhammad I, Ullah S, *et al.* Plant hormones; role in adventitious roots formation in medicinally valuable compounds in extinct plant species: A review. Pak J Agric Sci 2021; 58(3): 799-812.

[40]   Shafi A, Hassan F, Zahoor I, Majeed U, Khanday FA. Biodiversity, Management and Sustainable Use of Medicinal and Aromatic Plant ResourcesInt j med aromat plants. Springer 2021; pp. 85-111.

[41]   Sithole MZ, Agholor IA. Delineation and dimension of deforestation. 2021.

[42]   Xia B, Chu R, Yu H, Zhang Z, Gu Y. *Ex-situ* Conservation in Botanical Gardens. Phytohortology: EDP Sciences. 2021; pp. 23-61.

[43]   Yang H, Tiersch TR. Concepts, history, principles, and application of germplasm cryopreservation technology. University of Florida Institute of Food and Agricultural Sciences Gainesville 2020.
[http://dx.doi.org/10.32473/edis-fa223-2020]

[44]   Agrawal A, Singh S, Malhotra EV, Meena D, Tyagi R. In vitro conservation and cryopreservation of clonally propagated horticultural species Conservation and utilization of horticultural genetic resources. Springer 2019; pp. 529-78.
[http://dx.doi.org/10.1007/978-981-13-3669-0_18]

[45]   Niazian M, Sadat-Noori SA, Tohidfar M, Galuszka P, Mortazavian SMM. Agrobacterium-mediated genetic transformation of ajowan (Trachyspermum ammi (L.) Sprague): an important industrial medicinal plant. Ind Crops Prod 2019; 132: 29-40.
[http://dx.doi.org/10.1016/j.indcrop.2019.02.005]

[46]   Yin M, Hong S. Cryopreservation of Dendrobium candidum Wall. ex Lindl. protocorm-like bodies by encapsulation-vitrification. Plant Cell Tissue Organ Cult 2009; 98(2): 179-85.
[http://dx.doi.org/10.1007/s11240-009-9550-x]

[47]   Yang X, Popova E, Shukla MR, Saxena PK. Root cryopreservation to biobank medicinal plants: A case study for Hypericum perforatum L. Cell Dev Biol Plant 2019; 55(4): 392-402.
[http://dx.doi.org/10.1007/s11627-019-09999-x]

[48]   Xue SH, Luo XJ, Wu ZH, Zhang HL, Wang XY. Cold storage and cryopreservation of hairy root cultures of medicinal plant Eruca sativa Mill., Astragalus membranaceus and Gentiana macrophylla

Pall. Plant Cell Tissue Organ Cult 2008; 92(3): 251-60.
[http://dx.doi.org/10.1007/s11240-007-9329-x]

[49]   Panis B, Nagel M, Van den houwe I. Challenges and prospects for the conservation of crop genetic resources in field genebanks, in *in vitro* collections and/or in liquid nitrogen. Plants 2020; 9(12): 1634.
[http://dx.doi.org/10.3390/plants9121634] [PMID: 33255385]

[50]   Magray MM, Wani KP, Chatto MA, Ummyiah HM. Synthetic seed technology. Int J Curr Microbiol Appl Sci 2017; 6(11): 662-74.
[http://dx.doi.org/10.20546/ijcmas.2017.611.079]

[51]   Maqsood M, Khusrau M, Mujib A, Kaloo ZA. Synthetic seed technology in some ornamental and medicinal plants: An overview. Propagat Genetic Manipulation of Plants. 2021; pp. 19-31.
[http://dx.doi.org/10.1007/978-981-15-7736-9_2]

[52]   Krishna Kumar G, Thomas TD. High frequency somatic embryogenesis and synthetic seed production in Clitoria ternatea Linn. Plant Cell Tissue Organ Cult 2012; 110(1): 141-51.
[http://dx.doi.org/10.1007/s11240-012-0138-5]

[53]   Fonseka DLCK, Wickramaarachchi WWUI, Madushani RPS. Synthetic seed production as a tool for the conservation and domestication of Celastrus paniculatus: A rare medicinal plant. Annu Res Rev Biol 2019; 1-8.
[http://dx.doi.org/10.9734/arrb/2019/v32i430092]

[54]   Soejarto D. Transcriptome Characterization, Sequencing, and Assembly of Medicinal Plants Relevant To Human Health. University of Illinois at Chicago 2011.

[55]   Lau W, Sattely ES. Six enzymes from mayapple that complete the biosynthetic pathway to the etoposide aglycone. Science 2015; 349(6253): 1224-8.
[http://dx.doi.org/10.1126/science.aac7202] [PMID: 26359402]

[56]   Xiao M, Zhang Y, Chen X, *et al.* Transcriptome analysis based on next-generation sequencing of non-model plants producing specialized metabolites of biotechnological interest. J Biotechnol 2013; 166(3): 122-34.
[http://dx.doi.org/10.1016/j.jbiotec.2013.04.004] [PMID: 23602801]

[57]   Verma M, Ghangal R, Sharma R, Sinha AK, Jain M. Transcriptome analysis of Catharanthus roseus for gene discovery and expression profiling. PLoS One 2014; 9(7): e103583.
[http://dx.doi.org/10.1371/journal.pone.0103583] [PMID: 25072156]

[58]   Aslam B, Basit M, Nisar MA, Khurshid M, Rasool MH. Proteomics: Technologies and their applications. J Chromatogr Sci 2017; 55(2): 182-96.
[http://dx.doi.org/10.1093/chromsci/bmw167] [PMID: 28087761]

[59]   Bantscheff M, Schirle M, Sweetman G, Rick J, Kuster B. Quantitative mass spectrometry in proteomics: A critical review. Anal Bioanal Chem 2007; 389(4): 1017-31.
[http://dx.doi.org/10.1007/s00216-007-1486-6] [PMID: 17668192]

[60]   Sengupta A, Narad P. Metabolomics Metabolomics omics approaches, technologies and applications. Springer 2018; pp. 75-97.
[http://dx.doi.org/10.1007/978-981-13-2925-8_5]

[61]   Gaj T, Gersbach CA, Barbas CF III. ZFN, TALEN, and CRISPR/Cas-based methods for genome engineering. Trends Biotechnol 2013; 31(7): 397-405.
[http://dx.doi.org/10.1016/j.tibtech.2013.04.004] [PMID: 23664777]

[62]   Ricroch AE, Martin-Laffon J, Rault B, Pallares VC, Kuntz M. Next biotechnological plants for addressing global challenges: The contribution of transgenesis and new breeding techniques. N Biotechnol 2022; 66: 25-35.
[http://dx.doi.org/10.1016/j.nbt.2021.09.001] [PMID: 34537403]

[63]   Carles M, Cheung MK, Moganti S, *et al.* A DNA microarray for the authentication of toxic traditional Chinese medicinal plants. Planta Med 2005; 71(6): 580-4.

[http://dx.doi.org/10.1055/s-2005-864166] [PMID: 15971136]

[64]   Li X, Ding X, Chu B, *et al.* Molecular authentication of Alisma orientale by PCR-RFLP and ARMS. Planta Med 2007; 73(1): 67-70.
[http://dx.doi.org/10.1055/s-2006-951746] [PMID: 17109255]

[65]   Hosokawa K, Hishida A, Nakamura I, Shibata T. The sequences of the spacer region between the atpF and atpA genes in the plastid genome allows discrimination among three varieties of medicinal Angelica. Planta Med 2006; 72(6): 570-1.
[http://dx.doi.org/10.1055/s-2005-916257] [PMID: 16773544]

[66]   Zerega NJC, Mori S, Lindqvist C, Zheng Q, Motley TJ. Using amplified fragment length polymorphisms (AFLP) to identify black cohosh (Actaea racemosa). Econ Bot 2002; 56(2): 154-64.
[http://dx.doi.org/10.1663/0013-0001(2002)056[0154:UAFLPA]2.0.CO;2]

[67]   Watanabe A, Araki S, Kobari S, *et al. In vitro* propagation, restriction fragment length polymorphism, and random amplified polymorphic DNA analyses of Angelica plants. Plant Cell Rep 1998; 18(3-4): 187-92.
[http://dx.doi.org/10.1007/s002990050554] [PMID: 30744218]

[68]   Chen G, Wang XL, Wong WS, Liu XD, Xia B, Li N. Application of 3' untranslated region (UTR) sequence-based amplified polymorphism analysis in the rapid authentication of Radix astragali. J Agric Food Chem 2005; 53(22): 8551-6.
[http://dx.doi.org/10.1021/jf051334g] [PMID: 16248552]

CHAPTER 8

Genetic Transformation in Medicinal Plants

Ummara Waheed[1,*], Sadia Shabir[1], Zahid Ishaq[2], Maria Khan[3], Saeed Rauf[4], Nadia Iqbal[5], Maria Siddique[6] and **Irum Shahzadi[7]**

[1] *Institute of Plant Breeding and Biotechnology, MNS University of Agriculture, Multan, Pakistan*

[2] *Department of Biochemistry, Nishtar Medical University, Multan, Pakistan*

[3] *Department of Education, University of Mianwali, Punjab, Pakistan*

[4] *Department of Plant Breeding and Genetics, College of Agriculture, University of Sargodha, Sargodha, Punjab, Pakistan*

[5] *Department of Biotechnology, Woman University Multan, Multan, Punjab, Pakistan*

[6] *Department of Environmental Science, COMSATS University Islamabad, Abbottabad Campus, Sargodha, Punjab Pakistan*

[7] *Department of Biotechnology, COMSATS University Islamabad, Abbottabad Campus, Pakistan*

Abstract: Secondary metabolites including terpenoids, terpenes and enzyme cofactor have significant importance in medicinal use. Extraction of plant-based compounds is quite challenging by conventional methods. Biotechnological methods like genetic engineering and *In Vitro* techniques, proteomics, genomics and biochemical pathways are being employed to serve the purpose. Different organic compounds including enzymes, recombinant proteins, vaccines, antibiotics and anticancer have been successfully extracted through the genetic transformation of tomato, rice, corn, soybean and *Nicotiana tabacum*. This report describes different biotechnological approaches with a special focus on tissue culture and genetic transformation methods for the investigation of medicinal plants and their important role in our economic industry.

Keywords: Genetic transformation, Metabolic engineering, Plant secondary metabolites, Tissue culture.

INTRODUCTION

Plant-based products have a significant importance in various walks of life including food, medicine, pharmacy and industry [1]. Their safe application, especially in pharmaceutical industry as compared to the other products, enhances

---

* **Corresponding author Ummara Waheed:**Institute of Plant Breeding and Biotechnology, MNS University of Agriculture, Multan, Pakistan; Tel: +00 92061-920-1681; Fax: +92-61-9201679; E-mail: ummara.waheed@mnsuam.edu.pk

Zulqurnain Khan, Azra Yasmin & Naila Safdar (Eds.)

their importance [2]. Though medicinal plants are very important but cultivated at a small scale in comparison with food crops like rice, wheat, sunflower, soybean and potato [3]. Medicinal plants are known to contain secondary metabolites like Terpenoids, alkaloids, saponin, alkaloids, and terpenes.

Their demand is increasing at the current time owing to their safe and effective utilization [4, 5]. Fast and accurate strategies are required for the production of secondary metabolites in bulk amounts [6]. Exploration of medicinal plants and various techniques like proteomics and metabolomics, *In Vitro* culturing and genetic transformation methods, have now become a central attention to scientists [6, 7].

Different *In Vitro* techniques like callus culture, meristem culture and suspension culture are being used to exploit the beneficial compounds from medicinal plants [8, 9] like monoclonal antibodies, recombinant antibodies, and hormones [10-15].

Callus Culture

Callus is an undifferentiated mass of cells and is one of the most important sources of the regeneration and multiplication of medicinal plants [11]. Callus culture is well known to be used for the extraction of secondary metabolites and different biologically active compounds like anthraquinone, rosmarinic acid, and baicalin through downstream processing [16, 17]. *Opuntia ficus-indica* (cactus pear) and *Beta vulgaris* (Red beet), *Leucophyllum frutescens*, and *Poliomintha glabrescens* are some of the examples through which secondary metabolites are being extracted from callus cultures [18, 19].

Suspension Culture

Suspension cultures are made through the callus, where callus culture is introduced into the liquid media subjected to shaking and transferred to the liquid phase in bioreactors. Suspension culture is a special technique for developing transgenic products as it comprises decontamination and simple downstream processing in an economical way. It is a well-established way for production of secondary metabolites as cells are allowed to grow in optimum environmental conditions [20]. Uniform quality and yields are obtained through suspensions including high valued compounds such as taxol, ginsenosides and resveratrol [20-23]. *Thalictrum rugosum*, *Vitis vinifera* and Taxa species are few examples for the extraction of medicinally important compounds [21].

Organ Culture

Organ culture is another *In Vitro* culturing technique that is used to bypass the limitations of the undifferentiated cultures of callus and suspension cultures. This technique involves the direct culturing of whole organs like roots, buds, trichomes and even seeds. They've established a useful biological strategy for investigating the biosynthesis of variety of bioactive chemicals, including nicotine and tropane alkaloids, ginsenosides, anthraquinones, artemisinin alkaloids, hyoscyamines and cannabinoids [24, 25]. Hairy root is one of the organ cultures that is also being utilized in conservation of endangered species [34]. Plant tissue culture is directly responsible for increasing the *In Vitro* synthesis of important plant chemicals using strategies such as optimization of nutrient media, biotransformation strategies, use of eliciting molecules and *Agrobacterium* mediated transformation [26, 27].

MOLECULAR ELUCIDATION OF MEDICINAL PLANTS FOR SECONDARY METABOLITES PRODUCTION

A wide range of biotechnological methods are available to explore secondary metabolism. Interpretation and metabolite engineering of pathways have been used to boost the supply of desired products or to create novel medicinally important products such as antidiabetic and anticancer compounds [28]. Molecular elucidation entails a number of methodologies and phases, that govern secondary metabolite pathways such as transcription factors and ending with the cloning of the regulatory genes [29].

Targeting specific biochemical pathways of metabolites, enzyme catalysts, biosynthetic reactions, genes encoding the biosynthetic enzymes and regulatory factors are all part of molecular elucidation.

Overexpression, gene silencing and mutations are used extensively to control production of certain metabolites in the plant cells [30]. Through quantitative trait loci and genome-wide association analyses, high-throughput sequencing technology is being utilized to discover responsible genes for certain traits by using reference genome (Fig. **1**).

Different strategies have been used to increase production of secondary metabolites' in medicinal plants [9, 10, 31, 32]. Downregulation or functional deletion of genes is one method for reducing the production of an undesired group of chemicals while increasing the concentration of desired secondary metabolites (Fig. **2**). The underlying idea behind these methods is to eliminate an enzymatic step in a pathway by lowering the quantity of the associated mRNA or protein [33].

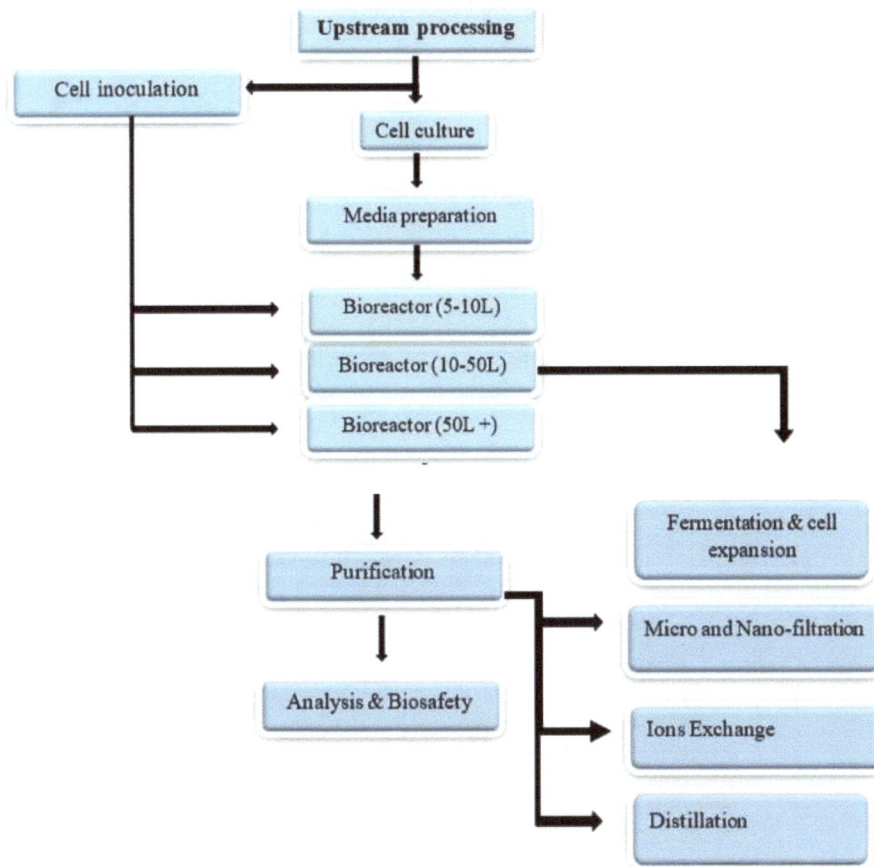

Fig. (1). Upstream processing involves the production and harvesting of medicinal products. It is initiated from cell inoculation and then cell expansion at a large scale. Final purified product was analyzed for biosafety measured.

METHODS OF GENETIC TRANSFORMATION

Medicinal plants have been subjected to genetic modification through two main methods.

Agrobacterium mediated transformation and direct methods of transformation (Fig. **3**). However, agrobacterium mediated transformation is the most preferred method due to the ease of use, stable gene delivery and economic importance.

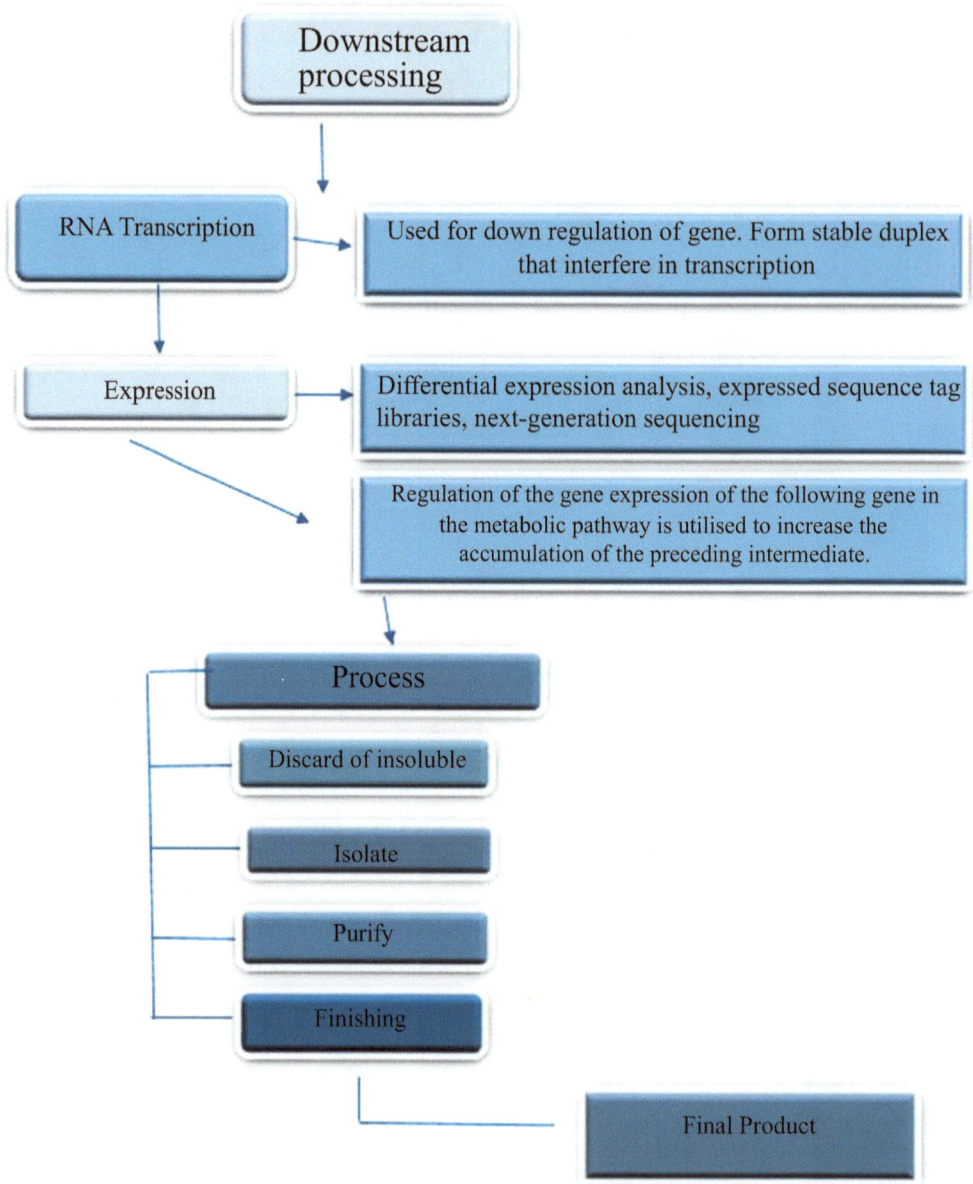

Fig. (2). Downstream processing involved in the molecular elucidation of genes. Downstream processing overcomes the production of unnecessary metabolites and enhances the expression of targeted product *via* molecular modifications.

Regulatory genes have been the best target for the generation of secondary metabolites. Transcription factors are a type of regulatory gene that binds to the promoter region of a gene and causes it to be upregulated in metabolic pathways. ORCA3, a transcription factor, for example, increases alkaloid production in *C. roseus* by upregulating TIA pathway genes which take part in jasmonate elicitor therapy. Anthocyanins production increases in maize suspension cells, caused by transcription factors C1 and R. Alternatively, in tomato, inhibition of the DET1 regulatory gene resulted in high concentration of apocarotenoid and flavonoid components.

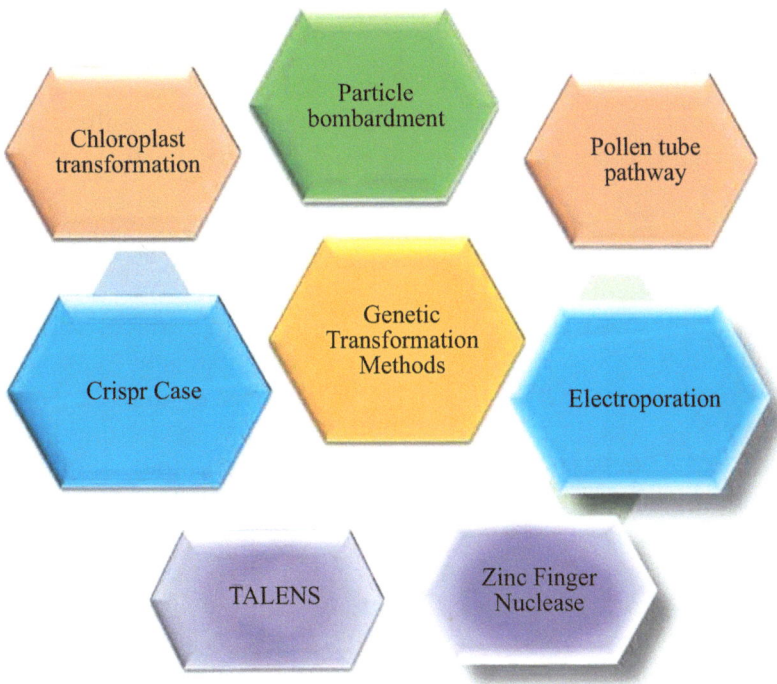

Fig. (3). Medicinal plants transformed by different genetic transformation techniques such as Electroporation, chloroplast transformation, pollen tube pathway and *via* recent techniques such as Zinc figure nuclease, Talens and Crisper Case.

Direct Genetic Transformation Methods

Direct genetic transformation methods include those that do not involve the vector to deliver gene of interest. Due to the ease of use, direct transformation methods are use for transformation of medicinal plants [33, 34]. PEG, Electroporation, biolistic methods are various direct transformation methods. Tobacco has been transformed through the Polyethylene Glycol (PEG) [35] while tobacco was also tested through the Electroporation methods [40, 50]. In some other techniques, model plants were transformed through Microinjection [36].

In the later years with the discovery of gene delivery through the gene gun, model crops along with some other economic crops like cotton were transformed through this method [36, 37].

Agrobacterium Mediated Genetic Transformation

Medicinal plants like many other food crops are being transformed through the naturally occurring soil borne bacteria like *Agrobacterium tumefaciens* and A. rh*izogenes*. Naturally, *Agrobacterium* adheres to the cell wall of the explant and stimulates *via* chemo attractants of the host plant [38]. Where it induces its virulent genes in the host plants through T-complex component. With stable gene transfer, it spreads its disease. Scientists engineered this organism by replacing the virulent genes with the gene of interest. Through Agrobacterium as a vector, many food and economic crops are being transformed [7, 39].

Compounds of great importance like enzymes, recombinant proteins, and vaccines are produced in bulk amount by genetic modification of tomato, rice, corn and soybean. Biotechnological methods like genetic transformation may prove to be an alternative and fast methods for manufacturing byproducts such as yeast, bacteria, and animals. Some examples are "Golden Rice" and "Golden banana" with the high level of carotene, polysterols, tocopherols and carotenoid from tomato, produced by genetic transformation. Root transformation by rhizogenic *Agrobacterium* is the best approach for secondary metabolite production, as large number of secondary metabolites are obtained from roots [40]. These metabolites are monoclonal antibodies, recombinant antibodies and hormones [14, 16, 27]. Hairy root structure of medicinal plants is an authentic and stead-fast method for secondary metabolites production because they are genetically and biosynthetically stable *In Vitro* cultures while suspension and callus cultures do not show stability [10, 20, 41, 42].

It can only deliver genes to a specific number of plant species, however limited for DNA-free editing or chloroplast or mitochondrial genome transformation [20, 21]. Transgenic hairy roots overexpressing genes of relevance have also increased secondary metabolite synthesis [31] (Table **1**).

It is a well-established fact that medicinal plants hold a central role in the production of important metabolites like alkaloids, saponins alkaloids, polyphenols, terpenoids, and nematocidal compounds naturally [54]. Modern biotechnological methods including tissue culturing and genetic transformation approaches have enhanced extraction of such important compounds [55, 56]. It is reported that more than 60 genera of medicinal plants have been transformed while many other plants are being grown through tissue culture techniques for the extraction of important compounds [57, 58].

Table 1. Expmaples of grobacterium mediated transfromation.

Gene	Function	Transformation Method	Uses	References
rolB	Increase Hairy Roots	*Agrobacterium*	Anti-inflammatory	[43]
badh	Enhance Drought and Salinity Tolerance in *Trachyspermum ammi* [L.] Sprague	*Agrobacterium*	Anti-cancer apoptosis, Stress-reliever, Antibacterial activity, Cardio protective	[44]
hGL	Gastric lipase production	*Agrobacterium*	Lipolysis	[45]
rHSA	Produce Recombinant human serum albumin	*Agrobacterium*	Carrier protein for steroids, thyroid hormones and fatty acids.	[46]
OsrHSA	Produce Recombinant human serum albumin	*Agrobacterium*	Carrier protein for steroids and thyroid hormones.	[47]
CYP82E4	Nicotine-fed teratomas converted nicotine to nornicotine	*Agrobacterium*	Control the amount of Nornicotine.	[48]
LSP1	Glycyrrhizin	*Agrobacterium*	Upregulated the expression of genes like geranyl diphosphate synthase, A reductase, farnesyl diphosphate synthase.	[29]
h6h	Hyoscyamine, 6b-hydroxyhyoscyamine	*Agrobacterium*	Responsible for hyoscyamine hydroxylation and epoxidation.	[49, 50]
TcPOT1.1	Cadaverine, putrescine, spermidine, spermine	*Agrobacterium*	Involved in polyamine transport pathway.	[51, 52]
TRII	Calystegine	*Agrobacterium*	Glycosidase inhibitors.	[53]

REFERENCES

[1]   Arora N, Pienkos PT, Pruthi V, Poluri KM, Guarnieri MT. Leveraging algal omics to reveal potential targets for augmenting TAG accumulation. Biotechnol Adv 2018; 36(4): 1274-92.
      [http://dx.doi.org/10.1016/j.biotechadv.2018.04.005] [PMID: 29678388]

[2]   Atanasov AG, Waltenberger B, Pferschy-Wenzig EM, *et al.* Discovery and resupply of pharmacologically active plant-derived natural products: A review. Biotechnol Adv 2015; 33(8): 1582-614.
      [http://dx.doi.org/10.1016/j.biotechadv.2015.08.001] [PMID: 26281720]

[3]   Cardoso JC, Oliveira MEBS, Cardoso FCI. Advances and challenges on the *in vitro* production of secondary metabolites from medicinal plants. Hortic Bras 2019; 37(2): 124-32.
      [http://dx.doi.org/10.1590/s0102-053620190201]

[4]   Castilho CVV, Neto JFF, Leitão SG, Barreto CS, Pinto SC, da Silva NCB. Anemia tomentosa var. anthriscifolia *in vitro* culture: Sporophyte development and volatile compound profile of an aromatic fern. Plant Cell Tissue Organ Cult 2018; 133(3): 311-23.

[http://dx.doi.org/10.1007/s11240-018-1383-z]

[5]   Chandran H, Meena M, Barupal T, Sharma K. Plant tissue culture as a perpetual source for production of industrially important bioactive compounds. Biotechnol Rep 2020; 26: e00450.
[http://dx.doi.org/10.1016/j.btre.2020.e00450] [PMID: 32373483]

[6]   Chandra H, Kumari P, Bontempi E, Yadav S. Medicinal plants: Treasure trove for green synthesis of metallic nanoparticles and their biomedical applications. Biocatal Agric Biotechnol 2020; 24: 101518.
[http://dx.doi.org/10.1016/j.bcab.2020.101518]

[7]   Chen J, Wang L, Chen J, *et al. Agrobacterium* tumefaciens-mediated transformation system for the important medicinal plant Dendrobium catenatum Lindl *in vitro*. Cell Dev Biol Plant 2018; 54(3): 228-39.
[http://dx.doi.org/10.1007/s11627-018-9903-4]

[8]   Cardillo AB, Talou JR, Giulietti AM. Expression of *Brugmansia candida Hyoscyamine 6beta-Hydroxylase gene in Saccharomyces cerevisiae* and its potential use as biocatalyst. Microb Cell Fact 2008; 7(1): 17.
[http://dx.doi.org/10.1186/1475-2859-7-17] [PMID: 18505565]

[9]   Dirar AI, Alsaadi DHM, Wada M, Mohamed MA, Watanabe T, Devkota HP. Effects of extraction solvents on total phenolic and flavonoid contents and biological activities of extracts from Sudanese medicinal plants. S Afr J Bot 2019; 120: 261-7.
[http://dx.doi.org/10.1016/j.sajb.2018.07.003]

[10]   Domingues RMA, Sousa GDA, Silva CM, Freire CSR, Silvestre AJD, Neto CP. High value triterpenic compounds from the outer barks of several Eucalyptus species cultivated in Brazil and in Portugal. Ind Crops Prod 2011; 33(1): 158-64.
[http://dx.doi.org/10.1016/j.indcrop.2010.10.006]

[11]   Efferth T. Biotechnology applications of plant callus cultures. Engineering 2019; 5(1): 50-9.
[http://dx.doi.org/10.1016/j.eng.2018.11.006]

[12]   Ele Ekouna JP, Boitel-Conti M, Lerouge P, Bardor M, Guerineau F. Enhanced production of recombinant human gastric lipase in turnip hairy roots. Plant Cell Tissue Organ Cult 2017; 131(3): 601-10.
[http://dx.doi.org/10.1007/s11240-017-1309-1]

[13]   Espinosa-Leal CA, Puente-Garza CA, García-Lara S. *in vitro* plant tissue culture: Means for production of biological active compounds. Planta 2018; 248(1): 1-18.
[http://dx.doi.org/10.1007/s00425-018-2910-1] [PMID: 29736623]

[14]   Figlan S, Makunga NP. Genetic transformation of the medicinal plant *Salvia runcinata* L. f. using *Agrobacterium rhizogenes*. S Afr J Bot 2017; 112: 193-202.
[http://dx.doi.org/10.1016/j.sajb.2017.05.029]

[15]   Georgiev V, Slavov A, Vasileva I, Pavlov A. Plant cell culture as emerging technology for production of active cosmetic ingredients. Eng Life Sci 2018; 18(11): 779-98.
[http://dx.doi.org/10.1002/elsc.201800066] [PMID: 32624872]

[16]   Goldstein DA, Thomas JA. Biopharmaceuticals derived from genetically modified plants. QJM 2004; 97(11): 705-16.
[http://dx.doi.org/10.1093/qjmed/hch121] [PMID: 15496527]

[17]   Hossan MS, Jindal H, Maisha S, *et al.* Antibacterial effects of 18 medicinal plants used by the Khyang tribe in Bangladesh. Pharm Biol 2018; 56(1): 201-8.
[http://dx.doi.org/10.1080/13880209.2018.1446030] [PMID: 29529970]

[18]   Soltani Howyzeh M, Sadat Noori SA, Shariati J V. Essential oil profiling of Ajowan (Trachyspermum ammi) industrial medicinal plant. Ind Crops Prod 2018; 119: 255-9.
[http://dx.doi.org/10.1016/j.indcrop.2018.04.022]

[19]   Hasne MP, Coppens I, Soysa R, Ullman B. A high-affinity putrescine-cadaverine transporter from

Trypanosoma cruzi. Mol Microbiol 2010; 76(1): 78-91.
[http://dx.doi.org/10.1111/j.1365-2958.2010.07081.x] [PMID: 20149109]

[20]   Isah T, Umar S, Mujib A, *et al*. Secondary metabolism of pharmaceuticals in the plant *in vitro* cultures: Strategies, approaches, and limitations to achieving higher yield. Plant Cell Tissue Organ Cult 2018; 132(2): 239-65.
[http://dx.doi.org/10.1007/s11240-017-1332-2]

[21]   Kim DII, Pedersen H, Chin CK. Stimulation of berberine production in Thalictrum rugosum suspension cultures in response to addition of cupric sulfate. Biotechnol Lett 1991; 13(3): 213-6.
[http://dx.doi.org/10.1007/BF01025820]

[22]   Klein TM, Wolf ED, Wu R, Sanford JC. High-velocity microprojectiles for delivering nucleic acids into living cells. Nature 1987; 327(6117): 70-3.
[http://dx.doi.org/10.1038/327070a0]

[23]   Kowalczyk T, Merecz-Sadowska A, Picot L, *et al*. Genetic manipulation and bioreactor culture of plants as a tool for industry and its applications. Molecules 2022; 27(3): 795.
[http://dx.doi.org/10.3390/molecules27030795] [PMID: 35164060]

[24]   Kren V, Martínková L. Glycosides in medicine: The role of glycosidic residue in biological activity. Curr Med Chem 2001; 8(11): 1303-28.
[http://dx.doi.org/10.2174/0929867013372193] [PMID: 11562268]

[25]   Kumar V, Roy BK. Population authentication of the traditional medicinal plant Cassia tora L. based on ISSR markers and FTIR analysis. Sci Rep 2018; 8(1): 10714.
[http://dx.doi.org/10.1038/s41598-018-29114-1] [PMID: 30013159]

[26]   Kuruppu AI, Paranagama P, Goonasekara CL. Medicinal plants commonly used against cancer in traditional medicine formulae in Sri Lanka. Saudi Pharm J 2019; 27(4): 565-73.
[http://dx.doi.org/10.1016/j.jsps.2019.02.004] [PMID: 31061626]

[27]   Lankatillake C, Huynh T, Dias DA. Understanding glycaemic control and current approaches for screening antidiabetic natural products from evidence-based medicinal plants. Plant Methods 2019; 15(1): 105.
[http://dx.doi.org/10.1186/s13007-019-0487-8] [PMID: 31516543]

[28]   Larquin PF. Uptake and integration of exogenous DNA in plants. In: Bajaj YPS, Ed. Plant protoplasts and genetic engineering II Biotechnology in agriculture and forestry. Berlin, Heidelberg, New York: Springer 1989; 9: pp. 54-74.
[http://dx.doi.org/10.1007/978-3-642-74454-9_3]

[29]   Liu ZB, Chen JG, Yin ZP, *et al*. Methyl jasmonate and salicylic acid elicitation increase content and yield of chlorogenic acid and its derivatives in gardenia jasminoides cell suspension cultures. Plant Cell Tissue Organ Cult 2018; 134(1): 79-93.
[http://dx.doi.org/10.1007/s11240-018-1401-1]

[30]   Máthé Á, Hassan F. Medicinal and aromatic plants of the world. Scientific, Production, Commercial and Utilization Aspects. Netherlands: springer 2015; 1.
[http://dx.doi.org/10.1007/978-94-017-9810-5]

[31]   Matveeva TV, Otten L. Widespread occurrence of natural genetic transformation of plants by Agrobacterium. Plant Mol Biol 2019; 101(4-5): 415-37.
[http://dx.doi.org/10.1007/s11103-019-00913-y] [PMID: 31542868]

[32]   Matveeva TV, Sokornova SV. Mediated transformation of plants for improvement of yields of secondary metabolites, bioprocessing of plant *in vitro* systems. Springer 2018; pp. 161-202.

[33]   McCabe DE, Martinell BJ. Transformation of elite cotton cultivars *via* particle bombardment of meristems. Nat Biotechnol 1993; 11(5): 596-8.
[http://dx.doi.org/10.1038/nbt0593-596]

[34]   Mishra S, Bansal S, Sangwan RS, Sangwan NS. Genotype independent and efficient Agrobacterium-

mediated genetic transformation of the medicinal plant Withania somnifera Dunal. J Plant Biochem Biotechnol 2016; 25(2): 191-8.
[http://dx.doi.org/10.1007/s13562-015-0324-8]

[35]   Mohagheghzadeh A, Gholami A, Hemmati S, Dehshahri S. Bag culture: A method for root-root co-culture. Z Naturforsch C J Biosci 2008; 63(1-2): 157-60.
[http://dx.doi.org/10.1515/znc-2008-1-229] [PMID: 18386507]

[36]   Mohanta YK, Biswas K, Jena SK, Hashem A, Abd Allah EF, Mohanta TK. Anti-biofilm and antibacterial activities of silver nanoparticles synthesized by the reducing activity of phytoconstituents present in the Indian medicinal plants. Front Microbiol 2020; 11: 1143.
[http://dx.doi.org/10.3389/fmicb.2020.01143] [PMID: 32655511]

[37]   Monfort LEF, Bertolucci SKV, Lima AF, *et al.* Effects of plant growth regulators, different culture media and strength MS on production of volatile fraction composition in shoot cultures of Ocimum basilicum. Ind Crops Prod 2018; 116: 231-9.
[http://dx.doi.org/10.1016/j.indcrop.2018.02.075]

[38]   Mulat M, Pandita A, Khan F. Medicinal plant compounds for combating the multi-drug resistant pathogenic bacteria: A review. Curr Pharm Biotechnol 2019; 20(3): 183-96.
[http://dx.doi.org/10.2174/1872210513666190308133429] [PMID: 30854956]

[39]   Naqvi S, Farré G, Sanahuja G, Capell T, Zhu C, Christou P. When more is better: Multigene engineering in plants. Trends Plant Sci 2010; 15(1): 48-56.
[http://dx.doi.org/10.1016/j.tplants.2009.09.010] [PMID: 19853493]

[40]   Neumann E, Schaefer-Ridder M, Wang Y, Hofschneider PH. Gene transfer into mouse lyoma cells by electroporation in high electric fields. EMBO J 1982; 1(7): 841-5.
[http://dx.doi.org/10.1002/j.1460-2075.1982.tb01257.x] [PMID: 6329708]

[41]   Ng HS, Wang CC, Tan JS, Lan JCW. Primary recovery of recombinant human serum albumin from transgenic oryza sativa with a single-step aqueous biphasic system. J Taiwan Inst Chem Eng 2018; 84: 60-6.
[http://dx.doi.org/10.1016/j.jtice.2018.01.018]

[42]   Niazian M, Noori SAS, Galuszka P, Tohidfar M, Mortazavian SMM. Genetic stability of regenerated plants *via* indirect somatic embryogenesis and indirect shoot regeneration of Carum copticum L. Ind Crops Prod 2017; 97: 330-7.
[http://dx.doi.org/10.1016/j.indcrop.2016.12.044]

[43]   Niazian M. Application of genetics and biotechnology for improving medicinal plants. Planta 2019; 249(4): 953-73.
[http://dx.doi.org/10.1007/s00425-019-03099-1] [PMID: 30715560]

[44]   Niazian M, Sadat-Noori SA, Tohidfar M, Galuszka P, Mortazavian SMM. Agrobacterium-mediated genetic transformation of ajowan (Trachyspermum ammi (L.) Sprague): An important industrial medicinal plant. Ind Crops Prod 2019; 132: 29-40.
[http://dx.doi.org/10.1016/j.indcrop.2019.02.005]

[45]   Ochoa-Villarreal M, Howat S, Hong S, *et al.* Plant cell culture strategies for the production of natural products. BMB Rep 2016; 49(3): 149-58.
[http://dx.doi.org/10.5483/BMBRep.2016.49.3.264] [PMID: 26698871]

[46]   Ogbole OO, Segun PA, Fasinu PS. Antimicrobial and antiprotozoal activities of twenty-four nigerian medicinal plant extracts. S Afr J Bot 2018; 117: 240-6.
[http://dx.doi.org/10.1016/j.sajb.2018.05.028]

[47]   Rana KL, Kour D, Kaur T, Devi R, Negi C, Yadav AN. Endophytic fungi from medicinal plants: Biodiversity and biotechnological applications. Microbial endophytes. Elsevier 2020; pp. 273-305.

[48]   Riggs CD, Bates GW. Stable transformation of tobacco by electroporation: Evidence for plasmid concatenation. Proc Natl Acad Sci 1986; 83(15): 5602-6.

[http://dx.doi.org/10.1073/pnas.83.15.5602] [PMID: 3016708]

[49]   Roy A, Jauhari N, Bharadvaja N. Medicinal plants as a potential source of chemopreventive agents. Anticancer plants: Natural products and biotechnological implements. Springer 2018; pp. 109-39.

[50]   Salehi B, Ata A, V. Anil Kumar N, Sharopov F, Ramirez-Alarcon K, Ruiz-Ortega A. Antidiabetic potential of medicinal plants and their active components. Biomolecules 2019; 9(10): 551.
       [http://dx.doi.org/10.3390/biom9100551] [PMID: 31575072]

[51]   Saito K, Sudo H, Yamazaki M, *et al.* Feasible production of camptothecin by hairy root culture of ophiorrhiza pumila. Plant Cell Rep 2001; 20(3): 267-71.
       [http://dx.doi.org/10.1007/s002990100320]

[52]   Scholl Y, Höke D, Dräger B. Calystegines in calystegia sepium derive from the tropane alkaloid pathway. Phytochemistry 2001; 58(6): 883-9.
       [http://dx.doi.org/10.1016/S0031-9422(01)00362-4] [PMID: 11684185]

[53]   Shree P, Mishra P, Selvaraj C, *et al.* Targeting COVID-19 (SARS-CoV-2) main protease through active phytochemicals of ayurvedic medicinal plants – *Withania somnifera* (Ashwagandha), *Tinospora cordifolia* (Giloy) and *Ocimum sanctum* (Tulsi) – a molecular docking study. J Biomol Struct Dyn 2022; 40(1): 190-203.
       [http://dx.doi.org/10.1080/07391102.2020.1810778] [PMID: 32851919]

[54]   Sliwinska E. Flow cytometry – a modern method for exploring genome size and nuclear DNA synthesis in horticultural and medicinal plant species. Folia Hortic 2018; 30(1): 103-28.
       [http://dx.doi.org/10.2478/fhort-2018-0011]

[55]   Sun QY, Ding LW, Lomonossoff GP, *et al.* Improved expression and purification of recombinant human serum albumin from transgenic tobacco suspension culture. J Biotechnol 2011; 155(2): 164-72.
       [http://dx.doi.org/10.1016/j.jbiotec.2011.06.033] [PMID: 21762733]

[56]   Verpoorte R, Memelink J. Engineering secondary metabolite production in plants. Curr Opin Biotechnol 2002; 13(2): 181-7.
       [http://dx.doi.org/10.1016/S0958-1669(02)00308-7] [PMID: 11950573]

[57]   Wilson SA, Roberts SC. Metabolic engineering approaches for production of biochemicals in food and medicinal plants. Curr Opin Biotechnol 2014; 26: 174-82.
       [http://dx.doi.org/10.1016/j.copbio.2014.01.006] [PMID: 24556196]

[58]   Zarei M, Mohammadi S, Jabbari S, Shahidi S. Intracerebroventricular microinjection of kaempferol on memory retention of passive avoidance learning in rats: Involvement of cholinergic mechanism (s). IJ Neurosci 2019; 129(12): 1203-12.

CHAPTER 9

Genome Editing and its Applications in Plants

Saba Yaseen[1], Azara Yasmeen[2], Naila Safdar[2] and Zulqurnain Khan[3,*]

[1] *Institute of Plant Breeding and Biotechnology, MNS University of Agriculture, Multan, Pakistan*

[2] *Department of Biotechnology, Fatima Jinnah Women University, Rawalpindi, Pakistan*

[3] *Department of Biotechnology, Institute of Plant Breeding and Biotechnology (IPBB), Muhammad Nawaz Shareef University of Agriculture, Multan-60000, Pakistan*

Abstract: Regularly interspaced short palindromic repeats/cas9 system (CRISPR-Cas) is a well-developed and frequently used genome editing system, which comprises a Cas9 nuclease and a single-guided RNA (that is an RNA-guided technique). Cas9 recognizes and cuts a specific DNA sequence by base-pairing with it, generating double-strand breakage (DSBs) that initiate cellular DNA repair mechanisms that result in alterations in the DSB regions or adjacent. CRISPR/Cas9 technology has transformed genetic modifications since its inception, and it is now routinely used to improve the genomics of large numbers of crops. CRISPR/Cas system is used for targeted modifications to improve plant growth, yield and tolerance to biotic and abiotic stress along with developing transgene-free gene-edited crops. The limitations of using the CRISPR/Cas9 technology, as well as ways for enhancing its responsiveness, are also investigated. This chapter also describes the introduction of CRISPR-edited DNA-free plants, which may be more acceptable than some other genetically-engineered organisms. The prospective uses of the CRISPR/Cas9 technology, as well as conventional breeding possibilities, are highlighted in this introductory chapter.

Keywords: CRISPR/Cas9, Conventional breeding, Genome editing, Genetically-engineered organisms.

INTRODUCTION

Because of the rapid advancement of decoding methods, the genome of many key crops has already been sequenced and analyzed [1]. Depending on these facts, we may breed better crops, identify the activities of genes that control essential agricultural features, and generate high-yielding crops [2].

* **Corresponding author Zulqurnain Khan:** Department of Biotechnology, Institute of Plant Breeding and Biotechnology (IPBB), Muhammad Nawaz Shareef University of Agriculture, Multan-60000, Pakistan; E-mail: zulqunrain.khan@mnsuam.edu.pk

Zulqurnain Khan, Azra Yasmin & Naila Safdar (Eds.)

The function of a gene is frequently studied using mutants, although the most common method of obtaining such genetic variants in living animals is the use of chemicals or physical techniques, such as genetic alterations with T-DNA or gene silence using RNA interference [3].

The adoption of those technologies has significantly sped the separation and subsequent integration of gene products into the genome of improved varieties of significant crop plants. Due to the lack of selection methods, the changes produced by these processes are unanticipated and frequently lead to a significant impairment of gene function, making mutant screening and detection time-taking [4]. Given that the genomes of important agricultural plants contain a large number of valid gene data that has yet to be assessed, the development of flexible, efficient, and accurate methods for altering target genes is essential. Synthetic sequence-specific nucleases (SSNs) were used to generate double-strand breakage in specific genomic areas to achieve these criteria [5]. When base pairs are inserted, deleted, or replaced, DNA lesions are repaired, and the desired gene is created. This method is now typically referred to as "genetic modification." The main two repair mechanisms are non-homologous end joining (NHEJ) and homologous recombination (HR). In somatic cells, the error-prone NHEJ is used to restore DSBs rather than a more precise HR [6]. Since HR is a more complex and involved method that requires inserting both donor DNA and the kSSN expressing cassette into plant cells. An illustration of plant genomic alteration. The ZFN, TALEN, and the more modern improved type II CRISPR/Cas, notably the CRISPR/Cas9 system, are the most often utilized plant genome editing techniques [7] (Table **1**).

Table 1. Successful reports of gene Editing in various crop plants.

Crop Species	Gene Editor	Target Gene	DNA Repair Type	Target Trait	References
Maize	ZFNs	*ZmIPK1*	HR	Herbicide tolerant and phytate reduced maize	[21]
Maize	ZFNs	*ZmTLP*	HR	Trait stacking	[22]
Rice	ZFNs	*OsQQR*	HR	Trait stacking	[23]
Rice	TALENs	*OsSWEET14*	NHEJ	Bacterial blight resistance	[24]
Wheat	TALENs	*TaMLO*	NHEJ	Powdery mildew resistance	[25]
Maize	TALENs	*ZmGL2*	NHEJ	Reduced epicuticular wax in leaves	[26]
Sugarcane	TALENs	*COMT*	NHEJ	Improved cell wall composition	[27]

(Table 1) cont.....

Crop Species	Gene Editor	Target Gene	DNA Repair Type	Target Trait	References
Sugarcane	TALENs	*COMT*	NHEJ	Improved saccharification efficiency	[28]
Soybean	TALENs	*FAD2-1A, FAD2-1B*	NHEJ	High oleic acid contents	[29]
Soybean	TALENs	*FAD2-1A, FAD2-1B, FAD3A*	NHEJ	High oleic, low linoleic contents	[29]
Potato	TALENs	*VInv*	NHEJ	Minimizing reducing sugars	[30]
Rice	TALENs	*OsBADH2*	NHEJ	Fragrant rice	[31]
Maize	TALENs	*ZmMTL*	NHEJ	Induction of haploid plants	[32]
Brassica oleracea	TALENs	*FRIGIDA*	NHEJ	Flowering earlier	[33]
Tomato	TALENs	*ANT1*	HR	Purple tomatoes with high anthocyanin	[34]
Rice	CRISPR/Cas9	*LAZY1*	NHEJ	Tiller-spreading	[35]
Rice	CRISPR/Cas9	*Gn1a, GS3, DEP1*	NHEJ	Enhanced grain number, larger grain size and dense erect panicles	[35]
Wheat	CRISPR/Cas9	*GW2*	NHEJ	Increased grain weight and protein content	[35]
Camelina sativa	CRISPR/Cas9	*FAD2*	NHEJ	Decreased polyunsaturated fatty acids	[36]
Rice	CRISPR/Cas9	*SBEIIb*	NHEJ	High amylose content	[37]
Maize	CRISPR/Cas9	*Wx1*	NHEJ	High amylopectin content	[38]
Potato	CRISPR/Cas9	*Wx1*	NHEJ	High amylopectin content	[38]
Wheat	CRISPR/Cas9	*EDR1*	NHEJ	Powdery mildew resistance	[39]
Rice	CRISPR/Cas9	*OsERF922*	NHEJ	Enhanced rice blast resistance	[40]
Rice	CRISPR/Cas9	*OsSWEET13*	NHEJ	Bacterial blight resistance	[41]
Tomato	CRISPR/Cas9	*SlMLO1*	NHEJ	Powdery mildew resistance	[42]

The use of ZFNs and transcription activator-like effectors in plants was previously limited due to the complexity of the synthesis of proteins, which required a long process. ZFN and transcription activator-like effector systems are more difficult to build than clustered regularly interspaced short palindromic systems, which just require a Cas9 protein and a manufactured sgRNA for DNA binding [8]. The development of the sgRNA sequence allows for the acquisition of added gene targeting properties, and it is this advancement that allows the CRISPR/Cas9 mechanism to be used in the real world. CRISPR/Cas9 has gotten a lot of attention since the first three findings on CRISPR/Cas9-mediated plant genetic alteration [9]. The sections follow details of some of the various methods that researchers may use to find CRISPR-induced mutations. Plant genetic editing with CRISPR/Cas9 was first reported in Nature Biotech in August 2013, and CRISPR/Cas9 has attracted prominence and appeal among plant scientists since then. The design strategies for CRISPR/Cas9 mechanism, and its different screening process to detect CRISPR/Cas9-mediated mutations, are discussed in the following sections[10].

SYSTEM OF CRISPR/CAS9

CRISPR/Cas9 Action Mechanism

CRISPR is a DNA fragment that has non-contiguous ssDNA repeats, which are snippets of varied sequences. CRISPR-allied genes interconnected to CRISPR loci, were discovered in the genome of Escherichia coli in 1987 [11]. CAS genes are typically located close to a CRISPR locus. CRISPR's function, however, was unknown at the time of its discovery. The identification of gaps within CRISPRs created by viral genomes and plasmids was a landmark point in our understanding of CRISPR activity. According to this discovery, CRISPR/Cas could be implicated in adaptive immunity in prokaryotes. In 2007, investigators have found that the phage-resistant phenotypes of bacteria could be modified by adding and removing certain spacers, providing the first evidence that CRISPR/Cas is an adaptive immune system [12]. During the investigation of the gap uptake process, motifs related to spacer precursors (proto-spacers) from the DNA of an invading phage virus were found. These structures are modest lengths of di- or trinucleotides put directly after or one position after the proto-spacers to help identify certain proto-spacers and the orientation of spacers integrated into repeat arrays [13]. In 2008, researchers discovered that a CRISPR RNA precursor (pre-crRNA) generated from a CRISPR locus was cleaved into mature RNA molecules by a complex of CAS proteins within the repeats (crRNA). Each crRNA is a tiny guide RNA with a spacer bordered by short DNA repeats, allowing CAS proteins to initiate an antiviral response. These discoveries revealed the molecular underpinnings of CRISPR/Cas-mediated adaptive immunity in *Streptococcus thermophilus* [14]. This was proposed in 2011 that the Cas9 system-mediated immunity had three stages: adaptation, expression, and interference. During the adaption stage, small plasmid or virus-derived sequences are incorporated into the CRISPR locus as new spacers, whereas transcription and maturing of crRNAs occur during the expression phase. Finally, during the intervention stage, the crRNAs guide Cas proteins to remove the corresponding plasmid or viral targeted sequences that match the spacers[15]. Based on an evolution analysis of Cas proteins and CRISPR/Cas systems, a comprehensive taxonomy of CRISPR/Cas systems into three basic types, I, II, and III, was proposed. [16]. Different types has a distinct protein on which the defense system depends to complete the process. Several CAS proteins function as hallmark proteins in type I and III systems, while Cas9 is a large single trademark multifunctional protein that creates crRNA as well as chops the target DNA in type II systems [17]. The type II system is a much simpler structure than the type I and III systems, and it can be much more RNA (tracrRNA) was discovered in the expression stage of the type II framework shown in S. pyogenes tries to control the development of crRNAs by base-pairing with the repeating regions of pre-rRNA transcripts with the help of an RNase III and Cas9 enzyme [18]. In 2012, developed a type II system mechanism and created sgRNA by fusing crRNA and tracrRNA. The DNA target-binding region of the sgRNA, which is 20 bp long, can be changed with any sequence of interest. A simple demonstration of CRISPR/Cas9 system is given in Fig. (**1**).

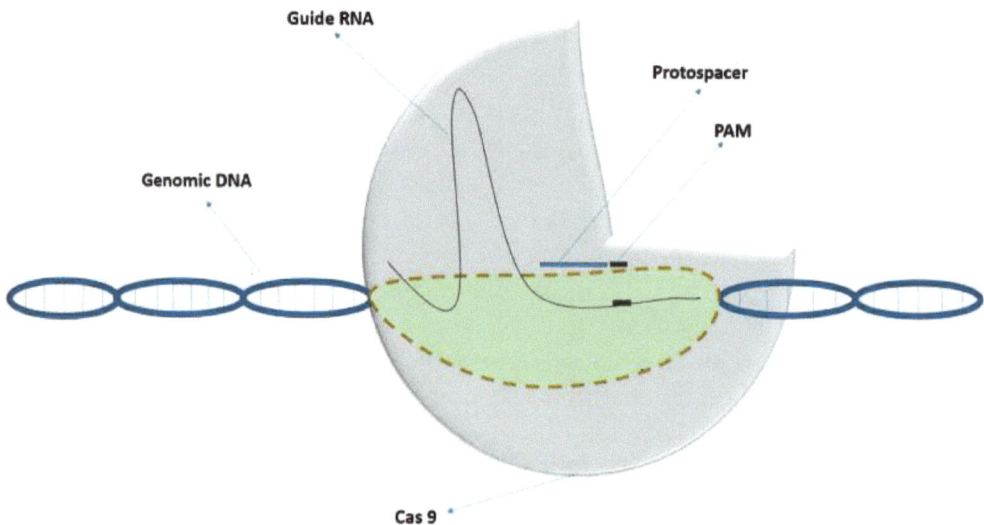

Fig. (1). CRISPR/Cas9 system for gene editing.

According to the findings, the Cas9 nuclease from *S. pyogenes* may be "programmed" with a sgRNA to generate DSBs three base pairs upstream of the PAM [19]. PAM nucleotide and position differed across CRISPR/Cas systems. The canonical PAM coupled with SpCas9 is 5'-NGG-3, while a DNA target sequence of 20 bp followed by a PAM is 20 NGG[20]. Cas9 nuclease has HNH and RuvC-like domains. Jinek has demonstrated that the HNH and RuvC-like domains may break complimentary and non-complementary DNA strands to the guide RNA. Because it is efficient, adaptable, and configurable, the approach has become the paradigm CRISPR/Cas9 system for editing genes *via* NHEJ/HR-mediated alterations [17].

CRISPR/CAS9 CONSTRUCTION

CRISPR/Cas9 technology was used effectively to change specific genes in a range of crop species. Some parameters, including sgRNA and cas9 overexpression, beta sheets of sgRNA and target DNA, cas9 codons, and focus DNA GC content, have been demonstrated to impact CRISPR/Cas9 editing effectiveness in plants [43]. As a consequence, while optimizing the CRISPR/Cas9 technique for high modifying effectiveness, several aspects must be examined [44]. Agrobacterium-mediated transformation may be used to introduce sgRNA and cas9 expression cassettes into targeted plants. Plant RNA polymerase III promoters, such as U3 and U6, which contain the translation initiation nucleotides "A" and "G," respectively, are often used to stimulate sgRNA transcription in plants [45]. The cas9 promoters maize ubiquitin and cauliflower mosaic virus 35S are the most often employed cas9 regulators. Plant-usage bias codons have also been introduced to the cas9 gene to allow for plant gene engineering. Because secondary structures in sgRNAs and target DNA may obstruct genetic alterations, selecting nucleotide sequence with a larger GC content may result in increased editing effectiveness [46].

It is a serious worry that genome editing technologies such as CRISPR/Cas9 might have unintended side effects [47]. The Cas9 nuclease is tolerant of certain mismatches between sgRNA and target DNA, leading to unwanted editing. An very specific 12-bp seeded code was utilized to build extremely accurate sgRNAs, which reduced the risk of off-target editing [48]. As the specificity, efficiency, and fidelity of Cas9 evolve, this method's off-target activity is also expected to decrease even more. For example, the high-fidelity nucleases SpCas9-HF1 and eSpCas9 lower the risk of incorrect DNA editing. If a single point mutation in the core of the Cas9 RuvC or HNH domain is made, the native SpCas9 is transformed into the Cas9 nickase [49]. It was utilised to induce paired single-strand breaks at the target region in order to increase the number of bases detected. Pair nicking was also shown to reduce off-target activity by 50-1000 times [50]. For non-homologous end mutagenesis, Cas9n fails in Arabidopsis plants as the modification frequency decreases by at least 740-fold, yet Cas9n's HR-mediated modification rate is the same as Cas9. The CRISPR/Cas9 mechanism has previously been revealed to be affected by temperature [51]. Compared to plants produced at 22 °C, Arabidopsis thaliana and citrus edited at 37 °C mutated more quickly. The key design criteria and approach employed in this work for on/off-target mutagenesis employing the CRISPR/Cas9 technology need to be studied further. To help future users get the most out of the CRISPR/Cas9 technology, bioinformatics experts created CRISPR-P 2.0 and CRISPR-GE, two new tools in the CRISPR toolbox. SgRNAs with the modest off-target functions may also be designed using these bioinformatic approaches [52].

CRISPR/CAS9-MEDIATED MUTATION SCREENING METHODS

Gene-scale mutant libraries have been generated utilizing CRISPR/cas9 for living cells, rice (*Oryza sativa*), and tomato cultivating using CRISPR/cas9-based variant databases for culture [53]. It is becoming increasingly common for scientists to analyse huge numbers of genetically modified variations, which includes detecting on-target edits and off-target changes, as well as disposing of gene transfer in progeny plants [54]. Numerous methods have been developed to deal with the time-consuming task of genetically altered recognition, including the restriction enzyme (RE) site deficit exam, surveying nuclease and T7 endonuclease I (T7EI) assays, polyacrylamide gel electrophoresis (PAGE)-based genotyping strategy, high-resolution melting evaluation (HRMA), and annealing at critical temperature PCR [55]. Based on chromosomal variances, these approaches have their own benefits and limitations. The aberrant change can be readily checked if it exhibits a noticeable knockdown phenotype. An albino phenotype caused by the absence of the phytoene deamination (PDS) gene was utilized to identify tobacco and rice variants in which the PDS gene had been changed using CRISPR/Cas9 and PDS-targeting sgRNA [56]. As a molecular marker in transposable elements, several antibiotic or herbicide resistance genes may also be used. The primary problem with phenotypic mutant testing is providing links between genetic variations and real plant traits. For both on-target and off-target alterations, high-throughput genotyping is the most efficient and accurate method [57]. Whole-genome scanning may also be used to detect genetically modified plants that have no transgenes. Whole-genome sequencing, on the other hand, is both expensive and time-consuming [58].

APPLICATIONS OF CRISPR/CAS9 IN CROPS

Research for Gene Functions

Numerous abiotic and biotic factors influence agriculture all around the globe. A growing global population, food instability, and environmental pollution have prompted farmers to explore for new ways to boost yields, quality, and resistance [59]. As a new method of improving agricultural varieties, CRISPR/Cas9 gene editing might be employed for functional genetic investigations. This technology might be used to improve a wide range of aspects of plant breeding in the future [60].

Genes Modifications for Yield Traits

Flowering in several agricultural plants is governed by seasonal variations in day duration, which may restrict the geographic distribution of cultivation for certain crops. By manipulating flowering alleles and their linkages, it is possible to control blooming timing [68]. The CRISPR/Cas9 approach for editing genes like FLOWER Genomic and SELF-PRUNING 5G has already resulted in considerable modifications in the blooming time of soybeans and tomatoes [69]. The thermo-sensitive genetically male sterility (TGMS) strain is one of the most often used male infertility strains in the two-line hybridization mating procedure. TMS5 is a line of thermos - responsive, genetically-engineered male-sterile mice [70]. CRISPR/Cas9 technique was demonstrated to be capable of accelerating high - yielding rice production by developing 11 new TGMS lines in just one year. The CRISPR/Cas9 approach proved successful in modifying four yield-enhancing genes: DEP1, Gn1a, IPA1, and GS3 (Table 2) .Dep1 and gn1a cultivars had more seeds per panicle, as well as higher grain products per panicle [71]. Using the CRISPR/Cas9 technology, researchers were able to alter profitability genes in farmed crop kinds with relative ease. With its erect leaflets, Zea mays can collect more sunlight and thrive at greater field densities than maize with a more typical leaf shape. Only recessive homozygotes with the upright leaf architecture exhibit the characteristics of LIGULELESS1 (LG1), which is a dominant gene [72]. To transgenic maize plants were utilized to produce a ZmLG1-targeting (sgRNA/Cas9] mutator, which was then transferred into six recipient lines of transgene maize through hybridization. This led to ZmLG1 genetic modification in the recipient lines, all of which had much lower leaf orientations than wild-type plants. Rather than using the standard method of delivering Cas9/sgRNA expression cassettes into target crops through transformations, this work shows a useful technique for genome editing [73]. Many main crop species' transformation and plant regeneration techniques are either unobtainable or inefficient and need lengthy and time-consuming tissue culture processes. New plant hormones known as strigolactones, limit shoot branching (tillering) and regulate root development to influence plant architecture [74]. Recent studies have shown that the CRISPR/Cas9 system may be utilized in rice to eliminate a kcy gene that regulates the manufacture of strigolactones. High tillering and a dwarf phenotype were common in the ccd7 mutant rice plants. Aside from that, certain CRISPR/Cas9 mutants have useful traits that may be exploited to breed and generate desired crops. Dwarfing and reduced fruit dehiscence in Brassica oleracea, valve-margin formation (*BnALC*) in polyploid oilseed rape, grain dormancy in barley (*Hordeum vulgare*), and chloroplast maturation in cotton (*GhCLA1*) are all connected to these genes (*Gossypium hirsutum*) [17].

Table 2. Applications of CRISPR in medicinal plants.

Species	Target Gene	Gene Description	Cas9/sgRNA Promoter	Results	Mutation Frequency	References
Salvia miltiorrhiza	SmCPS1	Committed diterpene synthase gene in tanshinone biosynthetic pathway	CaMV 35S/AtU6-26	8 heterozygous and 3 homozygous hairy root mutants	11.5% and 30.8% for the homozygous and chimeric mutants	[61]
Salvia miltiorrhiza	SmRAS	Rosmarinic acid synthase gene in phenolic acid biosynthetic pathway	CaMV 35S/AtU6-26, OsU3	5 biallelic, 2 heterozygous and 1 homozygous hairy root mutants	50%	[62]
Salvia miltiorrhiza	SmLACs	Laccase genes in phenolic acid and lignin biosynthetic pathway	AtUBQ/AtU6	15 single-locus crispr lines and 14 dual-locus crispr lines	90.6%	[63]
Salvia miltiorrhiza	SmbZIP2	Basic leucine zipper transcription factor; negative regulator in phenolic acid biosynthetic pathway	CaMV 35S/AtU6-26	SmbZIP2-deficient hairy roots	12%	[64]
Dendrobium officinale	C3H, C4H, 4CL, CCR, and IRX	Involved in lignocellulose biosynthetic pathway	MtHP, CVMV, MMV, PCISV, CaMV 35S /OsU3	DoLACs-deficient hairy roots	16.7%, 20%, 33.3%, 33.3% and 6.7% for C3H, C4H, 4CL, CCR and IRX	[65]
Dendrobium Chao Praya Smile	DOTFL1	Terminal flower 1 gene modulating flowering and inflorescence architecture	Ubi/OsU3, OsU6a	13 homozygous mutant plants	10.1%	[66]
Cannabis sativa	CsPDS1	Phytoene desaturase gene; Marker	CaMV 35S/AtU6	CsPDS1-deficient seedlings	2.5% and 51.6% for the homozygous and chimeric mutants	[67]

Gene Modifications to Improve the Quality of Products

Furthermore, by using CRISPR/ Cas9 system, a single dominant Waxy gene controlling amylose content was knocked out in two rice varieties, and the resulting mutants showed low amylose levels and elevated glutinosity. This research provides a simple and successful method for transforming a low-quality rice variety into a higher-quality one [75]. Furthermore, the GBSS gene, which encodes a granule-bound starch synthase, was damaged in tetraploid potato using CRISPR/Cas9. Only lines with mutations in all four GBSS alleles showed a decrease in GBSS enzymatic activity. These lines had a lower amylose concentration and a higher amylopectin/amylose ratio [76]. To improve polyploid crops, the capacity of CRISPR/Cas9 to concurrently mutate several genes simultaneously offers a straightforward and robust tool for the improvement of polyploid crops. A high concentration of cadmium in rice may provide a major health risk to both people and animals that use rice as a meal or feed. The CRISPR/Cas9 method has been used to create new Indica rice types with reduced Cd content by deleting the metal transporter gene OsNramp5 [77]. With CRISPR/Cas9 technology, it is now possible to produce safe rice from a Cd-contaminated paddy area using few resources. Healthy GABA is produced through a metabolic route termed the "gamma-Aminobutyric acid shunt". To enhance GABA concentration in tomatoes, researchers have recently changed a single gene or several genes at the same time using CRISPR/Cas9 technologies [78] (Table **2**). A further advantage of these genetically modified tomato plants is that they may be used as parental lines in the development of new hybrid tomato varieties.

Genome Modifications to Improve Abiotic Stress Resistance

Drought defense mechanisms in tomato plants are mediated by the mitogen-activated protein kin ases3 (SlMAPK3] gene, which was found using CRISPR/Cas9-based mutations. Rice variants produced using CRISPR/Cas9 were utilized to investigate the role of stress/ABA-activated protein kinase2 (SAPK2] in response to stress in rice [79]. To investigate the role of C-repeat conditional factors (CBFs) in Arabidopsis plant cold stress response, we employed CRISPR/Cas9 technology to create cbf1, 3 dual and cbf1, 2, 3 (CBFs) triple mutants. If a geneticist understood how genes work, he or she may be able to employ genetic markers to develop more resilient crops. Drought-resistant maize was developed using this technique (Shi *et al*) (Table **2**). The maize ARGOS8 gene, an ethylene response modulator is generated at low concentrations in maize [80]. To boost ARGOS8 production, GOS2 was inserted into the 50-untranslated domain of native ARGOS8 through CRISPR/Cas9, or the native ARGOS8 activator was changed to GOS2 *via* HR. When compared to their parent strain, these ARGOS8 varieties showed better grain yields even during a dry year. Gene editing utilizing the CRISPR/Cas9 system may reduce abiotic stressors, such as high heat, dehydration, salt, nutrient insufficiency, and high levels of toxic substances [81].

Genome Modification to Develop Biotic Stress Resistance

EDR1 has previously been identified as down regulation of powdery mildew susceptibility in Arabidopsis. Wheat has a high protein content (*Triticum aestivum*). Three EDR1 homologs were simultaneously knocked out using CRISPR/Cas9, resulting in Taedr1 plants that are more susceptible to powdery mildew [82]. Cas9-guided RNA-directed Cas9 knockouts for wheat and tomato improved their powdery mildew resistance by mutating a gene called MLO [83]. Citrus canker is caused by the bacterial pathogen Xanthomonas citri subsp. citri, which causes the canker susceptibility gene CsLOB1 to be expressed in susceptible plants. CRISPR/Cas9 was used to alter the CsLOB1 gene in grapefruit Duncan (Citrus paradisi Mac.), resulting in citrus variants that are resistant to canker. Many distinct hosts show recessive tolerance to the eIF (eukaryotic translation initiation factor) genes. CRISPR/Cas9 was successful in engineering virus-resistant cucumber and *Arabidopsis* plants by targeting genes. CRISPR/Cas9 was used to delete the eIF4G gene, which regulates the recessive RTSV susceptibility trait, to generate RTSV-resistant rice cultivars that are resistant to the rice tungro spherical virus (RTSV) [84]. In plants, ethylene-responsive factors (ERFs) play a role as regulators of various stress tolerance, as well as a role in the ethylene response process. Utilizing CRISPR/Cas9, researchers were able to increase rice's susceptibility to the blast fungi by targeting the OsERF922. Resistant to Botrytis cinerea has been found in mutant grapes caused by CRISPR/Cas9-induced loss of function of the VvWRKY52 gene [85]. When taken as a whole, these studies show that CRISPR/Cas9 may be used to enhance resistance to disease in essential agricultural species. The CRISPR/Cas9 system can break viral genomes as well as fight illness by changing the plant DNA. TMV and cucumber mosaic virus genomes may be edited using Francisella ovicidal nuclease (FnCas9] and a SgRNA specific for TMV or cucumber mosaic virus. Tobacco and *Arabidopsis* plants that expressed FnCas9/sgRNA have molecular resistance to the RNA viruses [86].

Gene Transcription or Translation Regulation

Multiple mechanisms exist for controlling the expression of a gene's product. Plant breeding relies heavily on manipulating gene expression to promote phenotypic variation [87]. In earlier investigations, cis-regulatory factors in the gene activation loop were linked to agricultural species of plants' development, domesticated, and selection. In tomatoes, CRISPR/Cas9 has recently been used to change the regulators of three genes associated with plant structure, inflorescence branching, and fruit size. Several promoter alterations indicated increasing variability in the trans-regulatory genes produced for each condition evaluated. When these transgenic crops are grown in the ground, they decreased plant height, modified the color of the leaves, and increased the tiller angle [88]. With the help of the CRISPR/Cas9 mechanism, plant uORFs have been altered to produce higher amounts of protein, which were then transcribed into four different variations. GDP-l-galactose phosphorylase codes one of its four terminals of a crucial enzyme in vitamin C synthesis (Lactuca sativa). Modifications in the LsGGP2 uORF resulted in increased ascorbate concentrations and increased tolerance to oxidative stress. Because of its potential to alter transcription and translational control, the CRISPR/Cas9 system has a lot of potential for improving crop output as well as understanding biological processes [17].

Mutant Libraries Construction

The task of critically analyzing the functions of all the genes in a plant genome that has been sequenced is a significant one. This problem can be solved by creating a genetic library that is saturated with mutants. To fine-tune the CRISPR/Cas9 system's ability to target certain genes, we changed the sgRNA's 20-bp target-binding region [89]. Genome-wide mutations and forward genetic testing may be carried out utilizing CRISPR/Cas9, which is both feasible and affordable. This discovery paved the way for the rising screening of plant mutant libraries using CRISPR/Cas9 in human cultured cells. When converting pooled sgRNA libraries into tomatoes, for instance, a variety of mutant strains were created. Using large-scale genetic screening and decoding, a homolog of an Arabidopsis boron outflow carrier gene and a gene related to immunity-associated leucine-rich repeat subclass II was swiftly revealed [90]. Additionally, two separate research groups have developed rice CRISPR/Cas9 mutant libraries, each of which has generated a substantial number of losses of function mutations by the transformation of sgRNA libraries.

Whenever these altered plants were being cultivated in the fields, they shortened the plants, modified the color of the leaves, and increased the tiller angle. In CRISPR/Cas9 mutant libraries, the association between phenotype and genotype was readily discovered using sgRNA, making this method practical for identifying mutant alleles. CRISPR/Cas9-based mutant libraries may be very useful in functional genomics research and crop improvement. Moreover, any effective agricultural plant transformation technique may be used to carry out genome-wide mutagenesis [48].

Generating Transgene-Free and Genome-Edited Crops

The ease, accuracy, and effectiveness of CRISPR/Cas9-induced genomic engineering and its capacity to make transgene-free, genetically engineered crops have all attracted great interest. It is possible to screen for mutant progeny plants that still carry the Cas9/sgRNA transgenes, even if they have been delivered into plants as transgenes *via* the CRISPR/Cas9 system. Plants that are genome-edited and devoid of transgenes are difficult to identify from those that have been mutated naturally[73]. Industrial use of CRISPR/Cas9 may be able to avoid the stringent biosafety regulations that are required for genetically engineered food crops. In the United States, biosafety regulations for anti-browning fungi Agaricus bisporus and waxy corn developed with CRISPR/Cas9 were met, among several instances [91].

An illustration of the process of creating crops free of transgenes and their effects on the genome. CRISPR/cas9 expression construct-carrying Agrobacterium plants produce transgenic plants in the T0 generation. It is necessary to collect the seeds after self-pollination for them to germinate. While non-transgenic plants that haven't been modified or even have lost their transgenes through recombination are eliminated in the T1 generation, edited plants are utilized to generate the T2 generation. It is usually impossible to completely modify a target gene in a single generation of a new organism [92]. Generating plants with the required alterations and no Cas9/sgRNA transgenes means you have a genomic sequence, and transgene-free plants in the T2 and following stages. Injection of *in vitro* pre-assembled CRISPR/Cas9 ribonucleoproteins (RNPs) into seedlings or embryonic cells may result in mutant or non-transgenic crops. To keep the modified crops free of transgenic, the RNPs technique was used, which prevents vector segments from being incorporated into the DNA [93].

CRISPR/CAS9 CHALLENGES AND CONCERNS

But even though crop breeding might benefit greatly from CRISPR/Cas9 technology, it still has certain limits. As a requirement to employ this technology, the small number of genes that influence critical agronomic variables poses a significant difficulty. Genomic sequencing data and good heritability materials for crop improvement are critically needed to be deciphered. Tissue culture-based plant reproduction is complicated by two added issues, namely a lack of efficient transformation methods. Biosafety issues might potentially impede the growth of CRISPR/Cas9.

CRISPR/Cas9-generated bioproduct's safety and off-target impacts are still hot topics of discussion. In contrast to animals and people, plants are more forgiving of off-target impacts. Modern technologies have made it possible to discover and eradicate plant mutations that have off-target consequences *via* segregation in subsequent crosses. The development of sgRNAs with high affinity for certain sequences, the selection of Cas9 with great accuracy, and the use of suitable experimental procedures may help limit off-target impacts. Another major problem is the present state of the market for genome-edited crops, which is bleak. The European Court of Justice (ECJ) has ruled that crops holding genomic sequence genes must adhere to certain strict regulations like other genetic modifications. EU investments in genome editing may suffer because of this action. Further research and improvements to the CRISPR/Cas9 genome engineering approach can help this technology play a vital role in the construction of sustainable farming methods that can keep a rapidly increasing global populace!

CONCLUSION

New options for plant genetic modification have been generated by the development of CRISPR/Cas9 technology. There are numerous desired aspects of CRISPR/genome Cas9's editing software for site-directed mutagenesis that are unreachable utilizing standard mutagenesis techniques.

Because RNA rather than proteins guides the Cas9 nuclease, CRISPR/Cas9 is an improvement over TALEN and ZFN. The CRISPR/Cas9 program's assembly techniques have improved with time, while efforts have been made to limit off-target consequences. There are also unique CRISPR/Cas9 design toolkits and several approaches for evaluating CRISPR/Cas9-induced variants being produced. CRISPR/Cas9 may be used to produce null alleles, new alleles, and ORFs by targeting coding sequences, regulatory regions, and ORFs.

If you want to change a single gene or a group of genes, a single transformation event is all that is required. Furthermore, variant libraries may be built using it. The use of CRISPR/Cas9 for plant genomes studies and sustainable agriculture, notably for production quality and biotic and abiotic stress moderation traits, became a popular method. With the ability to develop crops that have no transgenes, CRISPR/Cas9 is especially enticing. CRISPR/Cas9 system seems an effective technique for improving plant genetics at this moment. Despite being a successful new technology for gene manipulation, the widespread use of CRISPR/Cas9 is raising new questions and issues.

REFERENCES

[1]   Casquet J, Thebaud C. Gillespie RGJMer Chelex without boiling, a rapid and easy technique to obtain stable amplifiable DNA from small amounts of ethanol stored spiders 2012; 121: 136-41.

[2]   Simopoulos APJO-o-efarTse. Importance of the ratio of omega-6/omega-3 essential fatty acids: evolutionary aspects 2003; 1-22.

[3]   Guan S, Wang P, Liu H, Liu G, Ma Y. Zhao LJAJoB Production of high-amylose maize lines using RNA interference in sbe2a 2011; 1068: 15229-37.

[4]   Bell CC, Magor GW, Gillinder KR. Perkins ACJBg. A high-throughput screening strategy for detecting CRISPR-Cas9 induced mutations using next-generation sequencing. 2014; 151: 1-7.

[5]   Plant genome engineering with sequence-specific nucleases 2013; 64: 327-50.

[6]   Fattah F, Lee EH, Weisensel N, Wang Y, Lichter N. Hendrickson EAJPg. Ku regulates the non-homologous end joining pathway choice of DNA double-strand break repair in human somatic cells. 2010; 62: e1000855.

[7]   Bortesi L. Fischer RJBa The CRISPR/Cas9 system for plant genome editing and beyond 2015; 331: 41-52.

[8]   Sebo ZL, Lee HB, Peng Y, Guo YJF. A simplified and efficient germline-specific CRISPR/Cas9 system for Drosophila genomic engineering 2014; 81: 52-7.

[9]   Fiaz S, Ahmad S, Noor MA, *et al.* Applications of the CRISPR/Cas9 system for rice grain quality improvement: Perspectives and opportunities 2019; 204: 888.

[10]   Andersson M, Turesson H, Olsson N, *et al.* Genome editing in potato *via* CRISPR-Cas9 ribonucleoprotein delivery. 164(4): 378-84.2018;

[11]   Kohara Y, Akiyama K, Isono KJC. The physical map of the whole E. coli chromosome: application of a new strategy for rapid analysis and sorting of a large genomic library. 1987; 50(3): 495-508.

[12]   Marraffini LA, Sontheimer EJJNRG. CRISPR interference: RNA-directed adaptive immunity in bacteria and archaea. 2010; 11(3): 181-90.

[13]   Fischer S, Maier LK, Stoll B, *et al.* An archaeal immune system can detect multiple protospacer adjacent motifs (PAMs) to target invader DNA. J Biol Chem 2012; 287(40): 33351-63.
        [http://dx.doi.org/10.1074/jbc.M112.377002] [PMID: 22767603]

[14]   Sapranauskas R, Gasiunas G, Fremaux C, Barrangou R, Horvath P. Siksnys VJNar. The Streptococcus thermophilus CRISPR/Cas system provides immunity in *Escherichia coli*. 2011; 39(21): 9275-82.

[15]   Van der Oost J, Jore MM, Westra ER, Lundgren M. Brouns SJJTibs. CRISPR-based adaptive and heritable immunity in prokaryotes. 2009; 34(8): 401-7.

[16]   Koonin EV, Makarova KS. Zhang FJCoim. Diversity, classification and evolution of CRISPR-Cas systems. 2017; 37: 67-78.

[17]   Bao A, Burritt DJ, Chen H, Zhou X, Cao D. The CRISPR/Cas9 system and its applications in crop genome editing. 2019; 39(3): 321-36.

[18]   Court DL, Gan J, Liang Y-H, Shaw GX, Tropea JE, Costantino N. *et al.* RNase III: genetics and function; structure and mechanism 2013; 47: 405-31.

[19]   Le Rhun A, Escalera-Maurer A, Bratovič M. Charpentier EJRb. CRISPR-Cas in Streptococcus pyogenes. 2019; 16(4): 380-9.

[20]   Gupta R, Gupta D, Ahmed KT, *et al.* Modification of Cas9, gRNA and PAM: key to further regulate genome editing and its applications. 2021; 178: 85-98.
        [http://dx.doi.org/10.1016/bs.pmbts.2020.12.001]

[21]   Shukla VK. Precise Genome Modification in Maize Using Zinc-Finger Nucleases.

[22]   Haroon M, Zafar MM, Farooq MA, *et al.* Conventional breeding, molecular breeding and speed breeding. brave approaches to revamp the production of cereal crops 2020.

[23]   Prajapat RK, Mathur M, Upadhyay TK, Lal D, Maloo S, Sharma D. Genome Editing for Crop Improvement Crop Improvement. CRC Press 2021; pp. 111-23.

[24]   Hutin M, Sabot F, Ghesquière A, Koebnik R, Szurek B. A knowledge-based molecular screen

uncovers a broad-spectrum *OsSWEET14* resistance allele to bacterial blight from wild rice. Plant J 2015; 84(4): 694-703.
[http://dx.doi.org/10.1111/tpj.13042] [PMID: 26426417]

[25]   Li S, Lin D, Zhang Y, *et al.* Genome-edited powdery mildew resistance in wheat without growth penalties. Nature 2022; 602(7897): 455-60.
[http://dx.doi.org/10.1038/s41586-022-04395-9] [PMID: 35140403]

[26]   Lee K, Zhang Y, Kleinstiver BP, *et al.* Activities and specificities of CRISPR /Cas9 and Cas12a nucleases for targeted mutagenesis in maize. Plant Biotechnol J 2019; 17(2): 362-72.
[http://dx.doi.org/10.1111/pbi.12982] [PMID: 29972722]

[27]   Jung JH, Altpeter F. TALEN mediated targeted mutagenesis of the caffeic acid O-methyltransferase in highly polyploid sugarcane improves cell wall composition for production of bioethanol. Plant Mol Biol 2016; 92(1-2): 131-42.
[http://dx.doi.org/10.1007/s11103-016-0499-y] [PMID: 27306903]

[28]   Kun RS, Meng J, Salazar-Cerezo S, Mäkelä MR, de Vries RP, Garrigues S. CRISPR/Cas9 facilitates rapid generation of constitutive forms of transcription factors in Aspergillus niger through specific on-site genomic mutations resulting in increased saccharification of plant biomass. Enzyme Microb Technol 2020; 136: 109508.
[http://dx.doi.org/10.1016/j.enzmictec.2020.109508] [PMID: 32331715]

[29]   Demorest ZL, Coffman A, Baltes NJ, *et al.* Direct stacking of sequence-specific nuclease-induced mutations to produce high oleic and low linolenic soybean oil. BMC Plant Biol 2016; 16(1): 225.
[http://dx.doi.org/10.1186/s12870-016-0906-1] [PMID: 27733139]

[30]   Clasen BM, Stoddard TJ, Luo S, *et al.* Improving cold storage and processing traits in potato through targeted gene knockout. Plant Biotechnol J 2016; 14(1): 169-76.
[http://dx.doi.org/10.1111/pbi.12370] [PMID: 25846201]

[31]   Shan Q, Zhang Y, Chen K, Zhang K, Gao C. Creation of fragrant rice by targeted knockout of the *OsBADH2* gene using TALEN technology. Plant Biotechnol J 2015; 13(6): 791-800.
[http://dx.doi.org/10.1111/pbi.12312] [PMID: 25599829]

[32]   Chen G, Zhou Y, Kishchenko O, *et al.* Gene editing to facilitate hybrid crop production. Biotechnol Adv 2021; 46: 107676.
[http://dx.doi.org/10.1016/j.biotechadv.2020.107676] [PMID: 33285253]

[33]   Khan Z, Khan SH, Mubarik MS, Ahmad A. Targeted genome editing for cotton improvement. Past, Present and Future Trends in Cotton Breeding 2018; p. 11.

[34]   Sun Q, Zhang N, Wang J, *et al.* A label-free differential proteomics analysis reveals the effect of melatonin on promoting fruit ripening and anthocyanin accumulation upon postharvest in tomato. J Pineal Res 2016; 61(2): 138-53.
[http://dx.doi.org/10.1111/jpi.12315] [PMID: 26820691]

[35]   Romero FM, Gatica-Arias A. CRISPR/Cas9: Development and application in rice breeding. Rice Sci 2019; 26(5): 265-81.
[http://dx.doi.org/10.1016/j.rsci.2019.08.001]

[36]   Haun W, Coffman A, Clasen BM, *et al.* Improved soybean oil quality by targeted mutagenesis of the fatty acid desaturase 2 gene family. Plant Biotechnol J 2014; 12(7): 934-40.
[http://dx.doi.org/10.1111/pbi.12201] [PMID: 24851712]

[37]   Sun Y, Jiao G, Liu Z, *et al.* Generation of high-amylose rice through CRISPR/Cas9-mediated targeted mutagenesis of starch branching enzymes. Front Plant Sci 2017; 8: 298.
[http://dx.doi.org/10.3389/fpls.2017.00298] [PMID: 28326091]

[38]   Li J, Li Y, Ma L. CRISPR/Cas9-based genome editing and its applications for functional genomic analyses in plants. Small Methods 2019; 3(3): 1800473.
[http://dx.doi.org/10.1002/smtd.201800473]

[39]   Yin K, Qiu J-L. Genome editing for plant disease resistance: Applications and perspectives. Philosophical Transactions of the Royal Society B 2019; 374(1767).
[http://dx.doi.org/10.1098/rstb.2018.0322]

[40]   Wang F, Wang C, Liu P, *et al.* Enhanced rice blast resistance by CRISPR/Cas9-targeted mutagenesis of the ERF transcription factor gene OsERF922. PLoS One 2016; 11(4): e0154027.
[http://dx.doi.org/10.1371/journal.pone.0154027] [PMID: 27116122]

[41]   Zafar K, Khan MZ, Amin I, *et al.* Precise CRISPR-Cas9 mediated genome editing in super basmati rice for resistance against bacterial blight by targeting the major susceptibility gene. Front Plant Sci 2020; 11: 575.
[http://dx.doi.org/10.3389/fpls.2020.00575] [PMID: 32595655]

[42]   Prajapati A, Nain V. Screening of CRISPR/Cas9 gRNA for mimicking Powdery Mildew resistant MLO ol-2 mutant. Bioinformation 2021; 17(6): 637-45.
[http://dx.doi.org/10.6026/97320630017637] [PMID: 35173386]

[43]   Liang G, Zhang H, Lou D, *et al.* Yu DJSr. Selection of highly efficient sgRNAs for CRISPR/Cas9-based plant genome editing. 2016; 6(1): 1-8.

[44]   Wang W, Akhunova A, Chao S, Akhunov EJB. Optimizing multiplex CRISPR/Cas9-based genome editing for wheat. 2016; 051342.
[PMID: 051342]

[45]   James Faresse N, Canella D, Praz V, Michaud J, Romascano D. Hernandez NJPg. Genomic study of RNA polymerase II and III SNAPc-bound promoters reveals a gene transcribed by both enzymes and a broad use of common activators 2012; 8(11): e1003028.
[PMID: e1003028]

[46]   Krupkova O, Cambria E, Besse L, Besse A, Bowles R, Wuertz-Kozak KJJs. The potential of CRISPR/Cas9 genome editing for the study and treatment of intervertebral disc pathologies. 2018; 1(1): e1003.

[47]   Manghwar H, Li B, Ding X, Hussain A, *et al.* CRISPR/Cas systems in genome editing: Methodologies and tools for sgRNA design, off-target evaluation, and strategies to mitigate off-target effects. 2020; 7(6)

[48]   Meng X, Yu H, Zhang Y, *et al.* Construction of a genome-wide mutant library in rice using CRISPR/Cas9. 2017; 10(9): 1238-41.

[49]   Cebrian-Serrano A, Davies BJMG. CRISPR-Cas orthologues and variants: Optimizing the repertoire, specificity and delivery of genome engineering tools. 2017; 28(7): 247-61.
[http://dx.doi.org/10.1007/s00335-017-9697-4]

[50]   Kuscu C, Arslan S, Singh R, Thorpe J. Adli MJNb. Genome-wide analysis reveals characteristics of off-target sites bound by the Cas9 endonuclease. 2014; 32(7): 677-83.

[51]   LeBlanc C, Zhang F, Mendez J, *et al.* Increased efficiency of targeted mutagenesis by CRISPR/Cas9 in plants using heat stress. 2018; 8(9): 1589-99.

[52]   Hirzel M, Soulé R, Schneider S, Gedik B, Grimm RJACS. A catalog of stream processing optimizations. 2014; 46(4): 1-34.
[http://dx.doi.org/10.1145/2528412]

[53]   Yuan H, Cheung CM, Poolman MG, Hilbers PA, van Riel NAJTPJ. A genome-scale metabolic network reconstruction of tomato (Solanum lycopersicum L.) and its application to photorespiratory metabolism. 2016; 85(2): 289-304.

[54]   Mlynárová Lu, Keizer LP, Stiekema WJ, *et al.* Nap J-PJTPC. Approaching the lower limits of transgene variability. 1996; 8(9): 1589-99.

[55]   Ganopoulos I, Argiriou A, Tsaftaris AJFC. Microsatellite high resolution melting (SSR-HRM) analysis for authenticity testing of protected designation of origin (PDO) sweet cherry products. 2011;

22(3-4): 532-41.

[56]   Dickinson ME, Flenniken AM, Ji X, *et al.* High-throughput discovery of novel developmental phenotypes. 2016; 537(7621): 508-14.
[http://dx.doi.org/10.1038/nature19356]

[57]   Pattanayak V, Lin S, Guilinger JP, *et al.* Liu DRJNb. High-throughput profiling of off-target DNA cleavage reveals RNA-programmed Cas9 nuclease specificity. 31(9): 839-43.2013;

[58]   Joensen KG, Scheutz F, Lund O, *et al.* Real-time whole-genome sequencing for routine typing, surveillance, and outbreak detection of verotoxigenic *Escherichia coli*. 2014; 31(9): 839-43.
[http://dx.doi.org/10.1128/JCM.03617-13]

[59]   Popp J, Pető K. Nagy JJAfsd. Pesticide productivity and food security. RE:view 2013; 33(1): 243-55.

[60]   Gao CJNRMCB. The future of CRISPR technologies in agriculture. 2018; 19(5): 275-6.
[http://dx.doi.org/10.1038/nrm.2018.2]

[61]   Bai Z, Li W, Jia Y, *et al.* The ethylene response factor SmERF6 co-regulates the transcription of SmCPS1 and SmKSL1 and is involved in tanshinone biosynthesis in Salvia miltiorrhiza hairy roots. Planta 2018; 248(1): 243-55.
[http://dx.doi.org/10.1007/s00425-018-2884-z] [PMID: 29704055]

[62]   Di P, Zhang L, Chen J, *et al.* ^{13}C tracer reveals phenolic acids biosynthesis in hairy root cultures of Salvia miltiorrhiza. ACS Chem Biol 2013; 8(7): 1537-48.
[http://dx.doi.org/10.1021/cb3006962] [PMID: 23614461]

[63]   Li Q, Feng J, Chen L, *et al.* Genome-wide identification and characterization of Salvia miltiorrhiza laccases reveal potential targets for salvianolic acid B biosynthesis. Front Plant Sci 2019; 10: 435.
[http://dx.doi.org/10.3389/fpls.2019.00435] [PMID: 31024599]

[64]   Shi M, Du Z, Hua Q, Kai G. CRISPR/Cas9-mediated targeted mutagenesis of bZIP2 in Salvia miltiorrhiza leads to promoted phenolic acid biosynthesis. Ind Crops Prod 2021; 167: 113560.
[http://dx.doi.org/10.1016/j.indcrop.2021.113560]

[65]   Kui L, Chen H, Zhang W, *et al.* Building a genetic manipulation tool box for orchid biology: identification of constitutive promoters and application of CRISPR/Cas9 in the orchid, Dendrobium officinale. Front Plant Sci 2017; 7: 2036.
[http://dx.doi.org/10.3389/fpls.2016.02036] [PMID: 28127299]

[66]   Li Y, Zhang B, Wang Y, Gong X, Yu H. *DOTFL1* affects the floral transition in orchid *Dendrobium* Chao Praya Smile. Plant Physiol 2021; 186(4): 2021-36.
[http://dx.doi.org/10.1093/plphys/kiab200] [PMID: 33930147]

[67]   Zhang X, Xu G, Cheng C, *et al.* Establishment of an *Agrobacterium* -mediated genetic transformation and CRISPR/Cas9-mediated targeted mutagenesis in Hemp (*Cannabis Sativa* L.). Plant Biotechnol J 2021; 19(10): 1979-87.
[http://dx.doi.org/10.1111/pbi.13611] [PMID: 33960612]

[68]   Kramnik I, Demant P, Bloom BB, Eds. Susceptibility to tuberculosis as a complex genetic trait: Analysis using recombinant congenic strains of mice Novartis Foundation symposium. WILEY 1998.

[69]   Soyk S, Müller NA, Park SJ, *et al.* Variation in the flowering gene SELF PRUNING 5G promotes day-neutrality and early yield in tomato. 2017; 49(1): 162-8.

[70]   Zhou II, Zhou M, Yang Y, *et al.* RNase ZS1 processes UbL40 mRNAs and controls thermosensitive genic male sterility in rice. 2014; 5(1): 1-9.
[http://dx.doi.org/10.1038/ncomms5884]

[71]   Massel K, Lam Y, Wong A, *et al.* RNase ZS1 processes UbL40 mRNAs and controls thermosensitive genic male sterility in rice. 2021; 14(2): 510-8.

[72]   Li C, Liu C, Qi X, *et al.* RNA-guided Cas9 as an *in vivo* desired-target mutator in maize 2017; 15(12): 1566-76.

[73]   Wolt JD, Wang K, Yang BJPbj. The regulatory status of genome-edited crops. 2016; 14(2): 510-8.

[74]   Kohlen W, Charnikhova T, Lammers M, *et al.* The tomato carotenoid cleavage dioxygenase 8 (S 1 CCD 8) regulates rhizosphere signaling, plant architecture and affects reproductive development through strigolactone biosynthesis. 2012; 196(2): 535-47.

[75]   Sergio F. Newton IJJoAE. Occupancy as a measure of territory quality. 2003; 72(5): 857-65.

[76]   Gudmundsson M, Eliasson A-CJCP. Occupancy as a measure of territory quality. 1990; 72(5): 857-65.

[77]   Tang L, Mao B, Li Y, *et al.* Knockout of OsNramp5 using the CRISPR/Cas9 system produces low Cd-accumulating indica rice without compromising yield. 2017; 7(1): 1-12.

[78]   Li X, Wang Y, Chen S, *et al.* Lycopene is enriched in tomato fruit by CRISPR/Cas9-mediated multiplex genome editing. 2018; 8:993.

[79]   Lou D, Wang H, Liang G. Yu DJFips. 2017.

[80]   Shinwari ZK, Jan SA, Nakashima K, Yamaguchi-Shinozaki KJPBR. Genetic engineering approaches to understanding drought tolerance in plants. 2020; 14(2): 151-62.
[http://dx.doi.org/10.1007/s11816-020-00598-6]

[81]   Martignago D, Rico-Medina A, Blasco-Escámez D, Fontanet-Manzaneque JB. Caño-Delgado AIJFips. Drought resistance by engineering plant tissue-specific responses. 2020; 10:1676

[82]   Zhang Y, Bai Y, Wu G, *et al.* Simultaneous modification of three homoeologs of Ta EDR 1 by genome editing enhances powdery mildew resistance in wheat. 2017; 91(4): 714-24.

[83]   Arora L. Narula AJFips. Gene editing and crop improvement using CRISPR-Cas9 system. 2017; 8:1932.

[84]   Macovei A, Sevilla NR, Cantos C, *et al.* Novel alleles of rice eIF4G generated by CRISPR/Cas9-targeted mutagenesis confer resistance to Rice tungro spherical virus. 2018; 16(11): 1918-27.

[85]   Wang X, Tu M, Wang D, *et al.* CRISPR/Cas9-mediated efficient targeted mutagenesis in grape in the first generation 2018; 16(4): 844-55.

[86]   Makarova S, Khromov A, Spechenkova N, Taliansky M, Kalinina NJB. Application of the CRISPR/Cas system for generation of pathogen-resistant plants. 2018; 83(12): 1552-62.

[87]   Xu Y, Crouch JHJCs. Marker-assisted selection in plant breeding: From publications to practice. 2008; 48(2): 391-407.

[88]   Sun Y, Dong L, Zhang Y, *et al.* 3D genome architecture coordinates trans and cis regulation of differentially expressed ear and tassel genes in maize. 2020; 21(1): 1-25.

[89]   Tarasava K, Oh EJ, Eckert CA, Gill RTJBJ. CRISPR-enabled tools for engineering microbial genomes and phenotypes. 2018; 13(9): 1700586.
[http://dx.doi.org/10.1002/biot.201700586]

[90]   Jacobs TB, Zhang N, Patel D, Martin GBJPp. Generation of a collection of mutant tomato lines using pooled CRISPR libraries. 2017; 174(4): 2023-37.

[91]   Gao W, Xu W-T, Huang K-L, Guo M-z, Luo Y-BJFC. Risk analysis for genome editing-derived food safety in China. 2018; 84: 128-37.
[http://dx.doi.org/10.1016/j.foodcont.2017.07.032]

[92]   Basso MF, Duarte KE, Santiago TR, *et al.* Efficient genome editing and gene knockout in Setaria viridis with CRISPR/Cas9 directed gene editing by the non-homologous end-joining pathway. 2021; 38(2): 227-38.

[93]   Liang Z, Chen K, Li T, *et al.* Efficient DNA-free genome editing of bread wheat using CRISPR/Cas9 ribonucleoprotein complexes. 2017; 8(1): 1-5.

CHAPTER 10

Genome Editing in Medicinal Plants for Abiotic Stress Tolerance

Zarmeen Zafar[1], Furqan Ahmad[1], Shoaib ur Rehman[1], Saira Karimi[2], Umar Akram[1], Zareena Ali[3] and Zulqurnain Khan[4,*]

[1] *Institute of Plant Breeding and Biotechnology, MNS University of Agriculture, Multan, Pakistan*

[2] *Department of Biosciences, COMSATS University, Islamabad, Pakistan*

[3] *Department of Biochemistry and Biotechnology, The Women University, Multan, Pakistan*

[4] *Department of Biotechnology, Institute of Plant Breeding and Biotechnology (IPBB), Muhammad Nawaz Shareef University of Agriculture, Multan-60000, Pakistan*

Abstract: In the changing climate scenarios, living organisms have been facing several biotic and abiotic stresses. Abiotic stresses are the major factors posing huge threats to plants. Drought, heat, and salinity are bigger problems emerging in the world due to climate change. For adapting various climatic conditions, plants trigger several biochemical and molecular mechanisms. At the biochemical level, secondary metabolites play an important role in the survival of plants in uneven situations. Modulation of expression of genes and stress response elements is observed under stress. To cope with abiotic stresses in medicinal plants, the understanding of the biochemical and molecular mechanisms is very important. The use of modern biotechnological approaches along with conventional breeding may be helpful in developing tolerance against abiotic stresses in medicinal plants. This chapter highlights various abiotic stresses, their mechanisms and strategies to develop tolerance against these stresses.

Keywords: Abiotic stress, Adenine base editor, Biotechnological approaches, CRISPR/Cas, Drought, Homologous recombination, Light, Pentatricopeptide repeat proteins, RNA interference zinc finger nucleases (ZFNs), Site-directed sequence editing, Salt stress, Secondary metabolites, Temperature, Transcription activator-like effector nuclease(TALENS), Ultraviolet.

---

* **Corresponding author Zulqurnain Khan:** Department of Biotechnology, Institute of Plant Breeding and Biotechnology (IPBB), Muhammad Nawaz Shareef University of Agriculture, Multan-60000, Pakistan; E-mail: zulqurnain.khan@mnsuam.edu.pk

INTRODUCTION

Medicinal plants, being complicated organisms, are subjected to various physiological and chemical conditions that have a profound effect on plant production, hindering high performance and endangering their existence [1]. Ultraviolet (UV) radiation, drought, high salinity, heavy metals, temperature (high or low), *etc.* are important among the various variables and together they are referred to as abiotic stressors [2]. Any increase in the frequency of abiotic stress affects the plant from germination to maturity [3].

Medicinal plants are useful flora that has chemically active elements in many of their parts and produces a physiological reaction when utilized to cure a variety of diseases. Medicinal plants have been used in countries such as Sri Lanka, China, Thailand, India, Japan, Nepal, and Pakistan since time immemorial [4]. The earliest evidence of medicinal plants' use dates back to 2600 BC from Mesopotamian culture [5]. To meet the ever-growing demands of the pharmaceutical industry, some medicinal plants should now be grown commercially [6]. Plants are at risk of abiotic stressors such as salt, drought, high and low temperatures, heavy metal toxicity, high and low light, nutrient deficiencies, ozone, and UV-B radiation due to continuous changes in the environment. Abiotic stressors are linked to decreased crop productivity and quality [7]. Abiotic stress depletes agricultural production and identified as a major barrier to food security worldwide [8]. The genome-editing technique is used to create a long-term solution to abiotic stress. The adoption of genomic engineering methods for making specific changes in the genome of a plant to improve adaptability, increase yields, and allocate resilience to various stresses has facilitated technological advances. Some of the most important methods for modifying plant genomes include homologous recombination (HR), zinc finger nucleases (ZFNs), transcription activator-like effector nucleases (TALENs), pentatricopeptide repeat proteins (PPRs), the CRISPR/Cas9 system, RNA interferance (RNAi), cisgenesis, and intragenesis. CRISPR/Cas has revolutionized biological research and has greater use in agriculture than the other genome editing techniques. CRISPR / Cas is a powerful way to modify genomes to achieve desirable traits. The candidate genes must be chosen carefully to get the desired features.

Medicinal Plants and Secondary Metabolites (SMs)

In many regions of the world, traditional medicine has been discovered from plants for thousands of years. Plants are categorised as medicinal because they generate secondary metabolites that are advantageous to human health. Along with having medicinal value, these plant's secondary metabolites are crucial for plants to respond to environmental changes [9]. Different categories of medicinal plants may have different therapeutic effects. Alkaloids, terpenoids, and phenylpropanoids are among the second metabolites studied to improve therapeutic value. In cells, these chemicals are made up of series of enzyme-catalyzed processes. Secondary metabolites are chemical compounds unrelated to regular plant growth and development [10]. Natural pharmacologically active compounds with antimalarial, antidiabetic, hepatoprotective, antiulcer, anti-inflammatory, and antibacterial properties are among the secondary metabolites. When there are severe environmental conditions, they are produced in large quantities in various parts of the plant. Natural mutations have a significant impact on the control of several secondary metabolic processes. Stressful situations induce secondary metabolite's concentration to rise in medicinal plants because they have an impact on the genes engaged in secondary metabolic pathways.

Natural Factors that Affect Secondary Metabolite's Development Include

A local and temporal strategy carefully controls the integration and collection of SMs, which is also affected by biotic and abiotic environmental factors. Variations in SMs biosynthesis are caused by adverse abiotic circumstances, such as droughts or floods, excessive light or heat, poor soil quality, or the presence of harmful substances [11].

Light

An essential abiotic component needed by plants for photosynthesis, development, and the buildup of secondary metabolic products is exposure to photons of different durations and intensities [12]. High irradiance can occasionally promote plant growth. SMs production and different plant species have their own ideal light set (quality and quantity), capable of creating high yields for SMs [13].

Ultraviolet Light

In medicinal plants, exposure to ultraviolet (UV) light has a typical impact on SMs. For instance, the UV-B radiation of Chrysanthemum enhanced the content of flavonoids and phenolic acids [14]. While UV light exposure generally has a good impact on the production of SMs, excessive UV-B and UV-C radiation has a negative impact on plant growth, development, photosynthesis, and other critical functions [15]. Response of different medicinal plants under UV stress is described in Table **1**.

Table 1. Effect of light on various plant SMs content.

Metabolite Catagory	Name of Metabolite	Concentration Change	Environment Factor	Parts	Plant Species	References
Alkaloids	Alkaloids	Increase	30 & 50% bright sunlight	Entire plant	*Mahonia bodinieri*	[16]
				Root, stem	*Mahonia breviracema*	[16]
	Saponins,	Increase	UV-B	Roots and leaves	*Withania somnifera*	[17]
	Essential oil	Increase	Bright sunlight	Leaf	*Mahonia breviracema*	[16]
Phenols	Anthocyanins, lignin, tannins	Increase	UV-B	Roots and leaves	*Withania somnifera*	[17]
Others	Phytosterol	Increase	UV-B	Root and leaf	*Withania somnifera*	[17]
	Glucosinolate	Increase	UV	Leaf	*Nasturtium officinale*	[18]
	Hexadecanoic acid	Increase	50% bright sunlight	Leaf	*Mahonia bodinieri*	[16]

Temperature Stress

The temperature at which plants are present has a direct impact on their growth and development. Crop development and productivity can be negatively impacted by low and high temperatures [19]. SM production is frequently increased by high temperature stress [20], and some research have shown that SMs fall on plants under high-temperature stress [21]. Behavior of different medicinal plants under temperature stress is described in Table **2**.

Table 2. The impact of temperature variation on different plant SMs content.

Catagory of Metabolite	Name of Metabolite	Parts	Change in Concentration	Environment Factor	Plant Specie	References
Sesquiterpene lactone	Artemisinin	Whole plant	Increase	A transient pre-chilling treatment	*Artemisia annua*	[22]

(Table 2) cont.....

Catagory of Metabolite	Name of Metabolite	Parts	Change in Concentration	Environment Factor	Plant Specie	References
Phenols	Phenolics	Roots, leaves & flowers	Increase	High temperature	*Astragalus compactus*	[23]
	Anthocyanins	Whole plant	Decrease	High temperature	*Chrysanthemum*	[24]
Fatty acid	α-linolenic acid, Jasmonic acid	Leaves	Increase	Low temperature	*Camellia japonica*	[25]

Drought Stress

Drought stress decreases water absorption in humans and plants, which negatively impacts several bodily functions and can change SM biosynthesis [26]. Under extreme stress, plants, for instance, produce more flavonoids and phenolics [27]. In another research, the quality of important SMs such as rutin, quercetin, and betulinic acid in Hypericum brasiliense and artemisinin in Artemisia annua improved as a result of exposure to drought stress [11]. In order to employ SMs to increase plant tolerance to drought stress, further study is required to better understand the regulatory proteins and genes involved in plant production for SMs. The impact of drought stress on different plant SM concentrations is discussed in Table (3).

Table 3. The impact of drought stress on different plant SM concentrations.

Class of Metabolite	Name of Metabolite	Parts	Specie of Plant	References
Pentacyclic triterpenoid	Betulinic acid	-	*Hypericum brasiliense*	[28]
Sesquiterpene lactone	Artemisinin	Whole plant	*Artemisia*	[28]
Phenolic Compunds	Rutin and Quercetin	-	*Hypericum brasiliense*	[28]
	Total phenolic Compound	Leaf	Trachyspermum ammi	[29]
		Roots and Shoots	*Hypericum brasiliense*	[30]
		Leaf	*Labisia pumila*	[31]
	Anthocyanins	Leaf	*Labisia pumila*	[31]

Salt Stress

High-salt soils lead to nutrient imbalances, hyperosmotic stress, and a reduction in plant growth, photosynthesis, and nutrient absorption [32]. Plant SM concentration may change in response to osmotic stress brought on by salt or ion toxicity [33]. In Plantago ovata, salt stress causes plants to produce more alkaloids, tannins, phenolics, saponins, flavonoids, proline, and saponins [34]. There is little question that many regulatory genes, enzymes, transcription factors, and stress brought on by too much salt have an impact on the biosynthesis of SMs, which in turn causes instability in the creation or accumulation of certain SMs. Early transgenic efforts should concentrate on enhancing salt tolerance in threatened plant species and making it easier to introduce salt-tolerant genes into threatened medicinal plant species. The impact of changing soil salinity on the content of different plant SMs is given in Table **4**.

Table 4. The impact of changing soil salinity on the content of different plant SMs.

Metabolite Class	Metabolite Name	Environment Factor	Concentration Change	Parts	Plant Species	References
Flavonoid compounds	Flavonoid	NaCl	High	Root,Shoot	*Plantago ovata*	[35]
Alkaloids	Recinine	NaCl	High	Shoot	*Ricinus communis*	[16]
Phenolics	Tannin	Salinity	High	Whole plant	*Achillea fragratissima*	[36]
Essential Oils /Monoterpenes	Oil contents	NaCl	High	Shoot	*Ricinus communis*	[16]
		50 and 25 mM NaCl		Leaf	Coriandrum sativum	[37]
-	Oil contents	High salinity	Low	Leaf	Coriandrum sativum	[37]
-		100 Mm NaCl		Shoots	Origanum majorana	[38]
-	α-Pinene; [Z]-Myroxide	Salinity	Low	Leaf	Coriandrum sativum	[37]

Abiotic Stress Alters the Patterns of Gene Expression in Medicinal Plants

Environmental factors have a major role in determining the quantity and concentration of secondary metabolites in medicinal plants. The creation of secondary metabolites will be aided by studies on the impact of abiotic stress on medicinal plants using gene expression analyses. Withania somnifera is one of the most significant medicinal plants and possesses a wide range of therapeutic benefits, including sedative, narcotic, thyroid stimulation, anti-inflammatory, hypotonic, anti-stress, general tonic, diuretic, antimicrobial, and antitumor activities [39].*W. somnifera* voice analysis of several genetic groups was thought to demonstrate their enhanced responses to drought stress, including osmoregulation (1P5CS), detoxification (GST and SOD), signal transmission (STK and PSP), metabolism (AD and LD), and regulation (HSP, MYB, and WRKY) [40]. Rehmannia glutinosa rough roots are used to treat high blood pressure, diabetes, mood swings, and fever as well as to enhance performance. Hematopoietic, hepatic, immunostimulating, and tonic characteristics. The RghBNG genetic expression response to several abiotic stresses is investigated in this plant.

Genome Editing Tools

Because random insertion fails to eliminate the open reading frame (ORF) for genetically modified organisms, mutations that have limited functionality, adverse effects, or abnormal protein products are more likely. To solve such challenges in the genetic engineering of plants, new technologies are needed. The following are important planning techniques that can be used to identify different aspects of medicinal plants, as well as how these technologies might be enhanced to be used more effectively. Some important genome editing techniques have been described in Fig. (**1**).

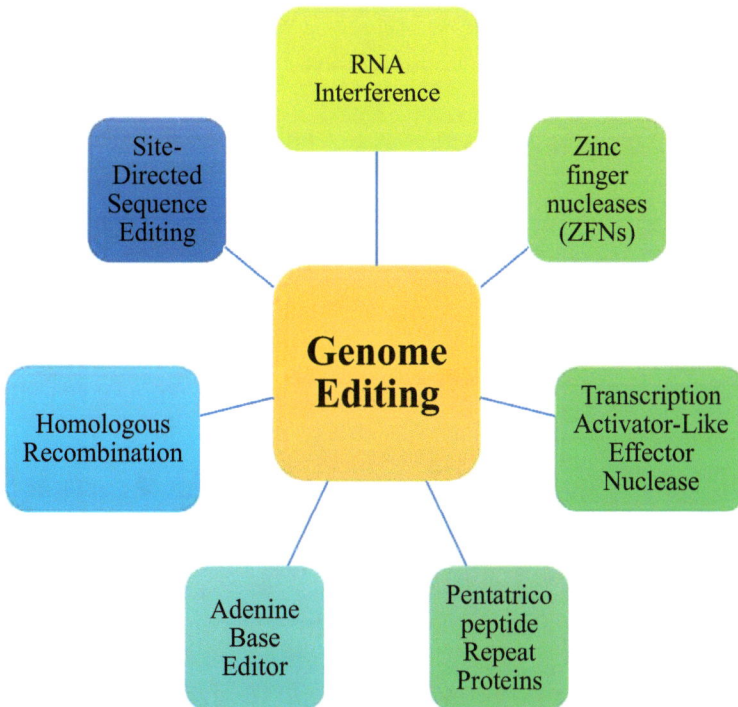

Fig. (1). Genome editing techniques in plants.

Homologous Recombination

Chromosome reunification (also known as homologous recombination, or HR) is a common and efficient method of genome engineering within the cell [41]. Because of this, genome editing technologies may reap the same benefits with little to no mistakes. The commencement of chromosomal double-stranded (DSB) breaks can be utilised to categorise the genome [42]. DSBs result in meiotic reunion during cell division. DSBs in eukaryotes are a great tool for genetic identification since they are highly conserved and may be started locally.

Zinc Finger Nucleases [ZFNs]

Custom DNA cleaving proteins called zinc finger nucleases (ZFNs) are used to separate DNA sequences into distinct regions [43]. ZFNs enable targeted genetic change by causing DNA double-strand breaks (DSBs), which enable genetic reunion through homologous recombination. A unique 6-bp hexamer sequence of DNA is identified by a set of two-finger modules in each ZFN's DNA binding domain. as a nuclease domain of FokI and a DNA-cleaving domain [44]. These domains can be connected to produce the zinc finger protein (ZFP) [44]. A unique genomic structure is created when DNA-cleaving and DNA-binding domains are joined. By combining DSBs, genetic sequence within the ZNF introduces double-stranded genetic sequences of DNA sequences, and irregular genome sequencing [44].

Transcription Activator-Like Effector Nuclease

In plants, TALENs (nuclease activator-like effector nucleases) have become a successful way for ZFNs to function as tools for genome programming [45]. Theoretically, TALENS and ZFNs both employ DSBs. Unspecific FokI endonucleases like TALENS are among the ZFNs. On the other hand, FokI domains for ZFNs attach to certain DNA binding domains based on activator-like effectors (TALEs) [45]. The Xanthomonas bacteria create TALE proteins that change the expression of genes in host plants [46].

Pentatricopeptide Repeat Protein

Compared to nuclear genomes, organellar genes have fewer developers, and organellar RNA has a longer half-life. The organelle transforms significant quantities of proteins binding to RNA to control the rate of gene expression since transcriptional regulation (RNA classification, RNA classification, RNA classification, and translation) is insufficient to control gene expression in organellar genomes [47]. The mediator protein present in the organellar genome that controls post-transcriptional regulation is replicated by PPR, or pentatricopeptide. PPRs are often recognised by the presence of a tandem repeat of 35-amino-acid motif [47].

Adenine Base Editor

Cytosine and 5-methylcytosine are automatically converted into uracil and thymine in cells due to hydrolytic deamination, which results in the transformation of C-G to T-A [48]. Each cell experiences these alterations 100–500 times every day [49]. However, these editing techniques are only capable of changing C-G to T-A. Target genomic loci are transformed from A-T base pairs to G-C base pairs, resulting in small nucleotide polymorphisms (SNPs). This challenge may be surmounted by the Adenine Foundation (ABEs) organizers. ABEs convert A-T to G-C in humans and bacteria, but they have not yet been adopted by plants [49]. Seventh-generation ABEs have significantly improved the precision and purity of A-T to G-C conversion in a genomic setting [49]. ABE employs tRNA adenosine deaminase related to CRISPR-Cas9, which is interestingly suppressed while converting A-T to G-C [49].

Interference of RNA

RNA (RNAi) is a kind of genetic post-transcriptional mutation that modifies the genetic composition in many ways [50]. RNAi can be brought on by small RNAs (siRNAs), piRNAs (PIWI-interacting RNAs), and microRNAs (miRNAs). Dual-stranded RNA is converted into single-stranded RNAs, which act as directives for the cleavage of DNA and compression translation. They are 20–30 nucleotide lengths that act as benchmarks for compression translation and DNA cleavage. SiRNA activity, which is based on sequential completion, reduces gene function [51]. Plant cells only make a small number of siRNAs, and the majority of siRNAs are viral. Transferring the chosen siRNA to the cell is crucial in order to study the function of any gene. On the other hand, plant cells have cell walls, which makes it difficult to transfer siRNA. A double-stranded region created by the reverse wrapping of hairpin RNA is utilised to deliver siRNA to plant cells.

Editing Site-Based Sequences

Pentatricopeptide proteins secreted in RNA editing are the source of the DYW domain (aspartate, tyrosine, and tryptophan tripeptide) and C-terminal glutamate (E) amino acids [52]. The homology and background of the protein nucleotide deaminase are those of DYW. The planning machine will replicate the programming enzyme Apolipoprotein B mRNA (APOBEC1) in mammals, which contains a sequence of cytidine deaminase, if the function of this domain is validated in plants [53].

Oligonucleotide-Directed Mutagenesis

Oligonucleotide-directed mutagenesis (ODM) is non-transgenic genome editing platform that modifies genomic DNA using aggregated oligonucleotides. One to a few nucleotides can be switched or inserted into the genes of live cells using ODM and plasmid as a model. In this method, the synthetic oligonucleotide is a single-stranded sequence with a small parallel / homologous parallel to a single duplex DNA strand. The cell phone repair system fixes different pairs, which leads to some flexibility.

Cisgenesis and Intragenesis

The most common techniques for gene or genome planning are genetic modification and transgenic production. Somatic hybridization allows genetic content from two genetically divided species to be merged. A genetic breakdown of interbreeding relatives known as cisgene has been made possible by advancements in genome sequencing technologies. The exchange of genes or DNA between members of the same species without any genetic interaction is known as cisgenesis or intragenesis [54]. The gene remains unchanged in cisgenesis, while the genetic component (promoter or regulator) is transferred to intragenesis.

CRISPR-Cas9

Cas9 is a please link to CRISPRs. CRISPR stands for short-term compound palindromic compounds that are usually compounded. The CRISPR program was first introduced in 1987 in Escherichia coli as a virus protection program. Later, they were discovered in 90% of archaeal genomes and 40% of subsequent bacterial genomes. Through a series of bioinformatics-based investigations, the different parts of CRISPR/Cas were identified, along with their roles in giving bacterial cells dynamic immunity. The CRISPR signature list, which consists of a series of 21–47 bp sequences (with a maximum of 32 bp — direct compounds) paired with 25–40 bp flexible sequences (spacers) and is repeated frequently across the bacterial genome, is discovered to be the main component of CRISPR loci. Nucleotides in duplicates are highly conserved among animal species, while spacers are similar in length but have a wide variety of shapes. The active CRISPR/Cas locus consists of two separate parts: the Cas operon and a series of short copies built of phage genomic sequences [55]. Most closely related CRISPR members were formerly written as One pre-processed RNA condensed to CRISPR RNAs (crRNAs), which control the target nucleic acid reduction activity of further Cas enzymes. A genetic nuclease sequence is made to recognise a certain genomic region, transmit it to a cell, and then divide the DNA into two strands (DSB). DNA modification techniques include nonhomologous end-joining

(NHEJ) and homology-targeted (HDR) repair work well for DSB. A homologous sequence based on a DSB or a "donor template" of DNA can be used as an HDR template to fine-tune the DSB. NHEJ mutations, on the other hand, tend to be flawed and often lead to implantation or removal near DNA fragmentation sites, leading to genetic mutations. NHEJ events prefer to modify DSBs in eukaryotic cells, providing a viable approach to plant genomics and agricultural development research. The most well-known and effective genetic engineering technique is CRISPR/Cas, a recently created system that recognises targeted DNA based on the Watson-Crick base between its RNA (s) and DNA guide. Its effectiveness was developed in Arabidopsis, rice, sorghum, and tobacco-based plant systems soon after it was first used to mutate animal genes and microbes [56]. CRISPR / Cas9 genetic targeting has become the standard technology used by thousands of laboratories worldwide since this discovery.

CRISPR / Cas9 Applications

Plants are now starting to employ CRISPR/Cas9 technology. Due to their application in creating genetic variations, genetically engineered CRISPR / Cas9 genes have become a standard tool in plant laboratories all over the world [57]. CRISPR / Cas9 is used to construct specific site genetic variables, multiplex gene-targeted variables, fragmentation, genetic modification, and targeted genetic insertion. Second-line metabolites from plants are interesting and are important research topics because of their chemical diversity, diverse biological functions, and therapeutic activity. Plant biology has long sought to understand the roots of an extensive range of plant metabolism. A thorough understanding of the biosynthesis route, comprising all the gaps and enzymes involved, is crucial for effective metabolic engineering because of the intricacy of the biosynthetic processes and controls involved in the creation of secondary metabolites [58].

Prioritizing the identification of the genes involved in the integrated or unstructured biosynthetic synthesis of a natural product is the goal of CRISPR/Cas9-mediated genome gene mutations in medicinal plants. The lack of knowledge about the biosynthetic genes for the medications used in medicine hinders access to the athletes who have improved.

The first step in arranging the CRISPR/Cas9-mediated plant genome in medicinal plants is to identify the genes involved in the manufacture of a natural product using integrated or simple biosynthetic techniques. Because it is impossible to create pharmaceuticals using genetically modified organisms, the biosynthetic genes of the currently used medications remain unknown. The CRISPR / Cas9 system, for example, is often used to reduce genes or deactivate their functions to limit the production of a specific set of chemicals while increasing the number of certain secondary metabolites. The basic idea behind these methods is to eliminate the enzymatic action along the way by reducing the amount of mRNA or protein associated with it [59].

CONCLUSION

Abiotic stress is a major factor in the growth and development of medicinal plants, especially in the face of growing climate-related challenges. With great concern about declining yields, it often fails to provide food security to the people of the world. In cases like these, it is important to use strategic strategies aimed at achieving agricultural sustainability to meet the growing need for alternative therapies. Genome planning has the potential to transform plant production around the world. Even though TALFNs and ZENs are commonly accustomed in plant research including HDR-based genetic modification, they nonetheless create undesirable effects and cause HDR-altered alleles. A major barrier to predictability is the lack of sufficient genetic data set for coping. Details of the various genome editing tools are required to address this problem. The existence of a DNA binding domain, on the other hand, is sufficient for limited predictions. CRISPR-Cas9 has the capability to alter biosynthetic pathways in heterologous plants using a genetically engineered and precisely controlled gene to maximize biopharmaceutical output. However, to have a greater impact on the biology of medicinal plants, much work needs to be done to amend CRISPR / Cas9 processes so that they can be easily used and readily available in practical and practical applications.

REFERENCES

[1]   Zhu JK. Abiotic stress signaling and responses in plants. Cell 2016; 167(2): 313-24.

[2]   Zafar SA, Zaidi SS e A, Gaba Y, Singla-Pareek SL, Dhankher OP, Li X. Engineering abiotic stress tolerance *via* CRISPR/Cas-mediated genome editing. J Exp Bot 2020; 71(2): 470-9.

[3]   Waqas MA, Kaya C, Riaz A, Farooq M, Nawaz I, Wilkes A. Potential mechanisms of abiotic stress tolerance in crop plants induced by thiourea. Front Plant Sci 2019; 10: 1336.

[4]   Ahmad SS, Husain SZ. Ethno medicinal survey of plants from salt range *Kallar Kahar* of Pakistan. Pak J Bot 2008; 40(3): 1005-11.

[5]   Gurib-Fakim A. Medicinal plants: Traditions of yesterday and drugs of tomorrow. Mol Aspects Med 2006; 27(1): 1-93.

[6]   Jaleel CA, Gopi R, Manivannan P, Panneerselvam R. Soil salinity alters the morphology in Catharanthus roseus and its effects on endogenous mineral constituents. EurAsian J Biosci 2008; 2(1): 18-25.

[7]   Smékalová V, Doskočilová A, Komis G, Šamaj J. Crosstalk between secondary messengers, hormones and MAPK modules during abiotic stress signalling in plants. Biotechnol Adv 2014; 32(1): 2-11.

[8]   Wang Y, Frei M. Stressed food–The impact of abiotic environmental stresses on crop quality. Agric Ecosyst Environ 2011; 141(3–4): 271-86.

[9]   Anand U, Jacobo-Herrera N, Altemimi A, Lakhssassi N. A comprehensive review on medicinal plants

as antimicrobial therapeutics: potential avenues of biocompatible drug discovery. Metabolites 2019; 9(11): 258.

[10]   Pagare S, Bhatia M, Tripathi N, Pagare S, Bansal YK. Secondary metabolites of plants and their role: Overview. Curr Trends Biotechnol Pharm 2015; 9(3): 293-304.

[11]   Verma N, Shukla S. Impact of various factors responsible for fluctuation in plant secondary metabolites. J Appl Res Med Aromat Plants 2015; 2(4): 105-13.

[12]   Zhang N, Sun Q, Zhang H, Cao Y, Weeda S, Ren S. Roles of melatonin in abiotic stress resistance in plants. J Exp Bot 2015; 66(3): 647-56.

[13]   Zhou M, Memelink J. Jasmonate: Responsive transcription factors regulating plant secondary metabolism. Biotechnol Adv 2016; 34(4): 441-9.

[14]   Li Y, Kong D, Fu Y, Sussman MR, Wu H. The effect of developmental and environmental factors on secondary metabolites in medicinal plants. Plant Physiol Biochem 2020; 148: 80-9.

[15]   Katerova Z, Todorova D, Sergiev I. Plant secondary metabolites and some plant growth regulators elicited by UV irradiation, light and/or shade.Medicinal Plants and Environmental Challenges. Springer 2017; pp. 97-121.

[16]   Li Y, Kong D, Fu Y, Sussman MR, Wu H. The effect of developmental and environmental factors on secondary metabolites in medicinal plants. Plant Physiol Biochem 2020; 148: 80-9.

[17]   Takshak S, Agrawal S áB. Secondary metabolites and phenylpropanoid pathway enzymes as influenced under supplemental ultraviolet-b radiation in withania somnifera dunal, an indigenous medicinal plant. J Photochem Photobiol B 2014; 140: 332-43.

[18]   Reifenrath K, Müller C. Species-specific and leaf-age dependent effects of ultraviolet radiation on two Brassicaceae. Phytochemistry 2007; 68(6): 875-85.

[19]   Khare S, Singh NB, Singh A, Hussain I, Niharika K, Yadav V. Plant secondary metabolites synthesis and their regulations under biotic and abiotic constraints. J Plant Biol 2020; 63(3): 203-16.

[20]   Punetha A, Kumar D, Suryavanshi P, Padalia RC, Thimmaiah VK. Environmental abiotic stress and secondary metabolites production in medicinal plants: A review. J Agric Sci Tarim Bilim Derg 2022; 28(3): 351-62.

[21]   Holková I, Bezáková L, Bilka F, Balažová A, Vanko M, Blanáriková V. Involvement of lipoxygenase in elicitor-stimulated sanguinarine accumulation in papaver somniferum suspension cultures. Plant Physiol Biochem 2010; 48(10–11): 887-92.

[22]   Yang RY, Feng LL, Yang XQ, Yin LL, Xu XL, Zeng QP. Quantitative transcript profiling reveals down-regulation of a sterol pathway relevant gene and overexpression of artemisinin biogenetic genes in transgenic artemisia annua plants. Planta Med 2008; 74(12): 1510-6.

[23]   Naghiloo S, Movafeghi A, Delazar A, Nazemiyeh H, Asnaashari S, Dadpour MR. Ontogenetic variation of volatiles and antioxidant activity in leaves of astragalus compactus lam.*Fabaceae.* EXCLI J 2012; 11: 436.

[24]   Nakayama M, Koshioka M, Shibata M, Hiradate S, Sugie H, Yamaguchi M atsu. Identification of cyanidin 3-O-(3 ", 6 "-O-dimalonyl-β-glucopyranoside) as a flower pigment of Chrysanthemum *Dendranthema grandiflorum.* Biosci Biotechnol Biochem 1997; 61(9): 1607-8.

[25]   Zhang C, Ding Z, Wu K, Yang L, Li Y, Yang Z. Suppression of jasmonic acid-mediated defense by viral-inducible microRNA319 facilitates virus infection in rice. Mol Plant 2016; 9(9): 1302-14.

[26]   Ogbe AA, Finnie JF, Van Staden J. The role of endophytes in secondary metabolites accumulation in medicinal plants under abiotic stress. S Afr J Bot 2020; 134: 126-34.

[27]   Azhar N, Hussain B, Ashraf MY, Abbasi KY. Water stress mediated changes in growth, physiology and secondary metabolites of desi ajwain *Trachyspermum ammi L.* Pak J Bot 2011; 43(1): 15-9.

[28]   Verma N, Shukla S. Impact of various factors responsible for fluctuation in plant secondary

metabolites. J Appl Res Med Aromat Plants 2015; 2(4): 105-13.

[29] Azhar N, Hussain B, Ashraf MY, Abbasi KY. Water stress mediated changes in growth, physiology and secondary metabolites of desi ajwain *Trachyspermum ammi L.*. Pak J Bot 2011; 43(1): 15-9.

[30] De Abreu IN, Mazzafera P. Effect of water and temperature stress on the content of active constituents of hypericum brasiliense choisy. Plant Physiol Biochem 2005; 43(3): 241-8.

[31] Jaafar HZ, Ibrahim MH, Karimi E. Phenolics and flavonoids compounds, phenylanine ammonia lyase and antioxidant activity responses to elevated co2 in labisia pumila *Myrisinaceae*. Molecules 2012; 17(6): 6331-47.

[32] Banerjee A, Roychoudhury A. Effect of salinity stress on growth and physiology of medicinal plants.Medicinal plants and environmental challenges. Springer 2017; pp. 177-88.

[33] Akula R, Ravishankar GA. Influence of abiotic stress signals on secondary metabolites in plants. Plant Signal Behav 2011; 6(11): 1720-31.

[34] Zargaran A, Borhani-Haghighi A, Faridi P, Daneshamouz S, Mohagheghzadeh A. A review on the management of migraine in the avicenna's canon of medicine. Neurol Sci 2016; 37(3): 471-8.

[35] Haghighi M, Afifipour Z, Mozafarian M. The effect of n-si on tomato seed germination under salinity levels 2012.

[36] Abd EL-Azim WM, Ahmed ST. Effect of salinity and cutting date on growth and chemical constituents of achillea fragratissima forssk, under ras sudr conditions. Res J Agric Biol Sci 2009; 5(6): 1121-9.

[37] Neffati M, Marzouk B. Changes in essential oil and fatty acid composition in coriander *coriandrum sativum l.* leaves under saline conditions. Ind Crops Prod 2008; 28(2): 137-42.

[38] Baatour O, Kaddour R, Aidi Wannes W, Lachaal M, Marzouk B. Salt effects on the growth, mineral nutrition, essential oil yield and composition of marjoram *origanum majorana*. Acta Physiol Plant 2010; 32(1): 45-51.

[39] Sathiyaseelan A, Shajahan A, Kalaichelvan PT, Kaviyarasan V. Fungal chitosan based nanocomposites sponges—an alternative medicine for wound dressing. Int J Biol Macromol 2017; 104: 1905-15.

[40] Singh R, Mishra A, Dhawan SS, Shirke PA, Gupta MM, Sharma A. Physiological performance, secondary metabolite and expression profiling of genes associated with drought tolerance in withania somnifera. Protoplasma 2015; 252(6): 1439-50.

[41] Siebert R, Puchta H. Efficient repair of genomic double-strand breaks by homologous recombination between directly repeated sequences in the plant genome. Plant Cell 2002; 14(5): 1121-31.

[42] Roy S. Maintenance of genome stability in plants: Repairing DNA double strand breaks and chromatin structure stability. Front Plant Sci 2014; 5: 487.

[43] Carroll D. Genome engineering with zinc-finger nucleases. Genetics 2011; 188(4): 773-82.

[44] Carlson DF, Fahrenkrug SC, Hackett PB. Targeting dna with fingers and talens. Mol Ther Nucleic Acids 2012; 1.

[45] Joung JK, Sander JD. TALENs: A widely applicable technology for targeted genome editing. Nat Rev Mol Cell Biol 2013; 14(1): 49-55.

[46] Knoop V, Staskawicz B, Bonas U. Expression of the avirulence gene avrbs3 from xanthomonas campestris pv. vesicatoria is not under the control of hrp genes and is independent of plant factors. J Bacteriol 1991; 173(22): 7142-50.

[47] Schmitz-Linneweber C, Small I. Pentatricopeptide repeat proteins: A socket set for organelle gene expression. Trends Plant Sci 2008; 13(12): 663-70.

[48] Li Z, Xiong X, Jian-Feng L. New cytosine base editor for plant genome editing. Sci China Life Sci

2018; 61(12): 1602.

[49] Gaudelli NM, Komor AC, Rees HA, Packer MS, Badran AH, Bryson DI. Programmable base editing of a• t to g• c in genomic dna without dna cleavage. Nature 2017; 551(7681): 464-71.

[50] Small I. RNAi for revealing and engineering plant gene functions. Curr Opin Biotechnol 2007; 18(2): 148-53.

[51] Birmingham A, Anderson E, Sullivan K, Reynolds A, Boese Q, Leake D. A protocol for designing sirnas with high functionality and specificity. Nat Protoc 2007; 2(9): 2068-78.

[52] Fujii S, Small I. The evolution of RNA editing and pentatricopeptide repeat genes. New Phytol 2011; 191(1): 37-47.

[53] Koito A, Ikeda T. Apolipoprotein B mRNA-editing, catalytic polypeptide cytidine deaminases and retroviral restriction. Wiley Interdiscip Rev RNA 2012; 3(4): 529-41.

[54] Hou H, Atlihan N, Lu ZX. New biotechnology enhances the application of cisgenesis in plant breeding. Front Plant Sci 2014; 5: 389.

[55] Jinek M, Chylinski K, Fonfara I, Hauer M, Doudna JA, Charpentier E. A programmable dual-RNA–guided DNA endonuclease in adaptive bacterial immunity. science. 2012; 337(6069): 816-21.

[56] Xie K, Yang Y. RNA-guided genome editing in plants using a CRISPR–Cas system. Mol Plant 2013; 6(6): 1975-83.

[57] Mondini L, Noorani A, Pagnotta MA. Assessing plant genetic diversity by molecular tools. Diversity 2009; 1(1): 19-35.

[58] Lin CS, Hsu CT, Yang LH, Lee LY, Fu JY, Cheng QW. Application of protoplast technology to crispr/cas9 mutagenesis: From single-cell mutation detection to mutant plant regeneration. Plant Biotechnol J 2018; 16(7): 1295-310.

[59] Wendt KE, Ungerer J, Cobb RE, Zhao H, Pakrasi HB. CRISPR/Cas9 mediated targeted mutagenesis of the fast growing cyanobacterium synechococcus elongatus UTEX 2973. Microb Cell Fact 2016; 15(1): 1-8.

CHAPTER 11

Genome Editing for Biotic Stress Resistance in Medicinal Plants

Muhammad Insaf¹, Muhammad Abu Bakar Saddique¹, Muhammad Ali Sher¹, Mahmood Alam Khan¹, Muhammad Usman², Saira Karimi³, Rana Binyamin⁴ and Zulqurnain Khan⁵,*

¹ *Institute of Plant Breeding and Biotechnology, MNS University of Agriculture, Multan, Pakistan*

² *Department of Food Science and Technology, MNS University of Agriculture, Multan, Pakistan*

³ *Department of Biosciences, COMSATS University, Islamabad, Pakistan*

⁴ *Institute of Plant Protection, MNS University of Agriculture, Multan, Pakistan*

⁵ *Department of Biotechnology, Institute of Plant Breeding and Biotechnology (IPBB), Muhammad Nawaz Shareef University of Agriculture, Multan-60000, Pakistan*

Abstract: Plants are continually subjected to a range of physical and biological stressors throughout their growth period. Insects and pests, like other biotic stressors, have created significant concerns about lower productivity, which jeopardizes agricultural production. Genome engineering, also known as genome editing, has emerged as a cutting-edge breeding technique capable of altering the genomes of plants, animals, microbes, and humans. Since ancient times, humans have used medicinal plants for food, medicine, and industrial purposes. Both traditional biotechnology and more recent next-generation sequencing (NGS) methods have been used successfully to improve natural chemicals derived from plants with medical potential. To modify the genome at the transcriptional level, protein-based editing approaches like zinc-finger nucleases (ZFNs) and transcription activator-like end nucleases (TALENs) were previously frequently employed. CRISPR/associated9 (Cas9) endonucleases are a powerful, resilient, and precise site-directed mutagenesis method in transcriptome gene editing. CRISPR/Cas9 genome editing employs specially created guide RNAs to detect a three-base pair protospacer adjacent motif (PAM) sequence situated downstream of the target DNA. The current review compiles current research published between 2010 and 2020 on the use of CRISPR/Cas9 genome-editing technologies in traditional medicines, describing significant innovations, difficulties, and prospects, as well as noting the technique's broader application in crop and lesser species. The CRISPR/Cas9 genome editing method has been utilised successfully in plants to boost agricultural productivity and stress tolerance.

* **Corresponding author Zulqurnain Khan:** Department of Biotechnology, Institute of Plant Breeding and Biotechnology (IPBB), Muhammad Nawaz Shareef University of Agriculture, Multan-60000, Pakistan; E-mail: zulqurnain.khan@mnsuam.edu.pk

Despite this, only a small number of medicinal plants have been altered using the CRISPR/Cas9 genome editing technique because to a lack of appropriate transformation and regeneration techniques, and also a lack of comprehensive genome and mRNA sequencing data. However, a variety of secondary metabolic activities in plants (*e.g.* alkaloids, terpenoids, flavonoids, phenolic acids, and saponin) altered lately using CRISPR/Cas-editing through knocking out, knocking in, and point mutations, modulation of gene expression, including targeted mutagenesis.

Keywords: Biotic stress, CRISPR/Cas9, Diseases, Insect/Pest, Stress resistance.

INTRODUCTION

Previously, zinc-finger nucleases (ZFNs) and transcription activator-like endonucleases (TALENs) were described as genome editing tools for manipulating the transcriptome at the transcriptional level [1]. ZFNs and TALENs both use a protein-based nuclease called Fok1 to knock out genes by inducing double-stranded breaks (DSBs) in the target DNA [2]. Additionally, tools have become less effective as a result of the presence of that nucleotide sequence and the ability of both proteins to dimer in the right orientation for the given spacer length [3].The clustered regularly interspaced short palindromic repeats (CRISPR/Cas9) technology provides several benefits over ZFNs and TALENs, including simplicity of use, ease of multiplexing, and capacity for large-scale library production [4].

The development of the CRISPR/Cas9 system does not need protein dimerization, and now allows very effective gene modification *via* the use of Tailored gRNAs that identify a three-base pair PAM region that is located downstream of the target gene [5]. CRISPR/Cas9 is an efficient and simple technique for gRNA genome editing. This technique is well-known for two significant benefits: I) the point mutation is entirely silenced, while RNAi dosage variation results in partial suppression; II) genomic modifications are durable and readily maintained in progeny, making them very useful in plant breeding [6].

Genome editing facilitated by CRISPR/Cas9 is used to generate site-specific breakage in DNA strands inside the gene, that are then repaired *via* the cell NHEJ or HDR machinery [7]. However, random addition or removal (indels) mediated by NHEJ mostly affects the open reading frame sequences that code for proteins or eliminates non-coding sequences cis-regulatory elements (promoters, regulators, *etc.*) [8]. Plants and plant-derived products have been utilized as a significant supplier of foods, medicines, health supplements, and also biomaterials for humans from time immemorial [9].

Natural compounds derived from plants have been identified as possible therapeutic targets for reducing human mortality and morbidity [10]. More than a few preclinical studies have taken conducted and proven the effectiveness of medicinal herbalism against a variety of disorders [11]. Numerous biotechnological treatments have been used to alter mechanisms in plants that lead to the synthesis of medicinal, nutritional, and industrial secondary metabolites [12]. Next-generation sequencing (NGS) methods have shown tremendous potential for evaluating the genetic diversity and metabolic pathways of medicinal plants. The breaks are repaired utilizing the host cell's repair process, either by a pathway of non-homologous end joining (NHEJ) or through Homologous recombination, which is a process that leads to the change of genes by utilizing a donor fragment. Genome editing uses four different kinds of nucleases to enhance plant resistance mechanisms that include mega nucleases ZFNs, TALENs, and the CRISPR–Cas9.

The production at an incredible rate of CRISPR/Cas9 research demonstrates that CRISPR technique requires lower financial resources with a greater percentage of success for gene amplification and manipulation than other current methods. CRISPR/Cas9 is a very effective tool for enhancing agronomic characteristics in plants [13]. *Streptococcus pyogenic* CRISPR/Cas9 system (SpCas9) has swiftly become indispensable in a variety of fields in plant study, and also in a variety of other fields of research. An sgRNA is used for this mechanism along with Cas9 to target DNA regions [14, 15].

The PAM element in a target gene reduces the possible target sequences. This constraint is irrelevant in the context of targeted mutagenesis [16]. CRISPR/Cas is a kind of bacterial defense mechanism that is particularly effective against bacteriophages, that may be used to modify specific genes in a variety of species. Diseases have posed a significant danger to our agriculture and manufacturing. Innovative biotechnological breeding procedures, like targeted genome modification, have also been touted as a good toolset for producing therapeutic plants with optimal secondary metabolite profiles [17]. The molecular mechanism of CRISPR-Cas9 editing tools as efficient genome editing approaches in herbal medicines, with an emphasis on recent innovations, limits, and possible future prospects are shown in Table **1**.

BIOTIC STRESSES IN PLANTS

Plants are susceptible to a large variety of pathogens that cause a variety of diseases. As a substitute, various herbicides, insecticides, and fungicides have been utilized, harming the environment directly and indirectly [29, 30]. Additionally, these compounds have an impact on the sequence of metabolic events in plants [31]. Thus, CRISPR/Cas9 technique is used to confer resistance to plants against diseases such as bacteria, viruses, fungus, nematodes, insects, and weeds. Biotic stressors, such as viral, fungal, and bacterial infections, account for between 20% and 40% of worldwide agricultural production losses [32]. By imparting pathogen resistance on host plants, we may mitigate diseases influence on crop growth and productivity, therefore solving the global food security dilemma. Advances in genome editing techniques have created new avenues for improving crop resilience. In recent years, the CRISPR/Cas system has been used to address a variety of agricultural difficulties, including achieving increased resistance to environmental stresses [33]. CRISPR/Cas technologies have been primarily used to combat viral infection; attempts are made to eradicate it. Increased Resistance against bacteria and fungi illness has recently been discovered.

Table 1. Diseases in medicinal plants.

Medicinal Plant	Disease	References
Daturainnoxia	Inflorescence and fruit rot	[18]
Menthaspp	Rust	[19]
Chlorophytumborivilianum	Macrophomina leaf spot	[20]
Cymbopoganflexuosus	Colletotrichum leaf spot	[21]
Cymbopogonmartinii	Ellisiella blight	[22]
Withaniasomnifera	Myrothecium leaf spot	[23]
Papaversomniferum	Damping-off	[24]
Pogostemoncablin	Alternaria blight	[25]
Hyoscyamusmuticus	Xanthomonas blight	[26]
Zingiber Officinale	Rhizome rot	[27]
Hibiscus moschatus	Leaf rot	[28]

CRISPR Cas for Insect Resistance

It is difficult to build insect resistance in plants using the CRISPR/Cas9 approach since there is no direction inside the genomic region of the insect, and additionally, there is no technique for embryonic chemical deposition. Whenever the gene for potato protease inhibitor II (pinII) was introduced into Arabidopsis using CRISPR/Cas9, the plant developed resistance to Lepidopteran bug disease [34] (Fig. 1). An aberration in the larval developmental phases was observed after the injection of Cas9 mRNA and Slabd-A-specific single guideRNA (sgRNA) into the lepidopteran pests *Spodoptera litura* [35]. Under the promoter PR1 and an expressing transgene Cry1Ac, developed from *Bacillus thuringiensis*, recombinant peanut gave protection against *Spodoptera litura* [36]. When it comes to developing aphid resistant in transgenic plants, the PR1 promoter is often regarded as the most effective promoter [37]. Under the control of the PR-1a promoter, the *Cry1Ab* gene confers resistance to the broccoli plant in response to *Plutella xylostella* bug [38].

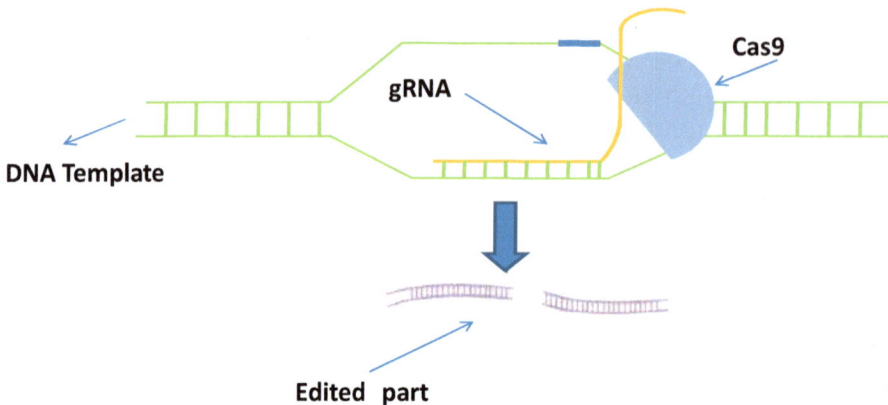

Fig. (1). DNA modification by CRISPR.

Virus Resistance *via* CRISPR/Cas

Arboviruses pose a significant danger to a wide variety of economic importance cereal and specially foods. According to the characteristics of their genomes, these viruses can be broken down into six distinct categories dsDNA viruses, which do not include any plant viruses, ssDNA reverse-transcribing viruses dsRNA, negative sense ssRNA and positive sense single-stranded RNA (ssRNAC) viruses. The majority of research on viral resistance using CRISPR-edited crops has focused on ssDNA Gemini-virus genes [39]. Plant viruses pose a significant danger to a wide variety of commercially significant staple and specialty crops. They are divided into six primary categories depending on the organization of the genome dsDNA viruses (which do not include plant viruses), single-stranded DNA (ssDNA), reverse-transcription Solitary RNA (ssRNA), twofold RNA (dsRNA), single-stranded Mrna (sRNA), as well as solitary RNA plus viruses are all examples of viral types [40]. The majority of research employing the CRISPR-edited plants for pathogenic resistance has focused on the genomes of ssDNA Gemini viruses [39].

Gemini Miridae a family of plant viruses that infect numerous significant plant families, including *Cucurbitaceae, Rubiaceae, Solanaceae, Malvaceae,* and *Fabaceae* are some of the families of plants [41]. Numerous fungi affect pathogenic invasions of MAPs, infecting aerial, foliage, and subterranean plant components. Powdery mildews are a kind of fungal disease that mostly affects leaves and young stems, eventually covering the whole surface of the plant-developing portions. Ruston, on the other hand, is another airborne disease that attacks leaves, stems, and branches, including fruit. They form pustules on the upper and lower surfaces of leaves. With symptoms starting with leaf spot and advancing to shrinkage and dissociation of the lethal zone from the surrounding healthy tissues with a variety of lesion patterns, the second group of infections, known as fungal pathogens, causes diseases like leaf spot and blight [42]. Blight results in the abrupt death of the plant or one of its conspicuous parts, such as leaves, twigs, or blossoms [42].

CRISPR/Cas Tolerance to Fungal Infections

Pathogens are responsible for a wide variety of diseases, including mould, rust, and rot. Not only do these diseases result in significant yearly output losses all around the globe, but they also lower the overall quality of both the crops that are produced. In addition, mycotoxigenic fungi pose a significant threat owing to the fact that they produce secondary metabolites known as *mycotoxins*. *Mycotoxins* are known for causing serious medical problems in both humans and livestock once they have been exposed to tainted food. Also on the basis of the existing understanding of molecular pathways engaged in the association between plant and pathogen, a number of different tactics have been developed to increase the plant varieties resistance for fungal attack by using Genes and potential candidates for the gene. Products that are involved in plants' fungal resistance have also been identified, and in the present day, these proteins are excellent candidates for editing using the CRISPR/Cas9 technique.

Resistance to Bacteria Through CRISPR/Cas

Only a few hundreds of the millions of types of bacteria that may be on earth are responsible for yield loss, which frequently manifests as various signs of diseases [43]. Controlling phytopathogenic bacteria may be challenging for a number of reasons, the most significant of which are the absence of appropriate agricultural agents and the prevalence of silent infections that go undiagnosed. The three primary elements of microbiological plant control are the application of genetic resistance, agronomic techniques, and biocontrol agents. This makes it possible to prevent and control the virus [44]. It is possible to divide phytopathogenic bacteria into two categories: crop-specific microbes, including *Clavibacter michiganensis* that is a species responsible for the bacterial ring rot that affects tomatoes, and polyphagous-specific bacteria, including *C. michiganensis*, that is a species responsible for disease in a variety of monocot and dicot plants.

CRISPR FOR DISEASE RESISTANCE IN MEDICINAL PLANTS

It is challenging to create insect resistance in plants using the CRISPR/Cas9 method since there is no assistance within the insect's genomic region and no such embryonic microinjection mechanism. When Arabidopsis was transduced by utilizing potato protease II [pin] gene using CRISPR/Cas9, it developed resistance against *Lepidopterous* insect pathogens [34]. Cas9 mRNA with Slabd-A-specific sguide RNA were injected into the cells (sgRNA) and caused developmental anomalies in the *lepidopteran* pest [35]. *Transgenic* peanuts with the promoter PR1 as well as the transgene Cry1Ac, derived from *Bacillus thuringiensis*, conferred resistance to *Spodopteralitura* [36]. PR1 promoter is often regarded as the most effective promoter into transgenic organisms for aphid resistant [37]. Under the promoter PR-1a, the Cry1Ab confers resistance to broccoli against the bug Plutellaxylostella [38].

Applications of Genome Editing in Medicinal Plants

This CRISPR-Cas9 technology is very versatile in medicine as this approach is a breakthrough in combating fatal diseases such as cancer, viral infections, genetic disorders, and novel pathogen identification. This approach has been effectively employed in mice to fix abnormalities in monogenic disorders, which causes Duchenne muscular dystrophy (DMD) [45]. HIV-1 provirus has been disrupted using CRISPR-Cas9 technology, which has already been applied in other applications such as replication [46], mankind papilloma virus replication [47], and hepatitis B viruses replication [48]. CRISPR-Cas9 has also been designed for use on intended living thing genetic digestive system disorders [49]. Additional findings show significant potential, primarily in the fight against anticancer therapy [50], for both Hutchinson-Gilford Pompe disease and the progeria phenotype [51]. CRISPR-Cas9 is currently being effectively employed in human cells to replace endogenous antibodies that have antibodies that are resistant to the Respiratory Syncytial Virus, the HIV; both influenza virus as well as the *Epstein-Barr* virus are contagious [52].

CRISPR-Cas9 enables quicker genetic alteration than other approaches due to simplicity, efficiency, cheap cost, and the ability to target many genes. It may also be used to genetically modify previously unsuitable plants. This gives an unparalleled potential for plant breeding as well as ecologically friendly farming [53]. CRISPR-Cas9 has enabled remarkable genetic manipulation to improve metabolic processes, biotic stress (bacteria and fungi, virus) tolerance, abiotic stresses, nutrient quality, crop yield, and weed control. For example, thermosensitive ploid male sterile in sorghum [54], maize, and wheat increased nutritional traits [55], enhanced disease tolerance or resistance [56], and [57] herbicide resistance [58]. The starch gene contained in potato granules was eliminated using CRISPR-Cas9, which led to the development of potato plants with features such as the ability to synthesise the amylopectin carbohydrate [59].

Future of CRISPR-edited Medicinal Plants

CRISPR-Cas and its tools are excellent for precise gene editing in medicinal plants, enabling gene knockout/knock-in, point mutation, gene regulation, and alteration of gene region. As a result, basic gene function studies and genetic engineering in medicinal herbs have been awarded additional support [60]. A growing number of important functional genes may be used as gene editing targets as a result of the success of high quality generation sequencing of medicinal plants. The advancement in omics technology will contribute to the availability of more abundant target genes for the effective use of gene-editing technology in medicinal plants, and advancement of synthetic biological and medical research regarding biosynthetic pathways of bioactive substances, genetic increase in the quality of pharmacological plants growth, and germplasm invention. While CRISPR-Cas has enormous potential for modifying the genes of medicinal plants, the method still has significant constraints. Considering the importance of functional genomics in gene editing, more research is needed to acquire high-quality genetic data on medicinal plants utilising cutting-edge sequencing methods. Furthermore, injecting cells with CRISPR-Cas9 reagents and regenerating medicinal plants remain difficult. As a result, robust transformation methods are needed, particularly some that employ carbon nanotube-based DNA nanostructures, cell-penetrating proteins, and plant viruses that really can efficiently diffuse into the pharmacological plant cell wall without requiring aid from machinery or inflicting injury on the tissue. Finally, developing an optimum cultivar entails altering several quantitative features and editing individual genes may not result in substantial phenotypic change; therefore, effective multiplexing genome editing approaches for herbal medicines are required.

Regulatory Aspects of CRISPR-edited Medicinal Plants

Concerns about the scientific, biosafety, and social ramifications of plant genetic manipulation remain unaddressed. They are primarily concerned with the selection of the target gene region, the construction of gRNA, off-target effects, and the delivery technique. The main concern is the possibility of unwanted genetic changes in plants as a consequence of off-target mutations [61]. During the DNA repair process, pieces of the CRISPR-Cas9 system may be digested into DNA and enter into predicted and/or unanticipated genomic sites [62]. However, transgenic integration and the possibility of off mutation can be avoided by using *in vitro* synthesised CRISPR-Cas9 ribonucleo-proteins [63].

Although this technology has been used for a variety of crops, it still has some limitations, such as limited stability, high prices, and a high degree of technical complexity that must be addressed [63]. Genetically engineered plants are those whose genomes have been altered in ways that would not have happened normally [64]. By contrast, genetic manipulation refers to DNA alterations that are comparable to those that may occur naturally (deletions, nucleotide substitutions, and insertions) or that might occur through traditional plant breeding [65]. The Cartagena Protocol on Biosafety established the framework for regulating the release and international trade of alive GMOs. In contrast, the manufacturing, usage, and regulation of genetically altered plants have taken distinct paths. While some countries actively discourage use and production, others freely cultivate and consume them [66]. The control of genome-edited organisms is governed by two frameworks. Certain governments control the manufacturing process, while others limit the features of the finished product [67]. While some countries have created biosafety laws, such as those governing genome-edited organisms, or have declared this deregulation, the majority of countries are still not doing so [67]. Crop genome editing regulation raises several issues, including market access and effectively addressing public concerns about the technology's biological protection without inhibiting its progress [68]. Transgene-free crops with changed genomes are currently available, similar to naturally occurring genetic variations. As a consequence, the dissemination of genome-modified crops or their derivatives may avoid the severe biosafety criteria that apply to transgenic plants [69]. In March of each year, the USDA issue a study concluding that genetic modification is sometimes equivalent to conventional breeding and hence does not need regulatory oversight under the American regulatory framework [70].

CONCLUSION

The CRISPR/Cas9 technique has many appealing characteristics, including high efficiency, ease of use, flexibility, and the ability to perform multiplexed modifications; as a result, it has emerged as the technology for modifying the genome with the most promise and a bright future in producing desired mutations in plants. Because of a lack of adequate sequence data for several MAPs, the CRISPR-Cas method has currently been implemented in only a few herbal medicines, but researchers are confident that additional research will fully exploit the potential of using CRISPR-Cas technology in genetic engineering other medicinal plants to recognise the genes and proteases contributing to the biosynthetic pathway of various secondary metabolites. According to recent research, CRISPR/Cas9 technology is fast emerging into the ultimate molecular tool for genetic modification, and CRISPR-Cas9 technology has altered the potential of gene editing in plants, especially medicinal plants. Multiplex genome editing, which comprises the targeted deletion of a few genes or even the simultaneous up- or down-regulation of several genes, may result in the formation of advantageous agronomic characteristics in target plants. With a larger toolset of CRISPR technique, researchers may soon be able to change complex characteristics . Thus, CRISPR-Cas9 can alter the biosynthetic pathway in heterologous medicinal plants by utilising an artificially created and precisely regulated genetic circuit to improve biopharmaceutical output.

REFERENCES

[1]   Chen K, Gao C. Targeted gene mutation in plants. Somatic Genome Manipulation. Springer 2015; pp. 253-72.
[http://dx.doi.org/10.1007/978-1-4939-2389-2_12]

[2]   Gaj T, Gersbach CA, Barbas CF III. ZFN, TALEN, and CRISPR/Cas-based methods for genome engineering. Trends Biotechnol 2013; 31(7): 397-405.
[http://dx.doi.org/10.1016/j.tibtech.2013.04.004] [PMID: 23664777]

[3]   Händel EM, Gellhaus K, Khan K, *et al.* Versatile and efficient genome editing in human cells by combining zinc-finger nucleases with adeno-associated viral vectors. Hum Gene Ther 2012; 23(3): 321-9.
[http://dx.doi.org/10.1089/hum.2011.140] [PMID: 21980922]

[4]   Sharma S, Kaur R, Singh A. Recent advances in CRISPR/Cas mediated genome editing for crop improvement. Plant Biotechnol Rep 2017; 11(4): 193-207.
[http://dx.doi.org/10.1007/s11816-017-0446-7]

[5]   Charpentier E, Doudna JA. Rewriting a genome. Nature 2013; 495(7439): 50-1.
[http://dx.doi.org/10.1038/495050a] [PMID: 23467164]

[6]   Barrangou R, Birmingham A, Wiemann S, Beijersbergen RL, Hornung V, Smith AB. Advances in CRISPR-Cas9 genome engineering: Lessons learned from RNA interference. Nucleic Acids Res 2015; 43(7): 3407-19.
[http://dx.doi.org/10.1093/nar/gkv226] [PMID: 25800748]

[7]   Shalem O, Sanjana NE, Zhang F. High-throughput functional genomics using CRISPR–Cas9. Nat Rev Genet 2015; 16(5): 299-311.
[http://dx.doi.org/10.1038/nrg3899] [PMID: 25854182]

[8]   Liu D, Hu R, Palla KJ, Tuskan GA, Yang X. Advances and perspectives on the use of CRISPR/Cas9 systems in plant genomics research. Curr Opin Plant Biol 2016; 30: 70-7.
[http://dx.doi.org/10.1016/j.pbi.2016.01.007] [PMID: 26896588]

[9]   Dey A, Nandy S, Mukherjee A, Modak BK. Sustainable utilization of medicinal plants and conservation strategies practiced by the aboriginals of Purulia district, India: A case study on therapeutics used against some tropical otorhinolaryngologic and ophthalmic disorders. Environ Dev Sustain 2021; 23(4): 5576-613.
[http://dx.doi.org/10.1007/s10668-020-00833-8]

[10]   Dey A, Gorai P, Mukherjee A, Dhan R, Modak BK. Ethnobiological treatments of neurological conditions in the Chota Nagpur Plateau, India. J Ethnopharmacol 2017; 198: 33-44.
[http://dx.doi.org/10.1016/j.jep.2016.12.040] [PMID: 28017696]

[11]   Anand U, Nandy S, Mundhra A, Das N, Pandey DK, Dey A. A review on antimicrobial botanicals,

phytochemicals and natural resistance modifying agents from Apocynaceae family: Possible therapeutic approaches against multidrug resistance in pathogenic microorganisms. Drug Resist Updat 2020; 51: 100695.
[http://dx.doi.org/10.1016/j.drup.2020.100695] [PMID: 32442892]

[12]   Dey A, Bhattacharya R, Mukherjee A, Pandey DK. Natural products against Alzheimer's disease: Pharmaco-therapeutics and biotechnological interventions. Biotechnol Adv 2017; 35(2): 178-216.
[http://dx.doi.org/10.1016/j.biotechadv.2016.12.005] [PMID: 28043897]

[13]   Mohanta TK, Bashir T, Hashem A, Abd Allah EF, Bae H. Genome editing tools in plants. Genes 2017; 8(12): 399.
[http://dx.doi.org/10.3390/genes8120399] [PMID: 29257124]

[14]   Chen L, Li W, Katin-Grazzini L, *et al.* A method for the production and expedient screening of CRISPR/Cas9-mediated non-transgenic mutant plants. Hortic Res 2018; 5(1): 13.
[http://dx.doi.org/10.1038/s41438-018-0023-4] [PMID: 29531752]

[15]   Ding D, Chen K, Chen Y, Li H, Xie K. Engineering introns to express RNA guides for Cas9-and Cpf1-mediated multiplex genome editing. Mol Plant 2018; 11(4): 542-52.
[http://dx.doi.org/10.1016/j.molp.2018.02.005] [PMID: 29462720]

[16]   Aglawe SB, Barbadikar KM, Mangrauthia SK, Madhav MS. New breeding technique "genome editing" for crop improvement: applications, potentials and challenges. 3 Biotech 2018; 8(8): 1-20.

[17]   Niazian M. Application of genetics and biotechnology for improving medicinal plants. Planta 2019; 249(4): 953-73.
[http://dx.doi.org/10.1007/s00425-019-03099-1] [PMID: 30715560]

[18]   Shukla RS, Alam M, Sattar A, Singh HN. First report of Rhizopus stolonifer causing inflorescence and fruit rot of Rauvolfia serpentina in India. Bull OEPP 2006; 36(1): 11-3.
[http://dx.doi.org/10.1111/j.1365-2338.2006.00938.x]

[19]   Kalra A, Singh HB, Pandey R, Samad A, Patra NK, Kumar S. Diseases in mint: Causal organisms, distribution, and control measures. J Herbs Spices Med Plants 2005; 11(1-2): 71-91.
[http://dx.doi.org/10.1300/J044v11n01_03]

[20]   Dadwal V, Savitri B. New report of a leaf spot disease of *Chlorophytum borivillianum* caused by *Macrophomina phaseolina* from India. J Mycol Plant Pathol 2012; 42(3): 397-8.

[21]   Thakur R, Husain A. A new leaf spot disease of lemon grass. Indian Phytopathol 1975; 28: 100-2.

[22]   Gupta M, Kumar S, Pandey R, Shukla R, Khaliq A, Kalra A. Leaf blight disease and its effect on essential oil content of palmarosa. Curr Res Med Aromat Plants 2000; 22(1B): 504-5.

[23]   Shivanna M, Parashurama T, Achar KS, Vasanthakumari M. Fungal foliar diseases in Withania somnifera and its effect on secondary metabolites Plant Biosystems-An I J Dealing with all Asp Plant Biol 2014; 148(5): 907-16.

[24]   Alam M, Sattar A, CHaURASIA H, Janardhanan K. Damping-off, a new disease of opium poppy caused by phythium dissotocum. Indian Phytopathol 1996; 49(1): 94-7.

[25]   Furukawa T, Kishi K. Alternaria leaf spot on three species of pelargonium caused by alternaria alternata in japan. J Gen Plant Pathol 2001; 67(4): 268-72.
[http://dx.doi.org/10.1007/PL00013028]

[26]   Sattar A, Alam M. Bacterial leaf blight disease of Egyptian henbane caused by Xanthomonas campestrispv. campestris. Plant Pathogenic Bacteria Chennai: Udayam Offset Pub 1998; 404-6.

[27]   Stirling GR, Turaganivalu U, Stirling AM, Lomavatu MF, Smith MK. Rhizome rot of ginger (*Zingiber officinale*) caused by *Pythium myriotylum* in Fiji and Australia. Australas Plant Pathol 2009; 38(5): 453-60.
[http://dx.doi.org/10.1071/AP09023]

[28]   Shukla R, Sattar A, Kumar S, Singh K, Husain A. Phytophthora leaf blight of hibiscus moschatus-new

record from India. Indian Phytopathology 1981; 34.

[29] Noll ME. The control of stomoxys calcitrans (stable flies) with essential oils. Sch Biol Sci. University of Bristol 2020.

[30] Isman MB, Grieneisen ML. Botanical insecticide research: Many publications, limited useful data. Trends Plant Sci 2014; 19(3): 140-5.
[http://dx.doi.org/10.1016/j.tplants.2013.11.005] [PMID: 24332226]

[31] Aktar W, Sengupta D, Chowdhury A. Impact of pesticides use in agriculture: Their benefits and hazards. Interdiscip Toxicol 2009; 2(1): 1-12.
[http://dx.doi.org/10.2478/v10102-009-0001-7] [PMID: 21217838]

[32] Savary S, Ficke A, Aubertot J-N, Hollier C. Crop losses due to diseases and their implications for global food production losses and food security. Springer 2012; pp. 519-37.

[33] Arora L, Narula A. Gene editing and crop improvement using CRISPR-Cas9 system. Front Plant Sci 2017; 8: 1932.
[http://dx.doi.org/10.3389/fpls.2017.01932] [PMID: 29167680]

[34] Bu QY, Wu L, Yang SH, Wan JM. Cloning of a potato proteinase inhibitor gene PINII-2x from diploid potato [Solanum phurejia L.] and transgenic investigation of its potential to confer insect resistance in rice. J Integr Plant Biol 2006; 48(6): 732-9.
[http://dx.doi.org/10.1111/j.1744-7909.2006.00258.x]

[35] Bi HL, Xu J, Tan AJ, Huang YP. CRISPR/Cas9-mediated targeted gene mutagenesis in *Spodoptera litura*. Insect Sci 2016; 23(3): 469-77.
[http://dx.doi.org/10.1111/1744-7917.12341] [PMID: 27061764]

[36] Zhu-Salzman K, Salzman RA, Ahn JE, Koiwa H. Transcriptional regulation of sorghum defense determinants against a phloem-feeding aphid. Plant Physiol 2004; 134(1): 420-31.
[http://dx.doi.org/10.1104/pp.103.028324] [PMID: 14701914]

[37] War AR, Paulraj MG, Ahmad T, *et al.* Mechanisms of plant defense against insect herbivores. Plant Signal Behav 2012; 7(10): 1306-20.
[http://dx.doi.org/10.4161/psb.21663] [PMID: 22895106]

[38] Cao J, Shelton AM, Earle ED. Gene expression and insect resistance in transgenic broccoli containing a Bacillus thuringiensis cry1Ab gene with the chemically inducible PR-1a promoter. Mol Breed 2001; 8(3): 207-16.
[http://dx.doi.org/10.1023/A:1013734923291]

[39] Ali Z, Abulfaraj A, Idris A, Ali S, Tashkandi M, Mahfouz MM. CRISPR/Cas9-mediated viral interference in plants. Genome Biol 2015; 16(1): 238.
[http://dx.doi.org/10.1186/s13059-015-0799-6] [PMID: 26556628]

[40] Roossinck MJ, Martin DP, Roumagnac P. Plant virus metagenomics: Advances in virus discovery. Phytopathology 2015; 105(6): 716-27.
[http://dx.doi.org/10.1094/PHYTO-12-14-0356-RVW] [PMID: 26056847]

[41] Zaidi SSA, Tashkandi M, Mansoor S, Mahfouz MM. Engineering plant immunity: Using CRISPR/Cas9 to generate virus resistance. Front Plant Sci 2016; 7: 1673.
[http://dx.doi.org/10.3389/fpls.2016.01673] [PMID: 27877187]

[42] Singh A, Gupta R, Saikia SK, Pant A, Pandey R. Diseases of medicinal and aromatic plants, their biological impact and management. Plant Genet Resour 2016; 14(4): 370-83.
[http://dx.doi.org/10.1017/S1479262116000307]

[43] Schloss PD, Handelsman J. Status of the microbial census. Microbiol Mol Biol Rev 2004; 68(4): 686-91.
[http://dx.doi.org/10.1128/MMBR.68.4.686-691.2004] [PMID: 15590780]

[44] Kerr A. Biological control of crown gall. Australas Plant Pathol 2016; 45(1): 15-8.

[http://dx.doi.org/10.1007/s13313-015-0389-9]

[45]   Long C, Amoasii L, Mireault AA, *et al.* Postnatal genome editing partially restores dystrophin expression in a mouse model of muscular dystrophy. Science 2016; 351(6271): 400-3.
       [http://dx.doi.org/10.1126/science.aad5725] [PMID: 26721683]

[46]   Ebina H, Misawa N, Kanemura Y, Koyanagi Y. Harnessing the CRISPR/Cas9 system to disrupt latent HIV-1 provirus. Sci Rep 2013; 3(1): 2510.
       [http://dx.doi.org/10.1038/srep02510] [PMID: 23974631]

[47]   Kennedy EM, Kornepati AVR, Goldstein M, *et al.* Inactivation of the human papillomavirus E6 or E7 gene in cervical carcinoma cells by using a bacterial CRISPR/Cas RNA-guided endonuclease. J Virol 2014; 88(20): 11965-72.
       [http://dx.doi.org/10.1128/JVI.01879-14] [PMID: 25100830]

[48]   Kennedy EM, Bassit LC, Mueller H, *et al.* Suppression of hepatitis B virus DNA accumulation in chronically infected cells using a bacterial CRISPR/Cas RNA-guided DNA endonuclease. Virology 2015; 476: 196-205.
       [http://dx.doi.org/10.1016/j.virol.2014.12.001] [PMID: 25553515]

[49]   Yang Y, Wang L, Bell P, *et al.* A dual AAV system enables the Cas9-mediated correction of a metabolic liver disease in newborn mice. Nat Biotechnol 2016; 34(3): 334-8.
       [http://dx.doi.org/10.1038/nbt.3469] [PMID: 26829317]

[50]   Chen M, Mao A, Xu M, Weng Q, Mao J, Ji J. CRISPR-Cas9 for cancer therapy: Opportunities and challenges. Cancer Lett 2019; 447: 48-55.
       [http://dx.doi.org/10.1016/j.canlet.2019.01.017] [PMID: 30684591]

[51]   Beyret E, Liao HK, Yamamoto M, *et al.* Single-dose CRISPR–Cas9 therapy extends lifespan of mice with Hutchinson–Gilford progeria syndrome. Nat Med 2019; 25(3): 419-22.
       [http://dx.doi.org/10.1038/s41591-019-0343-4] [PMID: 30778240]

[52]   Moffett HF, Harms CK, Fitzpatrick KS, Tooley MR, Boonyaratanakornkit J, Taylor JJ. B cells engineered to express pathogen-specific antibodies protect against infection. Sci Immunol 2019; 4(35): eaax0644.
       [http://dx.doi.org/10.1126/sciimmunol.aax0644] [PMID: 31101673]

[53]   Cong L, Ran FA, Cox D, *et al.* Multiplex genome engineering using CRISPR/Cas systems. Science 2013; 339(6121): 819-23.
       [http://dx.doi.org/10.1126/science.1231143] [PMID: 23287718]

[54]   Li J, Zhang H, Si X, Tian Y, Chen K, Liu J. Generation of thermosensitive male-sterile maize by targeted knockout of the ZmTMS5 gene. J Genet Genomics 2017; 44(9): 465-8.
       [http://dx.doi.org/10.1016/j.jgg.2017.02.002]

[55]   Okada A, Arndell T, Borisjuk N, *et al.* CRISPR/Cas9-mediated knockout of *Ms1* enables the rapid generation of male sterile hexaploid wheat lines for use in hybrid seed production. Plant Biotechnol J 2019; 17(10): 1905-13.
       [http://dx.doi.org/10.1111/pbi.13106] [PMID: 30839150]

[56]   Li A, Jia S, Yobi A, *et al.* Editing of an alpha-kafirin gene family increases, digestibility and protein quality in sorghum. Plant Physiol 2018; 177(4): 1425-38.
       [http://dx.doi.org/10.1104/pp.18.00200] [PMID: 29925584]

[57]   Zhang Y, Bai Y, Wu G, *et al.* Simultaneous modification of three homoeologs of *TaEDR1* by genome editing enhances powdery mildew resistance in wheat. Plant J 2017; 91(4): 714-24.
       [http://dx.doi.org/10.1111/tpj.13599] [PMID: 28502081]

[58]   Endo M, Mikami M, Toki S. Biallelic gene targeting in rice. Plant Physiol 2016; 170(2): 667-77.
       [http://dx.doi.org/10.1104/pp.15.01663] [PMID: 26668334]

[59]   Andersson M, Turesson H, Nicolia A, Fält AS, Samuelsson M, Hofvander P. Efficient targeted multiallelic mutagenesis in tetraploid potato (Solanum tuberosum) by transient CRISPR-Cas9

expression in protoplasts. Plant Cell Rep 2017; 36(1): 117-28.
[http://dx.doi.org/10.1007/s00299-016-2062-3] [PMID: 27699473]

[60]   Chen K, Wang Y, Zhang R, Zhang H, Gao C. CRISPR/Cas genome editing and precision plant breeding in agriculture. Annu Rev Plant Biol 2019; 70(1): 667-97.
[http://dx.doi.org/10.1146/annurev-arplant-050718-100049] [PMID: 30835493]

[61]   Liang Z, Chen K, Zhang Y, *et al.* Genome editing of bread wheat using biolistic delivery of CRISPR/Cas9 *in vitro* transcripts or ribonucleoproteins. Nat Protoc 2018; 13(3): 413-30.
[http://dx.doi.org/10.1038/nprot.2017.145] [PMID: 29388938]

[62]   Gorbunova V, Levy AA. Non-homologous DNA end joining in plant cells is associated with deletions and filler DNA insertions. Nucleic Acids Res 1997; 25(22): 4650-7.
[http://dx.doi.org/10.1093/nar/25.22.4650] [PMID: 9358178]

[63]   Malnoy M, Viola R, Jung MH, *et al.* DNA-free genetically edited grapevine and apple protoplast using CRISPR/Cas9 ribonucleoproteins. Front Plant Sci 2016; 7: 1904.
[http://dx.doi.org/10.3389/fpls.2016.01904] [PMID: 28066464]

[64]   Wang GP, Yu XD, Sun YW, Jones HD, Xia LQ. Generation of marker-and/or backbone-free transgenic wheat plants *via* Agrobacterium-mediated transformation. Front Plant Sci 2016; 7: 1324.
[http://dx.doi.org/10.3389/fpls.2016.01324] [PMID: 27708648]

[65]   Knott GJ, Doudna JA. CRISPR-Cas guides the future of genetic engineering. Science 2018; 361(6405): 866-9.
[http://dx.doi.org/10.1126/science.aat5011] [PMID: 30166482]

[66]   Garcia-Ruiz H. Susceptibility genes to plant viruses. Viruses 2018; 10(9): 484.
[http://dx.doi.org/10.3390/v10090484] [PMID: 30201857]

[67]   Eckerstorfer MF, Engelhard M, Heissenberger A, Simon S, Teichmann H. Plants developed by new genetic modification techniques—comparison of existing regulatory frameworks in the EU and non-EU countries. Front Bioeng Biotechnol 2019; 7: 26.
[http://dx.doi.org/10.3389/fbioe.2019.00026] [PMID: 30838207]

[68]   Kupferschmidt K. EU verdict on CRISPR crops dismays scientists. Science 2018; 361(6401): 435-6.
[http://dx.doi.org/10.1126/science.361.6401.435]

[69]   Tuteja N, Verma S, Sahoo RK, Raveendar S, Reddy INBL. Recent advances in development of marker-free transgenic plants: Regulation and biosafety concern. J Biosci 2012; 37(1): 167-97.
[http://dx.doi.org/10.1007/s12038-012-9187-5] [PMID: 22357214]

[70]   Waltz E. CRISPR-edited crops free to enter market, skip regulation. Nat Biotechnol 2016; 34(6): 582-3.
[http://dx.doi.org/10.1038/nbt0616-582] [PMID: 27281401]

SECTION 3: NANO-ENGINEERING OF MEDICINAL PLANTS

CHAPTER 12

Medicinal Plants: Traditional Trends to Modern Therapeutics

Naila Safdar[1,*], **Azra Yasmin**[1] and **Zulqurnain Khan**[2]

[1] *Department of Biotechnology, Fatima Jinnah Women University, Rawalpindi, Pakistan*

[2] *Department of Biotechnology, Institute of Plant Breeding and Biotechnology (IPBB), Muhammad Nawaz Shareef University of Agriculture, Multan-60000, Pakistan*

Abstract: Medicinal plant therapies are becoming more common, as more people seek natural cures and health approaches devoid of synthetic chemicals' adverse effects. The biological and pharmacological potential of plants is studied and utilized all around the globe for various purposes including the treatment of infections and diseases owing due to bioactive compounds in plants produced as a result of secondary metabolism. The study of medicinal plants is helpful in clinical trials to find pharmacologically useful chemicals, and this method has produced thousands of valued medicines. Opium, aspirin, quinine, and digoxin are some examples. Plants possess a large number of bioactive compounds. On the basis of their chemical structure, they are divided into four classes: alkaloids, flavonoids, tannins, and terpenes. Plants can now be turned into "factories" that create therapeutic proteins, vaccines, and many more products for use in the production of biotech pharmaceuticals, medications, and therapies. This chapter discusses the diversity and importance of medicinal plants in various sectors as well as highlights the successful drug products produced by the said entities and their future trends.

Keywords: Commercial value, medicinal plants, modern Phyto-therapeutics, plant-based drugs, secondary metabolites.

INTRODUCTION

Medicinal plants are defined as plants or herbs imparting therapeutic characteristics or that have a positive pharmacological effect on the human or animal body. Recently, there has been a lot of focus on using environmental friendly and bio-friendly solutions derived from plants for the inhibition and cure of many diseases. According to WHO, herbal remedies are used by 80% of individuals around the globe for various aspects of their healthcare. According to

---

* **Corresponding author Naila Safdar:** Department of Biotechnology, Fatima Jinnah Women University, Rawalpindi, Pakistan; E-mail: nailahussain@fjwu.edu.pk

Zulqurnain Khan, Azra Yasmin & Naila Safdar (Eds.)

WHO, some 21,000 species of plants are considered to have the potential for disease prevention and cure [1].

Numerous plants have been utilized for a variety of medicinal purposes from the beginning. Indian Vids, Unani Hakims, and the Mediterranean and European cultures have all used plants as medicine for over 4000 years, according to evidence. Traditional medical knowledge is still valued around the world. Increased therapeutic use of plants is a consequence of insufficient drug supply, out-pricing of medications, adverse consequences of most of the man-made compounds, as well as resistance development against medications that are being used for various infectious diseases [2].

DIVERSITY OF MEDICINAL PLANTS

Over 50,000 species of plants are employed for pharmaceuticals and healthcare goods, accounting for over one-tenth of all plant species. The distribution of therapeutic plants, on the other hand, isn't constant around various countries. For instance, India and China contain the most medicinal plants, with 7500 and 11,146 plant species, respectively, accompanied by the USA, Thailand, Philippines, and many other countries, with share of medicinal plants ranging from 6% in Malaysia to 16% in India compared to total species of plant. Fig. (**1**) shows the diversity of plant species around the world. Not only do some plant families have more medicinal plants than others, but they also have a higher proportion of vulnerable species [3].

Medicinal Plant Species

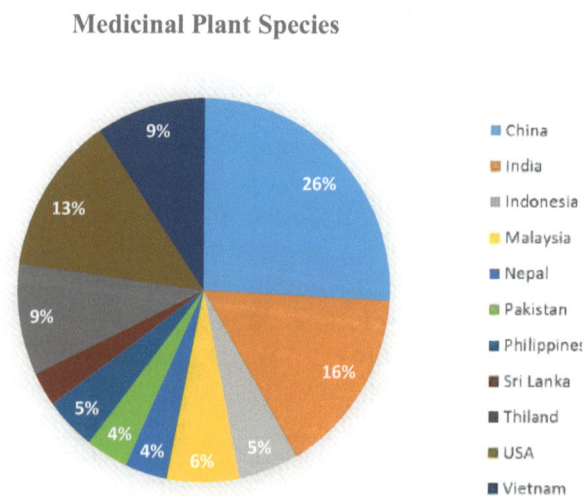

Fig. (1). Percentage of medicinal plant species present around the world.

Approximately seventy-five percent of the population gets their health care primarily from plants and plant extracts. Over one-third of all species of plants were used for medicinal purposes at some point. In developed countries such as America, plant pharmaceuticals are predicted to make up up to 25% of overall drugs, but in developing economies such as China and India, they might account for nearly 80%. As a consequence, the strategic value of medicinal plants in India exceeds that of the entire world. Two third of plant species being used in Western medicine are found in these nations, and the indigenous nation's primary care is based on traditional medicine [4].

MEDICINAL PLANTS IN PAKISTAN

In Pakistan, at least 12% of the flora is utilized medicinally. Baluchistan and Northern Pakistan is the hotspot of many medicinally important genera, including Allium, Cousinia, and Astragalus. Pakistan's plant hotspots are found over 13 natural regions, ranging from mangrove forests to alpine pastures [5]. Examples of some of the species having medicinal importance include *lantago ovata, Berberis lyceum, Papaver somniferum, Zingiber officinale, Foeniculum vulgare,* and *Cichorium intybus. Zingiber officinale* and *Foeniculum vulgare* are often used to treat gastrointestinal issues. *Berberis lyceum* is a valuable species used to treat diabetes, joint problems, and liver ailments. *Piper nigrum, Berberis lyceum, Plantago ovata, Butea monosperma, Rubia cordifolia, Zingiber officinale,* and *Nigella sativa* are among the other prominent species with a high usage value [6].

Traditional Plants

Plants as a source of medicines are thought to be safer, as they exhibit little or no adverse effects, than their synthetic counterparts. Medicinal plants are believed to contain a wide range of constituents that are employed to generate pharmacopoeial or synthetic medicines. Moreover, since some plants are considered as an essential source of nourishment, they are prescribed for various therapeutic applications [7]. In addition to medication, plants are used in organic dyeing, pest management, culinary, fragrance, herbal teas, as well as many other purposes. Certain plants and herbs are utilized in various parts of the world in order to prevent homes and other places free from flies, ants, fleas, and mice. Medicinal herbs have become key sources of biopharmaceuticals [8].

Herbal plants like black pepper, myrrh, cinnamon, ginseng, sandalwood, aloe, clover, barberry, and safflowers are known for treating sores, wounds, and sores. Basil leaves, chives, fennel seeds, cilantro, mint, apple, oregano, lemon balm, thyme, sage, and rosemary stand among the most useful medicinal herbs. Many plants are consumed to purify the blood, to help transform or improve a longstanding condition by getting rid of metabolic toxins. 'Blood cleansers' is another name for these. Certain plants and herbs enhance immunity by lowering the risk of illnesses like fever. Antibiotic characteristics can be found in some botanicals. Turmeric is commonly used as a home treatment for wounds and cuts. Traditional Indian medical practitioners offer antipyretic herbs such as black pepper, safflower, and sandalwood to relieve fever and inflammation induced by certain ailments. Antacids are found in marshmallow leaves and roots. These herbs help to keep the healthy stomach acid required for optimal digestion. Medicinal plants were reported to be used by Indian physicians to treat poisons from animals and snake bites. The aromatic characteristics of cardamom and coriander are well-known. Antiseptic plants such as aloe, turmeric, sandalwood, and khare khasak are extensively utilized and have significant medical potential. Certain cough syrups contain extracts of ginger and cloves. They're recognized for their expectorant properties, which help in the removal of mucus from the trachea, lungs, and bronchi. Chamomile, ajwain, calamus, basil, chrysanthemum, cardamom, coriander, peppermint, fennel, spearmint, ginger, cinnamon and turmeric are all beneficial for improving blood circulation. As a result, they're used to stimulate heart functioning. Disinfectant properties are found in certain therapeutic plants, which kill disease-causing microorganisms. Pathogenic bacteria that cause transferable infections are also inhibited by them. Mild tonics are made from aromatic herbs including goldenseal, aloe, chirayata, and barberry. Toxins in the blood are reduced by the bitter taste of such plants. They are also beneficial in the fight against infection. Certain herbs, such as cayenne, camphor, myrrh, and guggul can be utilized as stimulants to enhance a system's or organ's activity [8, 9].

Commercial Gains

Medicinal plants have traditionally been considered essential in the development of medications, especially against various cancers and infectious diseases, as well for many other therapeutic benefits cardiovascular disease (such as statins) and multiple sclerosis (for example, Gilenya). Medicinal plants are employed on an industrial scale to generate complete extracts and tinctures, refined extracts, and chemical compounds, in addition to their direct usage in traditional medicine. Refined extracts are usually more active than single components of a compound because they contain a whole blend of active elements [10].

Plant-based or natural product derived pharmaceuticals account for roughly 30% of all drugs on the global market. Clinical applicants currently under development have a comparable ratio. Although recombinant proteins and some peptides are becoming commercially popular, the dominance of molecules having low mass used in various disease treatment remains unchallenged, owing to better compliance and bioavailability. Plant-based therapeutic protein development could be a safer, more efficient, and cost-effective way to produce proteins [11].

Plants have a higher ability to assimilate genetic information and generate complex proteins that can be utilized to make more effective treatments as compared to conventional production methods. Traditional cell culture methods necessitate a commitment of both money and time. Because the technology is being developed in natural, renewable resources, plant-based pharmaceuticals have much-reduced facility and production costs. Because expansion is not confined to specialized manufacturing facilities, scaling output to satisfy expanded and diversified demand will be relatively simple. Plant-based protein manufacturing is the most cost-effective technique for providing patients with the benefits of increased and faster access to novel medicines for certain types of pharmaceuticals [12].

MODERN PLANT-BASED PHARMACEUTICS

Pharmaceuticals derived from medicinal plants have been successful because of innovative applications of biotechnology to modify the properties of plants, to produce vaccines, proteins, and other therapeutic products that are harvested by medical groups to treat many illnesses and life-threatening diseases like diabetes, cancer, heart disease, cystic fibrosis, Alzheimer's disease, and HIV. Medicinal plants are cost-effective, efficient, and safer alternatives as compared to other methods such as animal cell cultures or microbial fermentation to develop various therapeutic products [12].

In the medicinal plant industry, there are several operators and a wide variety of commercial operations. Which include global organizations that trade vast volumes of bulk natural resources around the world, as well as small-town firms that produce value-added plants and herbs for direct sale to local clients, and everything in between. At least 120 significant medications produced from plants are now in use in one or more countries around the world [13]. With the evolution of plant-manufactured pharmaceuticals, a new type of industrial manufacturing has emerged. Plants are being used by companies like ZEA Biosciences to generate cost-effective and scalable medicinal components. Avivin and glucuronidase, generated in transgenic maize plants, are among the very first products obtained as a result of molecular farming [13]. They are still utilized as reagents in research, and Merck's subsidiary MilliporeSigma continues to sell avidin. Many proteins used for diagnostic and technical purposes have been synthesized by using plants since then, some companies (such as Diamante, Leaf Expression Systems, and Agrenvec) have expanded their portfolios to include this protein category [14].

After phase I–III clinical trials, one biopharmaceutical product that has been tested in humans has achieved commercial authorization. Taliglucerase alfa is a recombinant human glucocerebrosidase developed by Protalix Biotherapeutics in cell suspension culture of *Daucus carota* [15]. In May 2012, the US Food and Drug Administration has approved taliglucerase alfa to be used in the form of injection for enzyme replacement rehabilitation for curing an adult having Gaucher disease, marketed by the name of Elelyso. In CHO cells, a different variant of the recombinant enzyme (imiglucerase) is generated [16, 17].

Protalix Biotherapeutics have used tobacco cells to develop a therapy for enzyme replacement intended for Fabry disease, which is in phase III [18]. Medicago, which is a large-scale Canadian business also operating in the United States, has also tested a vaccine derived from plants intended treating for seasonal influenza. Hemagglutinin (influenza) proteins were produced in plants of *N. benthamiana*, which resulted in enveloped VLPs (Virus-like particles) that look like wild-type virions of influenza but are devoid of viral RNA. To make quadrivalent virus-like particles (QVLP) for each season's formulation, monovalent VLPs containing hemagglutinins from various strains were combined. Both clinical investigations found that the QVLP vaccine that was plant-derived provided significant protection against influenza-related illnesses of respiratory system. As a result, ZMapp, which is a mixture of three monoclonal antibodies produced for experimental purpose to treat Ebola patients, was developed using transient expression in *N. benthamiana* [19].

Three monoclonal antibodies were developed individually by Mapp Biopharmaceutical in the United States and integrated into a product fit for direct therapeutic use in four to six months. ZMapp was tested on human beings, even before clinical trials, against the West African Ebola virus during the 2014 crisis period. Five out of the seven patients who received the medicine during experiment showed recovery, but each patient required 9 g of the mixture, which required between thirty and fifty kilograms of leaf tissue. In response to the demand produced by the Ebola epidemic, a collaboration of non-profit as well as governmental and corporate partners was created to develop antibodies specific to Ebola in CHO (Chinese Hamster Ovary) cells [20].

Dow AgroSciences LLC transformed tobacco to produce a vaccine that worked effectively against Newcastle disease in chickens. It was the first plant-based biopharmaceutical that was approved by both USDA as well as FDA in 2006 [21]. The vaccine which is a recombinant glycoprotein, hemagglutinin-neuraminidase (HN), with one out of two viral surface glycoproteins and is the primary cell surface antigen present in most of the domestic poultry and other avian species that elicits neutralizing antibodies and protective immunological response. The

HN protein that was isolated from cells of tobacco plants and injected into chickens as a crude extract, lowered the cost of production sufficiently for marketable viability. Although the HN protein that was unpurified provided complete protection in trials for virus encounters, Dow AgroSciences decided not to market it due to a strategic business choice [22].

Bioactive Metabolites

Bioactive chemicals are found in abundance in plants. Depending on their functional role, bioactive metabolites are classified as main or secondary metabolites [23]. Environmental factors such as abiotic or biotic stresses, various developmental states and tissue locations (fruits, flowers, leaves, or bark) cause plants and fungi to create hundreds of secondary metabolites. The most important ones include auxins, ethylene, abscisic acid, gibberellin, cytokinin, and polyamines [24].

Bioactive substances are extracted with a help of a variety of techniques such as maceration, soxhlet extraction, and steam distillation. To reduce the requirement of solvent while applying gentler conditions for extraction, various methods are employed which include the use of microwave, ultrasound, electric field, high pressure, or supercritical fluid [25, 26]. On the basis of metabolic origin, secondary metabolites are classified as nitrogen-containing compounds, phenolic compounds and terpenes [27].

Secondary metabolites produced by plants have been demonstrated to exhibit a variety of biological and therapeutic properties, providing a strong scientific foundation for the application of plants in traditional medical practice in many ancient societies. They include antiviral, antibacterial, and antifungal properties, and hence protect these plants against various infections. Moreover, they also contain crucial UV-absorbing chemicals, preventing major light damage to the leaves. Some medicinal herbs like forage grass have been found to have estrogenic characteristics and interact with animal fertility [28].

Groups of Bioactive Metabolites

Secondary plant metabolites can be grouped into various classes based on their chemical structures or metabolic origin (Fig. **2**). Alkaloids, flavonoids, tannins and terpenes are examples of plant secondary metabolites [28].

Alkaloids

Alkaloids are nitrogen-based heterocyclic molecules. Amino acids like tryptophan, tyrosine, and lysine are the fundamental metabolites from which they are produced. Alkaloids have been used as a laxative, cough suppressant, and sedative against snake venom, and inflammation for over 3000 years. Papaverine, a benzylisoquinoline alkaloid, has been demonstrated to inhibit many viruses, and indoquinoline alkaloids from *Cryptolepis sanguinolenta* have been proven to inhibit Gram negative bacteria as well as yeast.

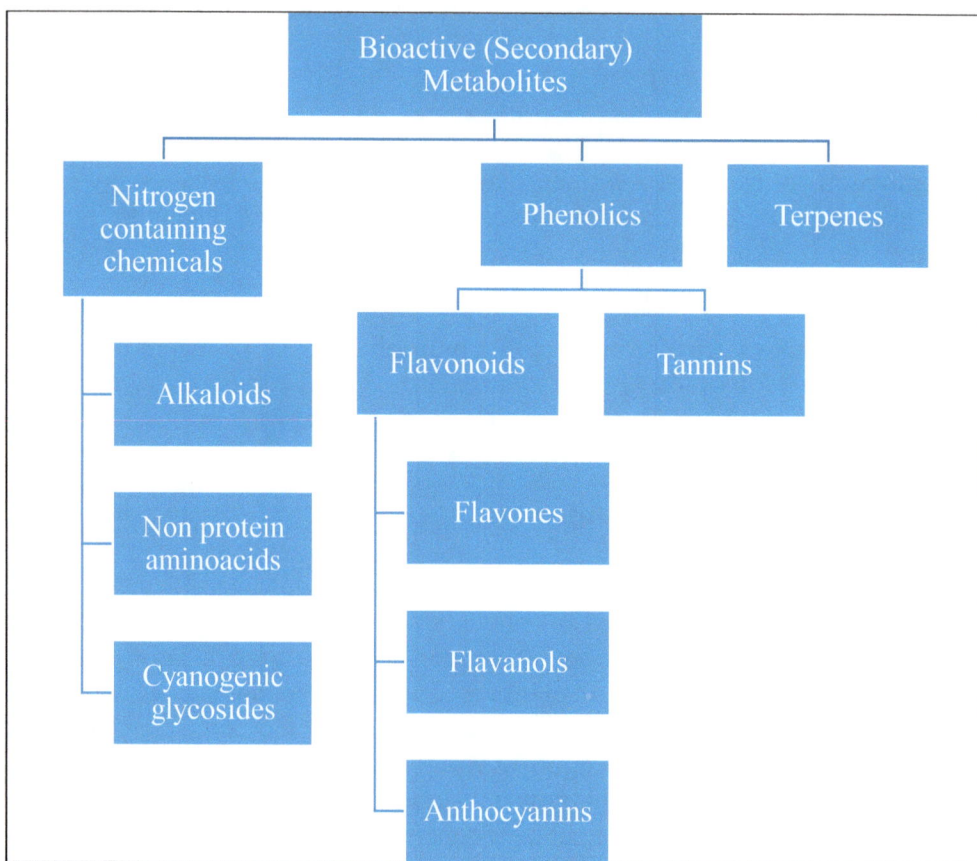

Fig. (2). Classification of plant secondary metabolites on the basis of their chemical structures.

Flavonoids

Flavonoids are polyphenolic compounds. They are present in almost all parts of plant like fruit and vegetable, seeds and nuts, as well as in stem, flower, they are also present in tea, honey, and resins. They are known for their therapeutic characteristics and are well known as antimicrobial, antispasmodic, anti-inflammatory, and anti-allergic agents as well as to enhance aquaresis. They are also effective in the regulation of hormones such as estrogens, androgens, and thyroid hormones.

Terpenes

Terpenes, also called terpenoids, are one of the most abundant as well as diversified classes of plant secondary metabolites. Tea, cannabis, sage, as well as citrus fruits are some of the examples of plants containing high concentrations of terpenes. They impart an intense odor that keeps away herbivores. The smell of their essential oils is helpful in making perfumes which include rose and lavender as a few examples. Monoterpenes are widely known because of their antiviral properties. Terpenes have the capability of anti-cancer as well as anti-diabetic action, which is particularly important since the prevalence of cancer and hyperglycemia is increasing in the modern age. One of the most commonly used terpenes is curcumin, which has antioxidant, anti-inflammatory, antiseptic, anti-cancer, and antiplasmodial, astringent and diuretic properties [29].

SUCCESS STORIES

Plants in Healthcare

Traditional medicinal plant knowledge has influenced modern pharmaceutics in the form of key medications. The use of whole plant or plant part extracts in traditional, complementary and alternative medicine is gaining worldwide acceptance. For example, the roots of *Rauwolfia serpentina* (Rauwolfia) are processed into capsules and tablets and used to treat hypertension and mental illnesses by interacting with the nervous system [30]. Cocaine is made from the leaves of *Erythroxylum coca* (*Papver somniferum*) and is sold under the brands Numbrino and Goprelto. Cocaine is a topical anesthetic used to provide local anaesthesia to mucosal membranes in the oral, vocal, and nasal canals that are accessible [31].

The leaf and root of *Atropa belladonna* (Belladonna) are used to treat hay fever or colds, arthritis pain, bronchospasms from asthma, haemorrhoids, nerve disorders, Parkinson's disease, indigestion, inflammatory bowel disease, and motion sickness. Belladonna is also found in topical ointments used to treat joint pain, sciatic nerve pain, and nerve pain in general. It is often used to treat mental illnesses, loss of control on muscular contraction, profuse sweating as well as asthma [32]. *Glycyrrhiza glabra* (Licorice) is a perennial herb used to cure a variety of ailments, including hyperdipsia, sexual debility, epilepsy, paralysis, fever, rheumatism, stomach ulcers, hemorrhagic diseases, skin infections, and jaundice [33].

Digoxin is a prescription medicine made from chemicals extracted from *Digitalis lanata* (foxglove). In the United States, *Digitalis lanata* is the most common source of digoxin. Foxglove is often used to treat fluid accumulation in the body, heart failure, and heart palpitations [34]. *Catharanthus roseus* or Periwinkle is widely used to prevent childhood leukemia as well as for treating cancer. It has been used to treat muscle discomfort, depression, as well as for healing wounds and wasp stings. Its benefits ranges from diabetes prevention to relieving stomach discomfort [35]. Vinblastine is a plant alkaloid sold under the trade name of Alkaban-AQ or Velban. It is used to treat Hodgkin as well as non-Hodgkin's lymphoma, breast, testicular, bladder, and lung cancers, soft tissue sarcoma, melanoma, Kaposi's sarcoma and choriocarcinoma. Germ cell tumor, histiocytosis and fibromatosis are also known to be treated by vinblastine [36].

The bark of *Cinchona officinalis*, often known as cinchona, is used to increase appetite, promote the flow of digestive juices, and treat bloating and other stomach issues. Varicose veins, leg cramps, and hemorrhoids are among the conditions for which it is prescribed. Cinchona is used to treat moderate swine flu, influenza, malaria, common cold, and fever. Some of the other uses include cancer, spleen enlargement, throat diseases, and muscle. Cinchona is used to numb discomfort, destroy germs, and act as an astringent in eye creams [37].

Pilocarpus jaborandi leaves are used to create medication. It was first utilized to make pilocarpine, a prescription drug approved by the FDA. Glaucoma is treated with pilocarpine eye drops. Pilocarpine pills are utilized to treat dry mouth caused by radiation therapy, as well as dry eyes and mouth caused by Sjogren syndrome [38]. Taxol is another example of anticancer plant alkaloid known for its effectiveness against ovarian, bladder, breast, prostate, esophageal, lung, melanoma, and many other cancers and solid tumors. Kaposi's sarcoma has also been treated with taxol [39].

Other than direct use of plants, many crops / medicinal plants have been modified for the mass production of many commercially important therapeutic products. This resulted in the engagement of modern day biotechnological tools following the production of commercial recombinant therapeutic goods.

Plant-based Vaccines

A large number of molecular farming studies proved the possibility of protein production in recombinant plants of specific species and established the usefulness of therapeutic protein products using cell-based or analytical assays [40]. Virus-like particles (VLPs) were generated as a result of the transient expression of structural proteins in Dengue virus and shortened non-structural RNA polymerase, which is RNA-dependent, in *N. benthamiana*, which elicited a significant antibody response in mice [41]. Plants are also utilized to make vaccines and many other therapeutic proteins for biopharma uses [42, 43].

Plants are particularly well suited to the manufacturing of animal vaccinations because crude extracts of plants can be directly fed to or inoculated into animals without the need for protein purification, lowering production costs dramatically. Because only those items will be feasible to be sold in a market that costs a few cents, this is critical for the development of animal vaccinations [44]. Piglets were vaccinated for enterotoxigenic *E. coli* or ETEC by introducing diets augmented with seeds of *Arabidopsis thaliana* possessing antibodies for anti-ETEC, and crude plant extracts of *N. benthamiana* leave selectively expressing the H5 virus of influenza managed to prevent the outbreak of avian influenza (H5N1) among chickens that were vaccinated [45, 46]

In early 2006, the world's earliest (injectable) plant-made vaccine was licensed. Dow AgroSciences LLC stated on January 31, 2006, that it had earned the United States Department of Agriculture's initial licensing for a plant-made vaccine. The vaccine was developed using a confined culture of plant-cell in production technique to treat viral Newcastle Disease (NDV). The plant-produced NDV vaccine was created using a cell line from transgenic plant of tobacco that was cultivated in a typical bioreactor as suspension culture. Bioreactor system can generate a large amount of vaccine in a matter of weeks using this technique. The resultant cells are extracted and treated to obtain a partially screened antigen that can be used in the final vaccine formulation. Birds that were subcutaneously injected with these vaccines got resistance for NDV infection. While the composition was never developed into a commercially marketable product, it did receive regulatory approval from the veterinary biologics centre of USDA, proving that the plant-based Vaccines might be created within the regulations that currently exists [45].

Plant-based Antibodies

Scientific studies had been conducted to investigate plant-derived antibodies or human protein replacements for diabetes, Gaucher's disease and, cystic fibrosis such as insulin, gastric lipase, and glucocerebrosidase [47, 48]. GMP (Good Manufacturing Practice) compliance is feasible by using cell suspension as well as hairy root cultures in bioreactors since productivity is possible inside a clean surrounding, comparable to other systems like the ones that are based upon microorganisms other animal cells. This is not possible for plants that are grown on hydroponic systems, rockwool blocks or in soil. As a result, regulatory agencies permit upstream manufacture in an environment that is non-GMP providing that the cleared plant tissue extracts are transported to a GMP environment in sterilized conditions for additional treatment [49]. Even so, because irregular growing circumstances and pollutants can impair product output and quality, the upstream of the farming stage should be properly managed to promote uniform growth of plants and prevent any interaction with bugs, pesticides, rodent excreta, and other harmful invasions. The establishment of these guidelines allowed for the synthesis of a monoclonal antibody, specific to HIV, in transgenic plants of tobacco to be used in phase I of clinical study in healthy females, upon giving a single vaginal dose of the antibody. This validated the plant-derived medicinal protein's safety and acceptability [50].

Human clinical trials have been conducted on two plant-based antibody products. CaroRxTM, the earliest antibody that has been clinically evaluated, was developed by Planet Biotechnology Inc. in the tobacco industry. CaroRxTM adheres to microbes that can result in dental decay, inhibiting the bacteria from adhering to the teeth. CaroRxTM is presently undergoing Phase II of clinical test in the US, as a part of US Food and Drug Administration approved Investigational New Drug (IDN) application.

Large Scale Biology Corporation (LSBC) announced the successful results of the first clinical trials carried out on humans, evaluating a cancer vaccination derived from plants in July 2008. Recombinant patient-specific idiotype vaccines against lymphoma of follicular B cells were expressed in tobacco by utilizing a temporary viral expression technique in plants. The vaccination was developed and processed within 3-4 months of receiving biopsy samples after which it was examined in Phase I of the safety and immunogenicity trial. Sixteen patients who were inoculated with their therapeutic antigen experienced no major side effects, although three fourth of them developed humoral or cellular immune responses, and half of them acquired responses specific to the antigen. Immune responses were observed to be directed toward the idiotype sequence itself rather than plant-specific glycan structures (glycosylation) in this situation. Bayer Pharmaceutics began clinical research of this plant derived vaccine antibody in December 2009, successfully presenting a protocol for Phase I study to the US Food and Drug Administration [51].

Plant-based Interferons and Insulins

Biolex Therapeutics, Inc. announced the very first discovery of a plant based therapeutic protein in a Phase II clinical trial with Locteron, an interferon alfa (IFN-) therapy for patients having chronic hepatitis C. Patients are currently being treated with IFN- in association with an antiviral medication weekly. Unfortunately, Ribavirin, an antiviral medicine, has a lot of negative consequences. In 2005, thirty two patients took part in Locteron Phase II clinical trial. In randomized double-blind research, IFN- generated from Lemna, an aquatic herb, was administered every two weeks in association with ribavirin. All of the hepatitis C patients (16/16) treated with 480 and 640 g dosages had an early virologic response. This is significant since early virologic responses in patients suffering from hepatitis C have shown to be a prerequisite for long-term virologic responses [52].

A phase I and II studies of insulin produced from safflower were also conducted by SemBioSys, which demonstrated that it had the same degree of safety as existing recombinant insulin. This insulin is made in lipid storage compartments, which makes it easier to be extracted. ORF Genetics creates numerous cytokines and growth factors in transgenic plants of barley to be used in cosmetics [53].

COMMERCIAL CHALLENGES

The limited variety of products obtained as a result of molecular farming in the market relative to the high amount of research suggests a tailback in commercial development, particularly in the pharma sector. The most acknowledged constraints of molecular farming are that the yield of plants is sometimes very low while the cost of downstream processing is high as well as the transformation into desired products is slow [54]. EU funded a project Pharma-Factory to solve this limitation by supplying tools like expression vectors, like *Agrobacterium*, and specific plant types, along with the accompanying methods. This could pave the way for a new approach to commercializing medicinal plants, in which the product, rather than the platform, receives its IP (Intellectual property). More work is needed in this area, particularly to translate plant-made medicinal products into easier clinics [55].

FUTURE TRENDS

The advantages of plant-based medications have been highlighted numerous times in the scientific literature, based on technological research and their benefits are greatly increasing enormously than those of other methods. Technology, on the other hand, is just beginning to be deployed because of initiatives by prominent pharma manufacturers with the resources and knowledge to interact with regulatory agencies and obtain market authorization, such as Pfizer with prGCD. After the arrangement for these initiatives is operational, plant-based protein production may become the favored form of upcoming protein synthesis. It's a good time to think about the technology's long-term potential now that plant-based pharmaceuticals are on the horizon. By evaluating at what might be accomplished now and planning for future advances, the pace that has been continuing to build up to this moment can be carried over into other areas [56].

Plant molecular farming still is a restricted medium for therapeutic product introduction to the general public. One of the most significant restrictions may be the large cell size of plants and the subsequently reduced amount of product biosynthesis facilities per unit of volume in comparison to microbial and other animal cells. Plant-based protein production, on the other hand, has several distinct advantages, including advanced modification technology for glycan chain, animal element and endotoxin less industrial production of allergens as well as the inexpensive synthesis of mucosal vaccines and other therapeutic products, since oral and topical administration eliminates the requirement for emulsions. The pace of synthesis is the major benefit of transient expression systems, which allows for the creation of recombinant proteins in a matter of time.

This technology is quite useful for development of vaccines and other diagnostic tools required in emergency situations, such as those for a new strains of influenza virus as well as SARS-CoV-2, required within a time of months or even weeks [57]. Protein synthesis capabilities become sparse in emergency scenarios when other treatments and diagnostics are unable to be discontinued or postponed in the face of a novel disease as demonstrated by the current COVID-19 pandemic. Plants can be grown although with studying the genome sequence of pathogens, so they'll be prepared to manufacture therapeutic products immediately when antigen sequences become available, allowing for the rapid filling of manufacturing gaps. As a result, many academic and corporate groups are developing tools for diagnosis and therapies against viral SARS-CoV-2 *via* transient expression in plants. Although it needs to see whether biological farming is capable to outperform conventional protein manufacturing techniques, the benefits of transient expression would allow plants to create a large space for recombinant proteins and other products as screening methods for infectious diseases [58].

Bioactive metabolites and products synthesized from plants have historically produced the bulk of novel medications. Natural product research has fallen in the pharmaceutical business during the last 15 years, owing to a focus on synthetic library screening with maximum throughput. A significant drop is observed in the approval of new drugs, as well as the potential loss of patent protection for critical drugs. Rapid genomic sequencing, combined with biosynthetic pathway alteration, may provide a massive resource for the discovery of pharmacological drugs in the future.

CONCLUDING REMARKS

As our lifestyles get more techno-savvy, we are drifting away from nature. We can't escape nature since we're a component of it. Medicinal plants are all-natural, have no negative consequences, are reasonably safe, eco-friendly, and widely available. Plants historically have been enough kind to treat ailments related to specific time periods. They must be planted, promoted, and preserved in order to save human life. Plant-=based products are becoming a sign of safety in comparison to artificially made medications. The naive dependence on synthetic substances, on the other hand, is receding, and people are moving back to natural ones in search of safety and stability. It's time to spread them out on a worldwide level.

REFERENCES

[1]   Sofowora A, Ogunbodede E, Onayade A. The role and place of medicinal plants in the strategies for disease prevention. Afr J Tradit Complement Altern Med 2013; 10(5): 210-29.
        [http://dx.doi.org/10.4314/ajtcam.v10i5.2] [PMID: 24311829]

[2]   Aziz MA, Adnan M, Khan AH, Shahat AA, Al-Said MS, Ullah R. Traditional uses of medicinal plants practiced by the indigenous communities at Mohmand Agency, FATA, Pakistan. J Ethnobiol Ethnomed 2018; 14(1): 2.
        [http://dx.doi.org/10.1186/s13002-017-0204-5] [PMID: 29316948]

[3]   Chen SL, Yu H, Luo HM, Wu Q, Li CF, Steinmetz A. Conservation and sustainable use of medicinal plants: Problems, progress, and prospects. Chin Med 2016; 11(1): 37.
        [http://dx.doi.org/10.1186/s13020-016-0108-7] [PMID: 27478496]

[4]   Thakur NS, Attar SK, Chauhan RS. Horti-medicinal agroforestry systems: A potential land use for commercial cultivation of medicinal and aromatic plants. Agro-forestry for Increased Production and Livelihood Security, Horti-Medicinal Agro-forestry Systems 2017; pp. 163-80.

[5]   Shinwari ZK, Qaiser M. Efforts on conservation and sustainable use of medicinal plants of Pakistan. Pak J Bot 2011; 43: 5-10.

[6]   Ahmad M, Zafar M, Shahzadi N, Yaseen G, Murphey TM, Sultana S. Ethnobotanical importance of medicinal plants traded in Herbal markets of Rawalpindi- Pakistan. J Herb Med 2018; 11: 78-89.
        [http://dx.doi.org/10.1016/j.hermed.2017.10.001]

[7]   ur Rehman F, Kalsoom M, Adnan M, *et al.* Importance of medicinal plants in human and plant pathology: A review int j phar. Biomed Res 2021; 8: 1-1.

[8]   Khan MA. Introduction and importance of medicinal plants and herbs national health portal india journal 2016.

[9]   Mohammed AH. Importance of Medicinal Plants Res Pharm Health Sci. 2019; 5: 124-5.

[10]   Grabley S, Sattler I. Natural products for lead identification: Nature is a valuable resource for providing tools mol methods drug discov. 2003; 87-107.

[11]   de Boer J, Aiking H. On the merits of plant-based proteins for global food security: Marrying macro and micro perspectives. Ecol Econ 2011; 70(7): 1259-65.
        [http://dx.doi.org/10.1016/j.ecolecon.2011.03.001]

[12]   Nagels B, Weterings K, Callewaert N, Van Damme EJM. Production of plant made pharmaceuticals: From plant host to functional protein. Crit Rev Plant Sci 2012; 31(2): 148-80.
        [http://dx.doi.org/10.1080/07352689.2011.616075]

[13]   Hood EE, Kusnadi A, Nikolov Z, Howard JA. Molecular farming of industrial proteins from transgenic maize chem higher plant bioeng. 1999; 127-47.

[14]   Santoni M, Ciardiello MA, Zampieri R, *et al.* Plant-made Bet v 1 for molecular diagnosis. Front Plant Sci 2019; 10: 1273.
        [http://dx.doi.org/10.3389/fpls.2019.01273] [PMID: 31649716]

[15]   Zimran A, Gonzalez-Rodriguez DE, Abrahamov A, *et al.* Long-term safety and efficacy of taliglucerase alfa in pediatric gaucher disease patients who were treatment-naïve or previously treated with imiglucerase. Blood Cells Mol Dis 2018; 68: 163-72.

[http://dx.doi.org/10.1016/j.bcmd.2016.10.005] [PMID: 27839981]

[16] Hollak CEM, vom Dahl S, Aerts JMFG, *et al.* Force majeure: Therapeutic measures in response to restricted supply of imiglucerase *Cerezyme* for patients with gaucher disease. Blood Cells Mol Dis 2010; 44(1): 41-7.
[http://dx.doi.org/10.1016/j.bcmd.2009.09.006] [PMID: 19804996]

[17] Zimran A, Wajnrajch M, Hernandez B, Pastores GM. Taliglucerase alfa: Safety and efficacy across 6 clinical studies in children and adults with Gaucher disease. Mol Genet Metab 2017; 120(1-2): S144.
[http://dx.doi.org/10.1016/j.ymgme.2016.11.384]

[18] Van der Veen SJ, Hollak CEM, van Kuilenburg ABP, Langeveld M. Developments in the treatment of fabry disease. J Inherit Metab Dis 2020; 43(5): 908-21.
[http://dx.doi.org/10.1002/jimd.12228] [PMID: 32083331]

[19] Pastores GM, Shankar SP, Petakov M, *et al.* Enzyme replacement therapy with taliglucerase alfa: 36 month safety and efficacy results in adult patients with gaucher disease previously treated with imiglucerase. Am J Hematol 2016; 91(7): 661-5.
[http://dx.doi.org/10.1002/ajh.24399] [PMID: 27102949]

[20] Pettit DK, Rogers RS, Arthur K, *et al.* CHO cell production and sequence improvement in the 13C6FR1 anti-Ebola antibody InMAbs. Taylor & Francis 2016; 8: pp. (2)347-57.

[21] Schillberg S, Raven N, Fischer R, Twyman R, Schiermeyer A. Molecular farming of pharmaceutical proteins using plant suspension cell and tissue cultures. Curr Pharm Des 2013; 19(31): 5531-42.
[http://dx.doi.org/10.2174/1381612811319310008] [PMID: 23394569]

[22] Schillberg S, Finnern R. Plant molecular farming for the production of valuable proteins: Critical evaluation of achievements and future challenges. J Plant Physiol 2021; 258-259: 153359.
[http://dx.doi.org/10.1016/j.jplph.2020.153359] [PMID: 33460995]

[23] Depuydt S, Van Praet S, Nelissen H, Vanholme B, Vereecke D. How plant hormones and their interactions affect cell growth mol cell biology growth different plant cells. 2016; 174.

[24] Pavlić B, Šojić B, Teslić N, Putnik P, Kovačević DB. Extraction of bioactive compounds and essential oils from herbs using green technologies in aromatic herbs in food. 2021; 233-62. Academic Press

[25] Azmir J, Zaidul ISM, Rahman MM, *et al.* Techniques for extraction of bioactive compounds from plant materials: A review. J Food Eng 2013; 117(4): 426-36.
[http://dx.doi.org/10.1016/j.jfoodeng.2013.01.014]

[26] Giacometti J, Bursać Kovačević D, Putnik P, *et al.* Extraction of bioactive compounds and essential oils from mediterranean herbs by conventional and green innovative techniques: A review. Food Res Int 2018; 113: 245-62.
[http://dx.doi.org/10.1016/j.foodres.2018.06.036] [PMID: 30195519]

[27] Das K, Gezici S. Secondary plant metabolites, their separation and identification, and role in human disease prevention Ann Phytomedicine 2018; 7: 13-24.

[28] Hussein RA. El-Anssary AA Plants secondary metabolites: The key drivers of the pharmacological actions of medicinal plants Herb Med 2019; 1: 11-30.

[29] Cox-Georgian D, Ramadoss N, Dona C, Basu C. Therapeutic and medicinal uses of terpenes.Medicinal Plants. Cham: Springer 2019; pp. 333-59.
[http://dx.doi.org/10.1007/978-3-030-31269-5_15]

[30] Shitiz K. Gupta SP Rauwolfia serpentina Himalayan Med Plants 2021; 111-49.

[31] Paniagua-Zambrana NY, Bussmann RW. Romero c erythroxylum coca lam e rythroxylaceae ethnobot andes 2020; 1-6.

[32] Bettermann H, Cysarz D, Portsteffen A, Kümmell HC. Bimodal dose-dependent effect on autonomic, cardiac control after oral administration of Atropa belladonna. Auton Neurosci 2001; 90(1-2): 132-7.
[http://dx.doi.org/10.1016/S1566-0702(01)00279-X] [PMID: 11485281]

[33] El-Saber Batiha G, Magdy Beshbishy A, El-Mleeh A, Abdel-Daim MM, Prasad Devkota H. Traditional uses, bioactive chemical constituents, and pharmacological and toxicological activities of Glycyrrhiza glabra L *Fabaceae*. Biomolecules 2020; 10(3): 1-19.
[http://dx.doi.org/10.3390/biom10030352] [PMID: 32106571]

[34] Kaul S, Ahmed M, Zargar K, Sharma P, Dhar MK. Prospecting endophytic fungal assemblage of Digitalis lanata Ehrh *foxglove* as a novel source of digoxin: A cardiac glycoside 3 Biotech 2013; 3(4): 335-40.

[35] Gajalakshmi S, Vijayalakshmi S, Devi RV. Pharmacological activities of *catharanthus roseus*: A perspective review. Int J Pharma Bio Sci 2013; 4: 431-9.

[36] Ichikawa M, Negoro R, Kawai K, Yamashita T, Takayama K, Mizuguchi H. Vinblastine treatment decreases the undifferentiated cell contamination of human ipsc-derived intestinal epithelial-like cells. Mol Ther Methods Clin Dev 2021; 20: 463-72.
[http://dx.doi.org/10.1016/j.omtm.2021.01.005] [PMID: 33614822]

[37] Bharadwaj KC, Gupta T, Singh RM. Alkaloid group of cinchona officinalis: Structural, synthetic, and medicinal aspects in synthesis of medicinal agents from plants 2018; 205-27. Elsevier

[38] Rocha JA, Andrade IM, Véras LMC, *et al.* Anthelmintic, antibacterial and cytotoxicity activity of imidazole alkaloids from pilocarpus microphyllus leaves. Phytother Res 2017; 31(4): 624-30.
[http://dx.doi.org/10.1002/ptr.5771] [PMID: 28111828]

[39] Druzhkova I, Lukina M, Dudenkova V, *et al.* Tracing of intracellular ph in cancer cells in response to taxol treatment. Cell Cycle 2021; 20(16): 1540-51.
[http://dx.doi.org/10.1080/15384101.2021.1949106] [PMID: 34308742]

[40] Bhusnure OG, Shinde MC, Vijayendra SS, Gholve SB, Giram PS, Birajdar MJ. Phytopharmaceuticals: An emerging platform for innovation and development of new drugs from botanicals. J Drug Deliv Ther 2019; 9: 1046-57.

[41] Ponndorf D, Meshcheriakova Y, Thuenemann EC, *et al.* Plant made dengue virus like particles produced by co-expression of structural and non-structural proteins induce a humoral immune response in mice. Plant Biotechnol J 2021; 19(4): 745-56.
[http://dx.doi.org/10.1111/pbi.13501] [PMID: 33099859]

[42] Clarke JL, Paruch L, Dobrica MO, *et al.* Lettuce-produced hepatitis c virus e1e2 heterodimer triggers immune responses in mice and antibody production after oral vaccination. Plant Biotechnol J 2017; 15(12): 1611-21.
[http://dx.doi.org/10.1111/pbi.12743] [PMID: 28419665]

[43] Lai H, Paul AM, Sun H, *et al.* A plant: Produced vaccine protects mice against lethal west nile virus infection without enhancing zika or dengue virus infectivity. Vaccine 2018; 36(14): 1846-52.
[http://dx.doi.org/10.1016/j.vaccine.2018.02.073] [PMID: 29490880]

[44] Topp E, Irwin R, McAllister T, *et al.* The case for plant-made veterinary immunotherapeutics. Biotechnol Adv 2016; 34(5): 597-604.
[http://dx.doi.org/10.1016/j.biotechadv.2016.02.007] [PMID: 26875776]

[45] Virdi V, Coddens A, De Buck S, *et al.* Orally fed seeds producing designer igas protect weaned piglets against enterotoxigenic *escherichia coli* infection. Proc Natl Acad Sci 2013; 110(29): 11809-14.
[http://dx.doi.org/10.1073/pnas.1301975110] [PMID: 23801763]

[46] Phan HT, Pham VT, Ho TT, *et al.* Immunization with plant-derived multimeric H5 hemagglutinins protect chicken against highly pathogenic avian influenza virus H5N1 Vaccines 2020; 8: 1-18.

[47] Yao J, Weng Y, Dickey A, Wang K. Plants as factories for human pharmaceuticals: Applications and challenges. Int J Mol Sci 2015; 16(12): 28549-65.
[http://dx.doi.org/10.3390/ijms161226122] [PMID: 26633378]

[48] Spiegel H, Stöger E, Twyman RM, Buyel JF. Current status and perspectives of the molecular farming

landscape molecular pharming: Applications, challenges and emerging areas 2018; 12: 3-23.

[49]   Sack M, Rademacher T, Spiegel H, *et al.* From gene to harvest: Insights into upstream process development for the GMP production of a monoclonal antibody in transgenic tobacco plants. Plant Biotechnol J 2015; 13(8): 1094-105.
[http://dx.doi.org/10.1111/pbi.12438] [PMID: 26214282]

[50]   Ma JKC, Drossard J, Lewis D, *et al.* Regulatory approval and a first-in-human phase i clinical trial of a monoclonal antibody produced in transgenic tobacco plants. Plant Biotechnol J 2015; 13(8): 1106-20.
[http://dx.doi.org/10.1111/pbi.12416] [PMID: 26147010]

[51]   Moon KB, Park JS, Park YI, *et al.* Development of systems for the production of plant-derived biopharmaceuticals. Plants 2019; 9(1): 30.
[http://dx.doi.org/10.3390/plants9010030] [PMID: 31878277]

[52]   Mizukami A, Caron AL, Picanço-Castro V, Swiech K. Platforms for recombinant therapeutic glycoprotein production recom glycoprotein product 2018; 1674: 1-14.

[53]   Thomas DR, Penney CA, Majumder A, Walmsley AM. Evolution of plant-made pharmaceuticals. Int J Mol Sci 2011; 12(5): 3220-36.
[http://dx.doi.org/10.3390/ijms12053220] [PMID: 21686181]

[54]   Shanmugaraj B, I Bulaon CJ, Phoolcharoen W. Plant molecular farming: A viable platform for recombinant biopharmaceutical production. Plants 2020; 9: 1-19.
[http://dx.doi.org/10.3390/plants9070842]

[55]   Schillberg S, Finnern R. Plant molecular farming for the production of valuable proteins: Critical evaluation of achievements and future challenges. J Plant Physiol 2021; 258-259: 153359.
[http://dx.doi.org/10.1016/j.jplph.2020.153359] [PMID: 33460995]

[56]   Capell T, Twyman RM, Armario-Najera V, Ma JKC, Schillberg S, Christou P. Potential applications of plant biotechnology against SARS-CoV-2. Trends Plant Sci 2020; 25(7): 635-43.
[http://dx.doi.org/10.1016/j.tplants.2020.04.009] [PMID: 32371057]

[57]   Rosales-Mendoza S, Márquez-Escobar VA, González-Ortega O, Nieto-Gómez R, Arévalo-Villalobos JI. What does plant-based vaccine technology offer to the fight against COVID-19? Vaccines 2020; 8(2): 183.
[http://dx.doi.org/10.3390/vaccines8020183] [PMID: 32295153]

[58]   Iranbakhsh a potential applications of plant biotechnology against sars-cov-2 Iran. J Biol 2021; 4: 89-98.

CHAPTER 13

Nanotechnology in Medicinal Plants

Shaghufta Perveen[1,*] and **Naila Safdar**[1]

¹ Department of Biotechnology, Fatima Jinnah Women University, Rawalpindi, Pakistan

Abstract: Nanoparticles have immense applications in plants from mass propagation to phyto-drug extraction and augmentation. Alongside, nanoparticles are also manifested as potential drug vehicles for carrying curative agents to the targeted tissues or part, accompanying control delivery of drugs to the infected site. Advancement in nanotechnology directed towards the transformation of metallo-drugs at the nanoscale brings new dimensions in therapeutics from the treatment of multidrug-resistant microbes to chemotherapies of tumors. With the nano-advancement, not only metals and their oxides are transformed at the nanoscale but also the potential phyto agents, proteins, and hormones are transformed into nanosized entities which change the entire fundamentals of therapeutic and curative practices. A lot of changes in medicine, drug delivery system and drug formulation as commenced just because of nanotechnology. The current chapter highlights nanotech advancements in the area of medicinal plant propagation, drug augmentation and extraction methodologies along with their limitations and future prospects.

Keywords: Nano-vehicles, Nanotized Herbs, Plant Therapeutics, Pharmacotherapy, Phyto-drugs.

INTRODUCTION

Plants are natural madicines for the treatment of numerous ailments and are a fundamental element of human societies throughout history. Medicinal plants not only account for the economic resilience of poor people in developing countries but also play a critical role in traditional as well as modern medicine [1]. Diversion of modern society towards natural products has been associated with low or no side effects which were obvious in conventional medication [2]. Complementary and alternative medicine (CAM) gained extensive scientific as well as commercial consideration. According to the World Health Organization, about 80% of people worldwide rely on herbal medicines to accomplish their primary healthcare needs. Like in Europe, 90% of wild resources are harnessed

---

* **Corresponding author Shaghufta Perveen:** Department of Biotechnology, Fatima Jinnah Women University, Rawalpindi, Pakistan; E-mail: shaghuftaperveen0346@gmail.com

Zulqurnain Khan, Azra Yasmin & Naila Safdar (Eds.)

for therapeutic purposes which account for the use of more than 1300 medicinal plant species. Similarly in the United States, about 79% of top prescriptions belong to natural resources [3]. The Himalayan region, parts of Bangladesh, Nepal, India, Afghanistan, Pakistan, Myanmar, Bhutan and China are acknowledged as a hotspot of medicinal plant diversity [4].

Presently, France, the UK, the USA, Italy and China are found to be the biggest global market for medicative plants. In Asia, Pakistan is ranked as 7th medicinal plant producer [5]. Around 600 species are reported for their uses as traditional medicine in Pakistan and more than 75% of the local population count on herbal medicines for most of their health issues. Substantial use of medicinal flora is in medicine cosmetics, dietary supplements and food manufacturing [6]. The majority of the indigenous population is still dependent on plant-based remedies. Other reasons for the increasing use of herbal medicines in all cultures are due to efficacy, unique targeting ability, minimal side effects and safety, which are questionable in the case of synthetic medicines [7]. This development directed towards an increase in demand for herbal medicines. Industrialized and developing countries scrutinize herbal products for the past 40 years. Moreover, herbs were redeemed with the advancement in technologies as a prospective source of new drug candidates. This entire condition led to the use of common as well as endangered medicinal plants.

PHYTOCHEMICALS IN PHARMACOTHERAPY

The plant produces diverse bioactive compounds with unusual chemical scaffolds. Bioactive compounds in medicinal plantsplay a key role in anti-analgesic, antifungal, antioxidant, antibacterial, anti-dysentery, antiviral, anti-malarial, anti-inflammatory, and anticancer activities. The polypharmacological effect of herbal constituents increased the activation of pro-drug, bioavailability, synergistic action and interference with cellular transport processes. These phytoconstituents have a lot of therapeutic effects *i.e.* against skin disorders, dyspepsia, diabetes, aphrodisiac, deflatulent, polyurea, jaundice, and haematological disorders. Bioactive compounds are plant secondary metabolites having both toxicological and pharmacological effects on animals and mankind. These bioactive compounds are derived from primary plant metabolites, mainly performing photosynthesis. Two main pathways are involved in the biosynthesis of these principal compounds *e.g.*, the mevalonic acid pathway and shikimic acid or aromatic amino acid pathway [8]. A diverse group of secondary metabolites are named as active compounds *i.e.*, alkaloids, oils, steroids, glycosides, enzymes, phenolics, resins and terpenoids. The presence of one or more active compounds is ascribed to the medicinal values of plant crude extracts/drugs.

Medicinal plant-based drugs normally belong to one of the following groups *i.e.* taurine, choline, fatty acids, polyphenols, prebiotics, caffeine, polysaccharides, carotenoids and phytoestrogens [9, 10]. Major classes of phytoconstituents with disease-preventing functions are anticancer, dietary fibre, immunity boosters, antioxidants, and neuro-pharmacological and detoxifying agents. Each class of these phytoconstituents has a wide array of chemicals with variant potencies. The function of phytochemicals can be decided by the type of functional group present on its side chain. Bioactive compounds of medicinal plants are endowed with boundless opportunities for new drug leads [11]. A substantial quantity of newly registered drugs have been derived from natural sources (plants) like paclitaxel and docetaxel from *Taxus brevifolia,* artemisinin from *Artemisia annua,* etoposide and teniposide from *Podophyllum* species [12]. About 1073 chemical entities had been approved from 1981-2010, which belonged to the small molecules group. Out of these 1073, 64% were either extracted or inspired by natural sources, while only 36% were purely synthetic [13]. Indispensable plant-based bioactive compounds which were notably used in pharmacotherapy are listed in Table **1**.

Table 1. Bioactive compounds used in pharmacotherapy.

Common Name	Scientific Name	Family	Drug Name	Therapeutics	References
Pacific yew tree	*Taxus brevifolia*	Taxaceae	Paclitaxel and its derivatives	Anticancerous	[14]
Periwinkle / Vinca rosa	*Catharanthus roseus*	Apocynaceae	Vinblastine and Vincristine	Anticancerous	[15]
Happy tree/cancer tree	*Camptotheca acuminata* Decne	Nyssaceae	camptothecin and its analogs	Anticancerous	[16]
Common snowdrop	*Galanthus nivalis* L	Amaryllidaceae	Galanthamine	Alzheimer's disease	[17]
Wormwood or sweet sagewort	*Artemisia annua* L	Asteraceae	Artemisinin	Antimalarial and anti-cancerous	[18]

Various compelling plant-based bioactive compounds *i.e.,* Podophyllotoxin, Paclitaxel, and Vinblastine were unable to meet the growing demands of the market. Because these species were slowly grown in their natural environment, some were even endangered due to over-exploitation. Moreover, the natural accumulation of these compounds is very low, over a long growth period. Besides this, they were also produced in very low concentrations in their natural green source [19, 20] The promising protective approach to avoid the extinction of these endangered medicinal herbs has been the cultivation of these plants under controlled conditions [21].

PHYTOCHEMICALS AUGMENTATION

The traditional method used for the cultivation of phytochemicals was field cultivation. But this system has various issues *i.e.* low yield and instability in phytoconstituents concentration due to variation in environmental, geographical and seasonal conditions. So this ancient method of field cultivation was replaced with plant cell, tissue and organ cultures. But these systems mostly failed to produce the desired quantity of product. So, researchers come up with the idea of elicitor application in tissue culture methods to augment the production of required phytochemicals. In this context, both biotic and abiotic elicitors were applied to achieve the targeted goal of phyto-drug augmentation. Besides this, hairy root culture was also sought to increase the production of desired phytoconstituents. Similarly, *in vitro* cultures were also explored to reconstruct the metabolic pathways of plants and to genetically engineer them. This search for a potential method for the augmentation of phyto-drug of interest put a step in the door of nanotechnology. This new invention has changed the focus of researchers from all around the world to explore this new technique for the enhancement of phytochemicals in herbal entities. The coupling of nanotechnology with micropropagation brings new dynamics to the field of phyto-augmentation *in vitro* cultures. This coupling was designed by keeping in mind that stress induction results in the activation of the defense system of plants which ultimately radiates the production of secondary metabolites (phytoconstituents). So, if stress will be imposed at the nanoscale then due to multi-dimensions and changes in surface plasmon resonance, different dimensions of stress will be induced that will ultimately radiate the production of secondary metabolites [22].

NANOPARTICLES IN PLANT THERAPEUTICS

Plant-mediated biosynthesis of nanoparticles is advantageous because of its simple methodology or protocol. Microbe-based nanoparticles were also synthesized and found to be a promising therapeutic tool but were not as advantageous as plants. Because the prior one required the multi-step isolation of potential microbe, specific conditions of culture preparation and subsequent subculturing as well as culture maintenance. Scaling up large-scale production of nanoparticles by using plants was a comparatively easy method [23, 24]. The main plant parts which were used as reducing and stabilizing agents for the biosynthesis of nanoparticles included stem, fruits and their outer coverings, leaves, barks, and roots [25]. To date, numerous metal salt solutions have been reported in the literature for the biosynthesis of nanoparticles *i.e.* titanium dioxide (TIO_2), zinc acetate dehydrate {Zn $(CH_3COO) \cdot 2H_2O$}, hexachloroplantic acid hexahydrate {$H_2PtCl_6 \cdot 6(H_2O)$}, silver nitrate ($AgNO_3$), copper (II) sulfate pentahydrate (CuSO$_4$.5H$_2$O), zinc oxide (ZnO), and hydrogen tetrachloroaurate (HAuCl$_4$).

Fruits and leaves of *Tribulus Terrestris* and leaves of *Podophyllum hexandrum* were reported for the green synthesis of silver nanoparticles. The biosynthesized nanoparticles reported strong antibacterial and anticancerous activity [26 - 28]. The strong antibacterial activity of spherical-shaped, 16 – 28 nm size silver nanoparticles against various isolates of multidrug resistance bacteria like *Bacillus subtilis*, *Staphylococcus aureus*, *Pseudomonas aeruginosa*, *streptococcus pyrogens* and *Escherichia coli* was reported [26, 27]. While the potential role of biosynthesized silver nanoparticles in caspase-mediated cell death and DNA damaging mediated inhibition of HeLa cell lines was also cited [28]. In another report, the antibacterial activity of cinnamon reduced and stabilized green silver nanoparticles [29]. The cytotoxic effect of *Caesalpinia pulcherrima*-mediated silver nanoparticles against HCT116 cell lines was also reported [30]. Besides them, silver and gold nanoparticles were also synthesized by using the leaves extract of *Mentha piperita* and *Mussaenda glabrata*. The biosynthesized nanoparticles have strong antimicrobial activity against human pathogens like *Staphylococcus aureus* and *E. coli* [31, 32]. Leaves of *Justicia wynaadensis* and *Berberis aristata* and the flower of *Nyctanthes arbortristis* were used for the biosynthesis of zinc oxide nanoparticles (ZnONPs). These biosynthesized zinc oxide nanoparticles have DNA-binding antifungal activity against many phytopathogens and antimitotic potential [33 - 35].

Similarly, the bactericidal activity of *Cissus quadrangular*-mediated biosynthesized titanium dioxide nanoparticles (TiO$_2$NPs) was reported against *Staphylococcus* and *E. coli* [36]. Medicinal plants *Cissus quadrangularis* and *Rosmarinus officinalis* L. were reported for the biosynthesis of copper oxide and iron nanoparticles respectively [37, 38]. The cytotoxic effect of iron nanoparticles on C26 and 4T1 cancer cell lines was cited [37]. While the antifungal activity of copper oxide nanoparticles against *Aspergillus flavus* and *Aspergillus niger* was reported [38]. Medicinal plant-mediated superficial biosynthesis of nanoparticles was considered economically viable, safe and less time-consuming than other biological organisms *i.e.* microbes [39, 40]. Another advantage was the increased antimicrobial potential of green synthesized nanoparticles than other nanoparticles. This increased antimicrobial potential was probably interrelated with the phytochemicals present in plants, which might be acting as stabilizing and capping agents during nano-synthesis [41].

NANOPARTICLES FOR PHYTO-ELICITATION

Nanoparticles can be applied by three means to augment the production of secondary metabolites in plants; (1) soil supplementation, (2) plant growth media supplementation and (3) foliar spray. For our convenience, we have categorized nanoparticles into three basic classes or elicitors to confer their elicitation potential upon secondary metabolite production in plants. Basic classes of nano-elicitors include:

1. Metallic nano-elicitors

2. Metal oxide nano-elicitors

3. Carbon-based nanomaterials

Metallic Nano-elicitors

Different metallic nanoparticles (Zn Au, Ag, Cu, Co) were explored to assess their effect on the production of plant secondary metabolites. The working proficiency of these nanoparticles depends on their characteristics, concentration applied, synthetic origin, time of exposure, and size.

Silver Nanoparticles (AgNPs) Mediated Elicitation

Among all types of nanoparticles, silver nanoparticles were most exploited concerning secondary metabolite elicitation in plants. Silver nanoparticles (30 mg/L) were effective in stimulating the production of lignans and neolignans in the cell suspension culture of *Linum usitatissimum* [42]. Approximately 3 mg/L concentration of silver nanoparticles was good enough to accumulate 2 times higher capsaicin in *Capsicum frutescens* cell suspension culture [43]. Silver nanoparticles' application at the exponential growth phase of *Corylus avellana* L. resulted in the stimulation of taxane yield. In the following experiment, a 5 ppm concentration was found enough to increase the percentage of taxol (378%) and baccatin III (163%) simultaneously in *Corylus avellana* L. cell suspension cultures [44]. In another study, a 500 mg/L concentration of silver nanoparticles was found good for phenolic production in *Brassica rapa* ssp seedlings [45]. However, 1 mg/L silver nanoparticles were ideal for diosgenin elicitation *in vitro* culture of *Trigonella foenum* graecum [46].

Silver nanoparticles induce reactive oxygen species (ROS) generation, which ultimately elicits the production of secondary metabolites. Exposure of *Brassica rapa* hairy root culture to silver nanoparticles caused induction of ROS along with enhanced production of phenolics, glucosinolates and flavonoids [45]. Similarly, a 40 mg/L concentration of silver nanoparticles caused ROS generation as well as essential oil augmentation in *Pelargonium graveolens* seedlings [47]. The shape and size of silver nanoparticles also play an important role in eliciting secondary metabolite concentration. Comparison between triangular (47 nm), spherical (8 nm) and decahedral (32 nm) silver nanoparticles at 50 mM concentration revealed that spherical shape nanoparticles enhanced the production of anthocyanin in *Arabidopsis thaliana* than triangular and decahedral-shaped nanostructures [43].

Elicitation Effect of Copper Nanoparticles (CuNPs)

Copper nanoparticles have both positive and negative effects on plant growth and secondary metabolite production. Copper nanoparticles (5 mM) were found effective in improving the concentration of total phenolics (2 fold) in *Verbena bipinnatifida* shoot culture [48]. Likewise, flavonoid (26.21%), lycopene (64.21%) and total phenolics (5.43%) production were increased in 50 mg/L copper nanoparticles treated shoot culture of *Solanum lycopersicum* [43]. Another report cited that 0.5 mg/L copper nanoparticles treatment resulted in a significant increase in the essential oil content of *Mentha longifolia* shoot cultures [49].

Effect of Bimetallic Nanoparticles on Elicitation

A study reported that a 1:3 ratio of bimetallic Ag: Au nanoparticles resulted in a significant improvement in the production of phenolics (23%) and flavonoids (4%) in *Prunella vlugaris* L callus culture [50]. Zn: Ag nanoparticles were effective at a ratio of 9:1 in elucidating the anolide content in *Withania somnifera*, which was grown under different conditions [51].

Metal Oxide Nano-elicitors

Different types of metal oxides were explored to assess their effect on plant secondary metabolite production. The most frequently used metal oxide elicitor in phytochemical studies included aluminium oxide (AL_2O_3), copper oxide (CuO), caesium oxide (CeO_2), zinc oxide (ZnO), cadmium oxide (CdO) and titanium oxide (TiO_2) nanoparticles.

Elicitation Potential of Copper Oxide Nanoparticles (CuONPs)

Treatment of *Brassica rapa* with copper oxide nanoparticles (CuONPs) resulted in a significant increase in anthocyanin (1.1 fold) and phenolic (1.3 fold) concentration along with stimulated ROS production. CuONP treatment with hairy root culture of *Brassica rapa* showed significant enhancement in glucosinolate and phenolic compound [45]. In another study, a significant increase in flavonoid and total phenolics was noted after 20 days of exposure of *Cichorium intybus* L. to CuONPs [43].

Elicitation Potential of Zinc Oxide Nanoparticles (ZnONPs)

Zinc oxide nano treatment (ZnONPs) has been positively regulating tropane alkaloids (1.2 fold) and phenolic (3.2 fold) compounds in *Hyoscyamus reticulate* hairy root culture [43]. A significant effect of ZnONPs (1 mg/L) on steviol glycoside accumulation in *Stevia rebaudiana* propagates was reported [52]. However, flavonoids and total phenolics were decreased in the same tissues under the same conditions. In another study, maximum accumulation of anthocyanins (100 mg/L) and phenolics (75 mg/L) was observed at 75 mg/L of ZnONPs. However, increased flavonoid (25 mg/L) contents were found at 25 mg/L concentration of ZnO NPs in *Lilium ledebourii* plantlets [53].

Elicitation Potential of Iron Oxide Nanoparticles (Fe_3O_4NPs)

Fe_3O_4NPs have been found to be more effective as elicitors than other metal oxide nanoparticles for the production of secondary metabolites. For example, the production of hyperforin has increased more than hypericin after exposure of *Hypericum perforatum* cell suspension culture to 100 ppb Fe_3O_4NPs [54]. In another study, foliar application of 30 ppb Fe_3O_4NPs significantly increased flavonoids and total phenolic contents in gamma-irradiated seeds of *Lepidum sativum* L [55]. Similarly, it was reported that 450 mg/L and 900 mg/L concentration of Fe_3O_4NPs for 24 hrs and 48 hrs respectively were ideal for the accumulation of hyoscyamine and scopolamine (5 fold each) in *Hyoscyamus reticulates* hairy root culture [56].

Elicitation Potential of Titanium Dioxide Nanoparticles (TiO_2NPs)

Aloin (118%) contents were significantly increased in *in vitro* cell suspension culture of *Aloe vera* treated with TiO_2NPs (Ghorbanpour *et al.*, 2015). In literature, the dose-dependent effect of TiO_2NPs (6 mg/L) was reported on the significant increase of coumaric acid, tannic acid and gallic acid in *Cicer arietinum* [57]. A comparative study was conducted to explore the effect of TiO_2NP and bulk titanium on *Hyoscyamus niger*. Experimental results are evident with the highest hyoscyamine and scopolamine content at 80 mg/L and 20 mg/L concentration of TiO_2NP. While bulk titanium has been ineffective at these concentrations [58].

Effect of Cerium Oxide Nanoparticles (CeO_2NPs) on Secondary Metabolites

Cerium oxide nanoparticles (CeO_2 NPs) were reported for a significant increase in carotenoids (26%) contents in *Solanum lycopersicum* hydroponic culture by treating with 0.1 mM concentration [59, 60]. Contrarily, CeO_2NPs and In_2O_3NPs at concentrations of 1000 mg/ L and 250 mg/L respectively induce oxidative stress followed by a 27% increase in phenolic compounds in *Arabidopsis thaliana* [60].

Silicon Dioxide Nanoparticles

Silicon dioxide nanoparticles (SiO_2NPs) caused an increase (2 fold) in the concentration of cirsimaritin, xanthhomicrol, rosmarinic acid and isokaempferide in *Dracocephalum kotschyi* hairy root culture after an exposure of 48 hrs [61]. But in another report, silicon dioxide nanoparticles (SiO_2NPs) were found to be less effective than titanium dioxide nanoparticles (TiO_2NPs) in augmenting the production of thymoquinone in *Nigella sativa* [43].

Carbon-based Nano-elicitation

Carbon-based nanomaterials are also explored for their potential as elicitors. In this context, besides single-walled (SWCNTs) and multi-walled carbon nano-tubes (MWCNTs), fullerene (Buckyball) was also applied in plant tissue culture to scrutinize their effect on secondary metabolite augmentation.

Elicitation Potential of Carbon Nano-tubes

Single-walled carbon nanotubes (SWCNTs) were electrifying the production of secondary metabolite both in the field and *in vitro* cultured plants. 500 mg/L concentration of SWCNTs spraying on greenhouse-generated *Tanacetum parthenium* resulted in a 2.5-fold increase in parthenolide contents [62]. Similarly, 0.002 g/L concentration of SWCNTs was enough for the significant increase in total tannins, phenolic and flavonoid content in *in vitro Simmondsia chinensis* shoot cultures [43].

Multi-walled carbon nano-tubes (MWCNTs) also have an inductive effect on the production of flavonoids, caffeic acid, phenolics and rosmarinic acid *in vitro Satureja khuzestanica* callus culture [63]. Similarly, a 10 mM concentration of fullerenes has a significant effect on the production of charantin, cucurbitacin-B, insulin and lycopene in *Momordica charantia* seedlings [64].

Chitosan Nanoparticles Mediated Elicitation

The biodegradability and non-toxicity of chitosan make it a promising entity for use in biomedicine and agriculture. Chitosan nanoparticles application was reported in the hydroponic culture of *Camellia sinensis* that caused enhanced accumulation of epigallocatechin gallate, gallic acid, epigallocatechin and epicatechin in *C. sinensis* [65]. Besides this, chitosan nanoparticles were also applied to increase the concentration of soluble alkaloids, phenol and proline in *Capsicum annuum* L. cultures [66].

NANOSIZED HERBAL MEDICINE

Seven metals *i.e.* iron, zinc, gold, tin, copper, lead and silver were used in the Ayurveda therapeutics system. These metal-based medicines were known as "Bhasma". Swarna, Bhasma is a gold ash also known as nano-size particles of gold metal. These metallic-medicines are prepared by repeating the process of metal grinding with herbal juices or some other specific matter and incineration [67].

Nanotechnology is also used to access the target cell and cell compartments for drug delivery [68]. Conversion of specific phytochemicals at the nanoscale has been employed to enhance their therapeutic values. The reduction of liquorice roots, milk veteh root, cassia twigs and seawort at the nanoscale resulted in increased anticancerous effects without damaging healthy cells of the body. Reduction of phytoconstituents at the nanoscale is advantageous because they have rapid curative effects, are ingested orally, only kill specific disease cell, and does not damage other cells/organs/tissues. These types of medicines are good in alleviating certain types of infections, like breast, skin, ovarian, cervical, pulmonary and liver cancer including metastatic lymphoma [67]. The antitumor activity of camptothecin has been improved by converting it into nano-size. Alkaloid camptothecin has overwhelmed broad-spectrum antitumor activity by targeting DNA topoisomerase [69].

Plant viruses were explored as a new type of nanoparticles for drug delivery in cancer therapy. Recently a common plant virus was modified to successfully deliver a drug in the specific cells of the human body with zero effect on surrounding tissues. As nano-viruses are a thousand times smaller than human hair width, so they were more effective in treating cancer with zero or no side effects. Among different types of nanoparticles, plant viruses are superior due to the ease of their construction, stability, target-oriented behavior and ability to cargo therapeutic agents. The stabilizing and target-oriented behavior of nanoparticles can be improved by attaching them to specific proteins. This function can be performed well by using viruses because their organization and structure are well-known. Encapsidation of 17 nm nanoparticles with 36 nm Red Clover Necrotic Mosaic Virus (RCNMV) resulted in the tethering of virion assembly to various nanoparticles and virions were assembled to form uniform sized virus nanoparticles of 33 nm diameters. This nano-virus was then used to deliver drugs by attaching them to their surface [70].

NANOTECHNOLOGY FOR THE EXTRACTION OF HERBAL COMPONENTS

An interesting and novel application of nanoparticles concerning medicinal plants is their use in phytoconstituents' extraction by following receptor theory. This technique is also known as ligand harvesting or ligand fishing. This technique is used for the purification of proteins. Beside protein purification, it can also be used to scrutinize and harvest specific bioactive components of complex biological/ botanical systems or specific extracts. In this technique, surface-modified nanoparticles were used for harvesting the metabolite of interest following the conjugation process without damaging the host cell. The nano-mediated extraction method is more advantageous than other techniques because of the spectrometric identification of isolates, host cell viability and exclusion of the organic solvent usage [71, 72]. Coupling of human serum albumin functionalized magnetic nanoparticles (HSAMNPs) with electrospray ionization mass spectrometry has been employed for the extraction of progeny II and III, gracillin and dioscin from *Dioscorea panthaica* [73]. Similarly, TiO_2NP conjugates with *Arabidopsis thaliana* flavonoids to harvest catechol rich flavonoids and enediol without affecting the cell viability of the foundation plant [72]. The schematic presentation of nanoparticles-mediated harvesting of secondary metabolites is shown in (Fig. **1**).

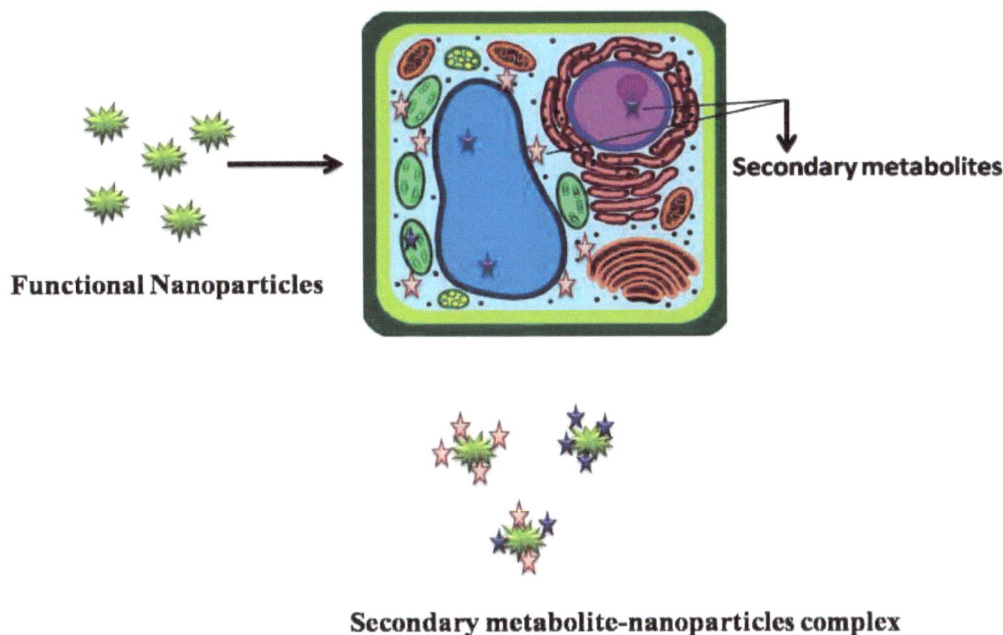

Functional Nanoparticles

Secondary metabolites

Secondary metabolite-nanoparticles complex

Fig. (1). Secondary metabolite-nano complex for harvesting phytoconstituents form intact plant.

NANOPARTICLES FOR DRUG DELIVERY

Nanoparticles received most attention among researchers because of their numerous therapeutic advantages, like administration routes, enhanced therapeutic effects and site-specific action [74]. Nanotechnology helps to extend the benefits of drug delivery systems by decreasing cumulative and toxic effects. Nano-incorporated drugs have increased the use of natural products in recent times. This method has increased the efficiency of natural products more than synthetic ones. Besides this, they also have lesser side effects than those caused by synthetic drugs [74]. Drug delivery system is also responsible for effective, reliable and safer means of drug transportation to target areas. For this purpose, they packed the drug in a way that their distribution is even at the selected administration route, prioritizing the best drug-receptor interaction and reducing its harmful effects [75].

Polymeric Nanoparticles

Polymeric nanoparticles are made up of biocompatible and biodegradable polymers, used for controlled/targeted delivery of drugs [76]. They can also be used to reduce the dose of therapeutics and increase phytoconstituent/ drug solubility as well as their absorption at the particular side. Nanoparticles are beneficial to use in blood since they are non-thrombogenic, have no effect on neutrophils, non-toxic, are stable, avoid the reticuloendothelial system, are noninflammatory and are nonimmunogenic [77]. Preference for nanomaterials depends on their reduced side effects, carrying more than one active constituent, sustained released system and increased residence time inside the body [77].

The size of polymeric nanoparticles can be in the range of 10 – 1,000 nm in diameter and exist either in the form of nanospheres (NSs) or nanocapsules (NCs). The difference between these two forms is that nanocapsules are oily cores and nanospheres are polymeric structures. Because of the difference in structure, active compounds loaded differently in nanocapsules active compounds adsorb on polymeric membranes or are dissolved in the oily core. However, in nanospheres, active compounds are either adsorbed or retained on the polymeric structure. Other types of polymeric nanoparticles are glycolic acid (PLGA) and poly---lactic acid (PLA). Some studies reported enhanced effects of encapsulation of drug delivery and bioavailability. The nano-encapsulation of *Phytolacca decandra* (Phytolaccaceae) extract with sodium arsenite and benzo pyrene enhances chemopreventive action and drug bioavailability in lung cancer and A549 cell lines [78]. Similarly, the loading of *Ocimum sanctum* L methanolic extract into sodium alginate chitosan nanoparticles enhanced their antimicrobial potential [79]. Nano-curcumin was formulated by loading curcumin with polymeric nanoparticles. This resulted in an increased efficacy against human pancreatic tumor cells. In another study, encapsulation of curcumin in PLGA nanospheres caused a decrease in the IC_{50} of prostate cancer cell lines. The hydrophobic property of Honokiol (HN) was altered by loading them with polymeric nanoparticles to enhance vascular administration [80]. The encapsulation of hydrophobically modified glycol chitosan with camptothecin enhanced antitumor activity against breast cancer cells at concentrations of 10 mg/kg and 30 mg/kg [74]. Similarly, the encapsulation of ethanolic extract of *polygala senega* with biodegradable PLGA enhanced the anticancerous effect against lung cancer cell line A549 [81].

Solid Lipid Nanoparticles and Nanostructured Lipid Carriers

Solid lipid nanoparticles (SLNs) are colloidal carriers of 50 - 1000 nm with high physiochemical stability and better protection against labile drug degradation [82]. SLNs were used to avoid the chemical degradation of drug molecules [82]. These nano lipid carriers seek attention as an alternative vehicle for colloidal drugs. These formulations are biodegradable and nontoxic and can be administered through dermal, oral, intravenous and pulmonary [83].

The bioavailability and absorption of quercetin in the intestine were increased five times by incorporating it with solid lipid nanoparticles. Dermal delivery and enhanced anti-inflammatory and antioxidant effect of quercetin were due to their incorporation with nanostructure lipid carrier [84]. Similarly, the oral bioavailability of curcumin was improved by incorporating it into SLN, which was composed of soy lecithin. These curcumin-loaded SLNs demonstrated extended drug release *in vitro* conditions [85]. In another study, incorporation of triptolide (TP) into SLNs caused an improved absorption and solubility into the skin. TPs are vine extracts which were suitable for the treatment of autoimmune diseases and inflammatory [74].

Liquid Crystal System

Liquid crystals (LC) are condensed structures, in-between between solid crystal and liquid isotopes. They might be ordered or disordered depending on their simplicity of efflux, might be cubic or hexagonal, and might be thermotropic (TLC) and lyotropic (LLC) based on general stipulations [86]. Liquid crystals can also be developed by vegetable oils because of their low molecular weight and viscosity. Another reason for the use of vegetable oil for the formulation of liquid crystals was their low occlusion, which facilitates both increased loading of the drug as well as enhanced dermal penetration capacity [87]. Interaction among the drug and its active target site has been promoted by liquid crystal. Besides this, they also facilitate the interaction between the cell membrane and active molecules to encourage their entry into the cell [74].

Different types of surfactants have been explored for the formulation of LC. Like silicone was used in combination with Cetearyl alcohol, essential oil, diacetyl phosphate, andiroba and ceteth-10 phosphate [74]. This system does not affect the stability of LC. Similarly, in another study, peach essential oil was used and increased physical stability of LC formulation [74]. Besides them, annatto oil from *Bixa orellana* and marigold oil were also used to develop an LC system by using hydrophilic/lipophilic balance. Recently natural products have been encouraged for the production of nanostructures with potential drug delivery capacity. In this way, new drug vehicles such as liquid crystal system were developed to promote indispensable factors such as the physical/ chemical stability of herbal compounds and bioavailability.

Liposome and Micro-emulsions

Liposomes are microscopic vesicles composed of one or more lipid bilayers. They are used for the encapsulation of hydrophilic substances. The primary structure of liposomes is based on sterols, antioxidants and phospholipids. Liposomes were classified as cationic, anionic, and neutral based on the charge present on them. They are also unilamellar, multilamellar and oligolamellare [74].

Liposomes were developed by using Cratylia mollis lectin (Cra) from *Cratylia mollis* Mart and investigated for their antitumor potential against Sarcoma-180 in Swiss mice. The results of experiments showed a 28% increase in tumor inhibition in Cra-loaded liposomes than a free form of Cra [74]. In another study, intranasal use of quercetin liposomes even at lower doses caused an improvement in anxiolytic and cognitive effects [74]. Similarly, silymarin hybrid liposomes have shown hepatoprotective activity against carbon tetrachloride in a dose dependent manner on buccal administration to albino rats [74]. The incorporation of camptothecin into PEGylate liposomes has caused an increase inthe stability and antitumor effect of camptothecin at 15 mg/kg. Sustainable delivery of breviscapine has been achieved by preparing breviscapine-loaded multivesicular liposome to reduce the frequency of injection and prolong drug circulation duration with subsequent improvement of patient compliance against breviscapine [74].

Another system that is used to improve the solubility, stability, bioavailability, targeted prolonged action and convey different degrees of hydrophilic/lipophilic activity of active constituents within the same formulation is microemulsions. The incorporation of triptolide with a microemulsion significantly suppressed carrageenan-induced rat paw edema with improved inflammatory effects [74]. In another study, *Syagrus romanzoffiana* fruit pulp extract was used to develop a nanoemulsion, which increased antioxidant properties of selected phyto-extracts [88].

NANOPARTICLES ENHANCE THE SOLUBILITY AND BIOAVAILABILITY OF PHYTO-DRUGS

The main obstruction that hinders the use of active phytochemicals in pharmaceutical formulations was their low solubility. This obstacle can be removed by using nanoparticles which solubilized these drugs and carried them to the target organ and body tissue. As nanoparticles have a high surface area to volume ratio, so they are capable of attracting more water molecules, which enhance the solubility of hydrophobic compounds [89]. Several nano-vesicles are used to enhance the solubility of phytochemicals including liposomes, polymer nanoparticles, micelles and solid lipid nanoparticles [90]. A few examples of phyto-drug whose bioavailability and solubility were increased by using nanoparticles are mentioned in Table **2**

Table 2. Increased bioavailability and solubility of phyto-drugs by nanotechnology.

Phyto-drugs	Green Source	Nature of Phyto-drug	Therapeutic Potential	Nanotechnique to Increase Solubility	References
Triptolide	*Tripterygium wilfordii*	Diterpenoid epoxide	Autoimmune diseases especially rheumatoid arthritis, leukemia, pancreatic cancer, psoriasis and polycystic kidney disease	Prepared as a biocompatible and biodegradable tripolide-loaded poly	[91]
Quercetin	onion, apples	Flavonoid	Antimicrobial, cardiovascular, antidiabetic, antioxidant, wound healing, anti-Alzheimer's, anti-inflammatory, and antiarthritic	Encapsulation in polymeric nanoparticles suspension	[89]
Tetrandrine	*Stephania tetrandra*	bis-benzylisoquinoline alkaloid	Antitumor activity and nonselective calcium blocker	Incorporated in solid lipid nanoparticles	[89]

(Table 2) cont.....

Phyto-drugs	Green Source	Nature of Phyto-drug	Therapeutic Potential	Nanotechnique to Increase Solubility	References
Cryptotanshinone	*Salvia miotiorrhiza* Bunge	quinoid diterpene	Cytotoxic, anti-parasitic, antioxidant, anti-inflammatory, anti-angiogenic and antibacterial.	Solid lipid nanoformulations	[92]
Hypericin	*Hypericum perforatum*	polycyclic aromatic dianthraquinone	Anti-inflammatory, anti-depressant, anti-viral and antibacterial	Hypericin-loaded solid lipid nanoparticles and suspension of hypericin polymeric nanoparticles	[90, 93]
Thymoquinone	*Nigella sativa*	Monoterpene	Antioxidant, analgesic, antipyretic, antiinflammatory and anticancer	Encapsulated with polymer nanoparticles, liposomes and cyclodextrin	[89, 94, 95]
Ampelopsin	*Ampelopsis grossedentata*	Flavonoid	Antihypertension, cough, antioxidant, antimicrobial, hepatoprotective, anti-inflammatory and anticancerous	Encapsulated in microemulsion	[96]

LIMITATIONS OF NANOTECHNOLOGY

Nanotechnology in plants is at its experimental stage and risk effects associated with it cannot be taken with ease.

1. Nanoformulations can access the body through the skin, lungs and digestive tract. This might cause activation of reactive oxygen species (ROS) which can damage cells and DNA. Another possible risk involves the crossing of blood-brain barrier. Nanoformulations are capable of becoming a part of blood-stream specifically in the case of nano-encapsulated or nano-coated drug delivery agents.

2. Nano-size dimensions can be a hazard for biological systems. Moreover, unforeseen interactions between nano-size chemicals and the body may lead to unpredictable results.

3. Nanoparticles as drug carriers may also interact with cellular organelles, enhance platelet aggregation and activate the blood clotting mechanism.

4. To use nanoparticles/nanomedicine in therapeutics like for cancer treatment, various legal, social and ethical aspects of nanomaterials at the community level need to be addressed. More medical trials are needed to validate nano's role in therapeutics.

5. Nanoparticles' size optimization is important for nano-herbal formulations. 20 nm to 30 nm-sized nanoparticles can be cleaned by renal excretion, whereas 200 nm or greater than this nano-size needs a removal mechanism by a mononuclear phagocytic system [97].

6. The surface properties of nanoparticles also play a critical role in their interaction with targeted cells and proteins. Nano-encapsulated herbal drugs get affected by nanoparticles' size, physiochemical properties and shape. For instance, inhalation of less than 100 nm nanostructures can cause oxidative stress and pulmonary inflammation [98]. Large aggregates of nanoparticles can be formed due to their instability, which might get confined in the lungs and pose serious health issues to patients.

FUTURE PROSPECTS

Though the nanomedicine market size is increasing globally and expected to extend with time still many constrains related to this field are unanswered including nano-biodistribution and Inefficient nano product development systems. Right now, the current investigation only focuses on the potential benefits of nanostructures and ignores their safety concerns under *in vivo* conditions. As development in nanotechnology has expanded its extensions to molecular research tools and diagnostics, there is a need to cater to the problem of primary healthcare facilities in poorer nations in a cost-effective and easy way. Moreover, there is a need to explore briefly the mechanism of nanotransport, biological barriers, adverse effect control, endocytosis and degradation pathways involved in nano-herbal therapies.

CONCLUDING REMARKS

The current chapter highlighted the use of nanotechnology in medicinal plant propagation, phyto-drug augmentation and extraction methodologies. Sparking light is shed on the potential application of nanoparticles in medicines specifically concerning controlling targeted drug delivery and transformation of proteins, hormones and phyto agents at nanoscale.

REFERENCES

[1]   Aziz MA, Adnan M, Khan AH, Shahat AA, Al-Said MS, Ullah R. Traditional uses of medicinal plants practiced by the indigenous communities at Mohmand Agency, FATA, Pakistan. J Ethnobiol Ethnomed 2018; 14(1): 2.

[http://dx.doi.org/10.1186/s13002-017-0204-5] [PMID: 29316948]

[2]   Xavier TF, Kannan M, Lija L, Auxillia A, Rose AKF, kumar SS. Ethnobotanical study of kani tribes in thoduhills of kerala, South India. J Ethnopharmacol 2014; 152(1): 78-90.
[http://dx.doi.org/10.1016/j.jep.2013.12.016] [PMID: 24393788]

[3]   Chen SL, Yu H, Luo HM, Wu Q, Li CF, Steinmetz A. Conservation and sustainable use of medicinal plants: Problems, progress, and prospects. Chin Med 2016; 11(1): 37.
[http://dx.doi.org/10.1186/s13020-016-0108-7] [PMID: 27478496]

[4]   Rashid N, Gbedomon RC, Ahmad M, Salako VK, Zafar M, Malik K. Traditional knowledge on herbal drinks among indigenous communities in Azad Jammu and Kashmir, Pakistan. J Ethnobiol Ethnomed 2018; 14(1): 16.
[http://dx.doi.org/10.1186/s13002-018-0217-8] [PMID: 29467005]

[5]   Kanwal H, Sherazi BA. Herbal medicine: Trend of practice, perspective, and limitations in Pakistan. Asian Pac J Health Sci 2017; 4(4): 6-8.
[http://dx.doi.org/10.21276/apjhs.2017.4.4.2]

[6]   Petrakou K, Iatrou G, Lamari FN. Ethnopharmacological survey of medicinal plants traded in herbal markets in the Peloponnisos, Greece. J Herb Med 2020; 19: 100305.
[http://dx.doi.org/10.1016/j.hermed.2019.100305]

[7]   Hasan MN, Azam NK, Ahmed MN, Hirashima A. A randomized ethnomedicinal survey of snakebite treatment in southwestern parts of Bangladesh. J Tradit Complement Med 2016; 6(4): 337-42.
[http://dx.doi.org/10.1016/j.jtcme.2015.03.007] [PMID: 27774416]

[8]   Ramawat KG, Dass S, Mathur M. The chemical diversity of bioactive molecules and therapeutic potential of medicinal plants. Herbal Drugs Ethnomedicine to Modern Medicine. Berlin, Heidelberg.: Springer 2009; pp. 7-32.
[http://dx.doi.org/10.1007/978-3-540-79116-4_2]

[9]   Chikezie PC, Ibegbulem CO, Mbagwu FN. Bioactive principles from medicinal plants. Res J Phytochem 2015; 9(3): 88-115.
[http://dx.doi.org/10.3923/rjphyto.2015.88.115]

[10]   Golmohamadi A, Möller G, Powers J, Nindo C. Effect of ultrasound frequency on antioxidant activity, total phenolic and anthocyanin content of red raspberry puree. Ultrason Sonochem 2013; 20(5): 1316-23.
[http://dx.doi.org/10.1016/j.ultsonch.2013.01.020] [PMID: 23507361]

[11]   Cragg GM, Newman DJ. Natural products: A continuing source of novel drug leads. Biochim Biophys Acta, Gen Subj 2013; 1830(6): 3670-95.
[http://dx.doi.org/10.1016/j.bbagen.2013.02.008] [PMID: 23428572]

[12]   Fadeyi SA, Fadeyi OO, Adejumo AA, Okoro C, Myles EL. *In vitro* anticancer screening of 24 locally used Nigerian medicinal plants. BMC Complement Altern Med 2013; 13(1): 79.
[http://dx.doi.org/10.1186/1472-6882-13-79] [PMID: 23565862]

[13]   Newman DJ, Cragg GM. Natural products as sources of new drugs over the 30 years from 1981 to 2010. J Nat Prod 2012; 75(3): 311-35.
[http://dx.doi.org/10.1021/np200906s] [PMID: 22316239]

[14]   Zein N, Aziz SW, El-Sayed AS, Sitohy B. Comparative cytotoxic and anticancer effect of Taxol derived from *Aspergillus terreus* and *Taxus brevifolia.* Biosci Res 2019; 16: 1500-9.

[15]   Moon SH, Pandurangan M, Kim DH, Venkatesh J, Patel RV, Mistry BM. A rich source of potential bioactive compounds with anticancer activities by *Catharanthus roseus* cambium meristematic stem cell cultures. J Ethnopharmacol 2018; 217: 107-17.
[http://dx.doi.org/10.1016/j.jep.2018.02.021] [PMID: 29452141]

[16]   Ma X, Song L, Yu W, *et al.* Growth, physiological, and biochemical responses of *Camptotheca acuminata* seedlings to different light environments. Front Plant Sci 2015; 6: 321.

[http://dx.doi.org/10.3389/fpls.2015.00321] [PMID: 26005446]

[17]   Kim JK, Park SU. Pharmacological aspects of galantamine for the treatment of Alzheimer's disease. EXCLI J 2017; 16: 35-9.
[http://dx.doi.org/10.17179/excli2016-820] [PMID: 28337117]

[18]   Xu C, Zhang H, Mu L, Yang X. Artemisinins as anticancer drugs: Novel therapeutic approaches, molecular mechanisms, and clinical trials. Front Pharmacol 2020; 11: 529881.
[http://dx.doi.org/10.3389/fphar.2020.529881] [PMID: 33117153]

[19]   Staniek A, Bouwmeester H, Fraser PD, *et al.* Natural products - learning chemistry from plants. Biotechnol J 2014; 9(3): 326-36.
[http://dx.doi.org/10.1002/biot.201300059] [PMID: 24677691]

[20]   Miralpeix B, Rischer H, Häkkinen S, *et al.* Metabolic engineering of plant secondary products: Which way forward? Curr Pharm Des 2013; 19(31): 5622-39.
[http://dx.doi.org/10.2174/1381612811319310016] [PMID: 23394556]

[21]   Tasheva K, Kosturkova G. Role of biotechnology for protection of endangered medicinal plants Environmental biotechnology-New approaches and prospective applications. Environmental Biotechnology. Intech Open 2013; pp. 235-86.
[http://dx.doi.org/10.5772/56068]

[22]   Fazili MA, Bashir I, Ahmad M, Yaqoob U, Geelani SN. *In vitro* strategies for the enhancement of secondary metabolite production in plants: A review. Bull Natl Res Cent 2022; 46(1): 35.
[http://dx.doi.org/10.1186/s42269-022-00717-z] [PMID: 35221660]

[23]   Jha AK, Prasad K, Prasad K, Kulkarni AR. Plant system: Nature's nanofactory. Colloids Surf B Biointerfaces 2009; 73(2): 219-23.
[http://dx.doi.org/10.1016/j.colsurfb.2009.05.018] [PMID: 19539452]

[24]   Bar H, Bhui DK, Sahoo GP, Sarkar P, De SP, Misra A. Green synthesis of silver nanoparticles using latex of Jatropha curcas. Colloids Surf A Physicochem Eng Asp 2009; 339(1-3): 134-9.
[http://dx.doi.org/10.1016/j.colsurfa.2009.02.008]

[25]   Akhtar MS, Panwar J, Yun YS. Biogenic synthesis of metallic nanoparticles by plant extracts. ACS Sustain Chem Eng 2013; 591-602.
[http://dx.doi.org/10.1021/sc300118u]

[26]   Gopinath V, Priyadarshini S, Venkatkumar G, Saravanan M, Ali D. Mubarak Ali D Tribulus terrestris leaf mediated biosynthesis of stable antibacterial silver nanoparticles. Pharm Nanotechnol 2015; 3(1): 26-34.
[http://dx.doi.org/10.2174/2211738503666150626160843]

[27]   Gopinath V, MubarakAli D, Priyadarshini S, Priyadharsshini NM, Thajuddin N, Velusamy P. Biosynthesis of silver nanoparticles from Tribulus terrestris and its antimicrobial activity: A novel biological approach. Colloids Surf B Biointerfaces 2012; 96: 69-74.
[http://dx.doi.org/10.1016/j.colsurfb.2012.03.023] [PMID: 22521683]

[28]   Jeyaraj M, Rajesh M, Arun R, *et al.* An investigation on the cytotoxicity and caspase-mediated apoptotic effect of biologically synthesized silver nanoparticles using Podophyllum hexandrum on human cervical carcinoma cells. Colloids Surf B Biointerfaces 2013; 102: 708-17.
[http://dx.doi.org/10.1016/j.colsurfb.2012.09.042] [PMID: 23117153]

[29]   Premkumar J, Sudhakar T, Dhakal A, Shrestha JB, Krishnakumar S, Balashanmugam P. Synthesis of silver nanoparticles (AgNPs) from cinnamon against bacterial pathogens. Biocatal Agric Biotechnol 2018; 15: 311-6.
[http://dx.doi.org/10.1016/j.bcab.2018.06.005]

[30]   Deepika S, Selvaraj CI, Roopan SM. Screening bioactivities of Caesalpinia pulcherrima L. swartz and cytotoxicity of extract synthesized silver nanoparticles on HCT116 cell line. Mater Sci Eng C 2020; 106: 110279.

[http://dx.doi.org/10.1016/j.msec.2019.110279] [PMID: 31753355]

[31]   Francis S, Joseph S, Koshy EP, Mathew B. Green synthesis and characterization of gold and silver nanoparticles using Mussaenda glabrata leaf extract and their environmental applications to dye degradation. Environ Sci Pollut Res Int 2017; 24(21): 17347-57.
[http://dx.doi.org/10.1007/s11356-017-9329-2] [PMID: 28589274]

[32]   MubarakAli D, Thajuddin N, Jeganathan K, Gunasekaran M. Plant extract mediated synthesis of silver and gold nanoparticles and its antibacterial activity against clinically isolated pathogens. Colloids Surf B Biointerfaces 2011; 85(2): 360-5.
[http://dx.doi.org/10.1016/j.colsurfb.2011.03.009] [PMID: 21466948]

[33]   Chandra H, Patel D, Kumari P, Jangwan JS, Yadav S. Phyto-mediated synthesis of zinc oxide nanoparticles of Berberis aristata: Characterization, antioxidant activity and antibacterial activity with special reference to urinary tract pathogens. Mater Sci Eng C 2019; 102: 212-20.
[http://dx.doi.org/10.1016/j.msec.2019.04.035] [PMID: 31146992]

[34]   Kumar NH, Andia JD, Manjunatha S, Murali M, Amruthesh KN, Jagannath S. Antimitotic and DNA-binding potential of biosynthesized ZnO-NPs from leaf extract of Justicia wynaadensis (Nees) Heyne-A medicinal herb. Biocatal Agric Biotechnol 2019; 18: 1-7.

[35]   Jamdagni P, Khatri P, Rana JS. Green synthesis of zinc oxide nanoparticles using flower extract of Nyctanthes arbor-tristis and their antifungal activity. J King Saud Univ Sci 2018; 30(2): 168-75.
[http://dx.doi.org/10.1016/j.jksus.2016.10.002]

[36]   Priyadarshini S, Mainal A, Sonsudin F, Yahya R, Alyousef AA, Mohammed A. Biosynthesis of TiO_2 nanoparticles and their superior antibacterial effect against human nosocomial bacterial pathogens. Res Chem Intermed 2020; 46(2): 1077-89.
[http://dx.doi.org/10.1007/s11164-019-03857-6]

[37]   Farshchi HK, Azizi M, Jaafari MR, Nemati SH, Fotovat A. Green synthesis of iron nanoparticles by Rosemary extract and cytotoxicity effect evaluation on cancer cell lines. Biocatal Agric Biotechnol 2018; 16: 54-62.
[http://dx.doi.org/10.1016/j.bcab.2018.07.017]

[38]   Devipriya D, Roopan SM. Cissus quadrangularis mediated ecofriendly synthesis of copper oxide nanoparticles and its antifungal studies against *Aspergillus niger, Aspergillus flavus.* Mater Sci Eng C 2017; 80: 38-44.
[http://dx.doi.org/10.1016/j.msec.2017.05.130] [PMID: 28866178]

[39]   Shah M, Fawcett D, Sharma S, Tripathy SK, Poinern GE. Green synthesis of metallic nanoparticles *via* biological entities. Materials 2015; 8(11): 7278-308.
[http://dx.doi.org/10.3390/ma8115377] [PMID: 28793638]

[40]   Mittal AK, Chisti Y, Banerjee UC. Synthesis of metallic nanoparticles using plant extracts. Biotechnol Adv 2013; 31(2): 346-56.
[http://dx.doi.org/10.1016/j.biotechadv.2013.01.003] [PMID: 23318667]

[41]   Roy A, Bulut O, Some S, Mandal AK, Yilmaz MD. Green synthesis of silver nanoparticles: Biomolecule-nanoparticle organizations targeting antimicrobial activity. RSC Advances 2019; 9(5): 2673-702.
[http://dx.doi.org/10.1039/C8RA08982E] [PMID: 35520490]

[42]   Zahir A, Nadeem M, Ahmad W, Giglioli-Guivarc'h N, Hano C, Abbasi BH. Chemogenic silver nanoparticles enhance lignans and neolignans in cell suspension cultures of *Linum usitatissimum* L. Plant Cell Tissue Organ Cult 2019; 136(3): 589-96.
[http://dx.doi.org/10.1007/s11240-018-01539-6]

[43]   Anjum S, Anjum I, Hano C, Kousar S. Advances in nanomaterials as novel elicitors of pharmacologically active plant specialized metabolites: Current status and future outlooks. RSC Advances 2019; 9(69): 40404-23.
[http://dx.doi.org/10.1039/C9RA08457F] [PMID: 35542657]

[44]   Jamshidi M, Ghanati F. Taxanes content and cytotoxicity of hazel cells extract after elicitation with silver nanoparticles. Plant Physiol Biochem 2017; 110: 178-84.
[http://dx.doi.org/10.1016/j.plaphy.2016.04.026] [PMID: 27112786]

[45]   Chung IM, Rekha K, Rajakumar G, Thiruvengadam M. Influence of silver nanoparticles on the enhancement and transcriptional changes of glucosinolates and phenolic compounds in genetically transformed root cultures of *Brassica rapa* ssp. rapa. Bioprocess Biosyst Eng 2018; 41(11): 1665-77.
[http://dx.doi.org/10.1007/s00449-018-1991-3] [PMID: 30056602]

[46]   Jasim B, Thomas R, Mathew J, Radhakrishnan EK. Plant growth and diosgenin enhancement effect of silver nanoparticles in Fenugreek (Trigonella foenum-graecum L.). Saudi Pharm J 2017; 25(3): 443-7.
[http://dx.doi.org/10.1016/j.jsps.2016.09.012] [PMID: 28344500]

[47]   Logeswari P, Silambarasan S, Abraham J. Synthesis of silver nanoparticles using plants extract and analysis of their antimicrobial property. J Saudi Chem Soc 2015; 19(3): 311-7.
[http://dx.doi.org/10.1016/j.jscs.2012.04.007]

[48]   Genady EA, Qaid EA, Fahmy AH. Copper sulfate nanoparticales *in vitro* applications on Verbena bipinnatifida Nutt Stimulating growth and total phenolic content increasments. Int J Pharm Res Allied Sci 2016; 5(196): 202.

[49]   Talankova-Sereda TE, Liapina KV, Shkopinskij EA, *et al.* The Influence of Cu и Co Nanoparticles on growth characteristics and biochemical structure of Mentha longifolia *in vitro*. InNanophysics, Nanophotonics, Surface Studies, and Applications. Cham: Springer 2016; pp. 427-36.

[50]   Fazal H, Abbasi BH, Ahmad N, Ali M. Elicitation of medicinally important antioxidant secondary metabolites with silver and gold nanoparticles in callus cultures of *Prunella vulgaris* L. Appl Biochem Biotechnol 2016; 180(6): 1076-92.
[http://dx.doi.org/10.1007/s12010-016-2153-1] [PMID: 27287999]

[51]   Fazal H, Abbasi BH, Ahmad N, *et al.* Sustainable production of biomass and industrially important secondary metabolites in cell cultures of selfheal (*Prunella vulgaris* L.) elicited by silver and gold nanoparticles. Artif Cells Nanomed Biotechnol 2019; 47(1): 2553-61.
[http://dx.doi.org/10.1080/21691401.2019.1625913] [PMID: 31213081]

[52]   Javed R, Usman M, Yücesan B, Zia M, Gürel E. Effect of zinc oxide (ZnO) nanoparticles on physiology and steviol glycosides production in micropropagated shoots of *Stevia rebaudiana* Bertoni. Plant Physiol Biochem 2017; 110: 94-9.
[http://dx.doi.org/10.1016/j.plaphy.2016.05.032] [PMID: 27246994]

[53]   Chamani E, Karimi Ghalehtaki S, Mohebodini M, Ghanbari A. The effect of Zinc oxide nano particles and Humic acid on morphological characters and secondary metabolite production in *Lilium ledebourii* Bioss. Iran J Genet Plant Breed 2015; 4: 11-9.

[54]   Shakya P, Marslin G, Siram K, Beerhues L, Franklin G. Elicitation as a tool to improve the profiles of high-value secondary metabolites and pharmacological properties of *Hypericum perforatum*. J Pharm Pharmacol 2018; 71(1): 70-82.
[http://dx.doi.org/10.1111/jphp.12743] [PMID: 28523644]

[55]   Ahamed TE, Ahamed ES. Synergy prospect low gamma irradiation doses incorporating elicitation with iron nanoparticles to hyper production biomass yield and bioactive secondary metabolites for cress, medicinal plant. J Plant Sci 2018; 6: 157-63.

[56]   Moharrami F, Hosseini B, Sharafi A, Farjaminezhad M. Enhanced production of hyoscyamine and scopolamine from genetically transformed root culture of *Hyoscyamus reticulatus* L. elicited by iron oxide nanoparticles *in vitro*. Cell Dev Biol Plant 2017; 53(2): 104-11.
[http://dx.doi.org/10.1007/s11627-017-9802-0] [PMID: 28553065]

[57]   Al-Oubaidi HK, Kasid NM. Increasing phenolic and flavonoids compounds of Cicer arietinum L from embryo explant using titanium dioxide nanoparticle *in vitro*. World J Pharm Res 2015; 4: 1791-9.

[58]   Ghorbanpour M, Hatami M, Hatami M. Activating antioxidant enzymes, hyoscyamine and

scopolamine biosynthesis of Hyoscyamus niger L. plants with nano-sized titanium dioxide and bulk application. Acta Agric Slov 2015; 105(1): 23-32.
[http://dx.doi.org/10.14720/aas.2015.105.1.03]

[59]   Hussain I, Singh NB, Singh A, Singh H, Singh SC, Yadav V. Exogenous application of phytosynthesized nanoceria to alleviate ferulic acid stress in *Solanum lycopersicum*. Sci Hortic 2017; 214: 158-64.
[http://dx.doi.org/10.1016/j.scienta.2016.11.032]

[60]   Ma C, Liu H, Guo H, *et al.* Defense mechanisms and nutrient displacement in Arabidopsis thaliana upon exposure to CeO_2 and In_2O_3 nanoparticles. Environ Sci Nano 2016; 3(6): 1369-79.
[http://dx.doi.org/10.1039/C6EN00189K]

[61]   Nourozi E, Hosseini B, Maleki R, Mandoulakani BA. Pharmaceutical important phenolic compounds overproduction and gene expression analysis in *Dracocephalum kotschyi* hairy roots elicited by SiO_2 nanoparticles. Ind Crops Prod 2019; 133: 435-46.
[http://dx.doi.org/10.1016/j.indcrop.2019.03.053]

[62]   Ahmadi SZ, Ghorbanpour M, Hadian J, Salehi-Arjmand H. Impact of foliar spray of spherical nano-carbon and salicylic acid on physiological traits and parthenolide content in two feverfew cultivars (*Tanacetum parthenium* Linn cv Pharmasaat and Jelitto). Faslnamah-i Giyahan-i Daruyi 2018; 17: 82-98.

[63]   Ghorbanpour M, Hadian J. Multi-walled carbon nanotubes stimulate callus induction, secondary metabolites biosynthesis and antioxidant capacity in medicinal plant *Satureja khuzestanica* grown *in vitro*. Carbon 2015; 94: 749-59.
[http://dx.doi.org/10.1016/j.carbon.2015.07.056]

[64]   Kole C, Kole P, Randunu KM, *et al.* Nanobiotechnology can boost crop production and quality: First evidence from increased plant biomass, fruit yield and phytomedicine content in bitter melon (Momordica charantia). BMC Biotechnol 2013; 13(1): 37.
[http://dx.doi.org/10.1186/1472-6750-13-37] [PMID: 23622112]

[65]   Chandra S, Chakraborty N, Dasgupta A, Sarkar J, Panda K, Acharya K. Chitosan nanoparticles: A positive modulator of innate immune responses in plants. Sci Rep 2015; 5(1): 15195.
[http://dx.doi.org/10.1038/srep15195] [PMID: 26471771]

[66]   Asgari-Targhi G, Iranbakhsh A, Ardebili ZO. Potential benefits and phytotoxicity of bulk and nano-chitosan on the growth, morphogenesis, physiology, and micropropagation of Capsicum annuum. Plant Physiol Biochem 2018; 127: 393-402.
[http://dx.doi.org/10.1016/j.plaphy.2018.04.013] [PMID: 29677682]

[67]   Pandey A, Pandey G. Usefulness of nanotechnology for herbal medicines. Plant Arch 2013; 13: 617-21.

[68]   De Jong WH, Borm PJ. Drug delivery and nanoparticles: Applications and hazards. Int J Nanomedicine 2008; 3(2): 133-49.
[http://dx.doi.org/10.2147/IJN.S596] [PMID: 18686775]

[69]   Cuong NV, Hsieh MF, Huang CM. Recent development in nano-sized dosage forms of plant alkaloid camptothecin-derived drugs recent pat anticancer drug discov. Recent Pat Anticancer Drug Discov 2009; 4(3): 254-61.
[http://dx.doi.org/10.2174/157489209789206913] [PMID: 19522697]

[70]   Lico C, Schoubben A, Baschieri S, Blasi P, Santi L. Nanoparticles in biomedicine: New insights from plant viruses. Curr Med Chem 2013; 20(28): 3471-87.
[http://dx.doi.org/10.2174/09298673113209990035] [PMID: 23745557]

[71]   Lala S. Nanoparticles as elicitors and harvesters of economically important secondary metabolites in higher plants: A review. IET Nanobiotechnol 2021; 15(1): 28-57.
[http://dx.doi.org/10.1049/nbt2.12005] [PMID: 34694730]

[72]   Kurepa J, Nakabayashi R, Paunesku T, *et al.* Direct isolation of flavonoids from plants using ultra small anatase TIO_2 nanoparticles. Plant J 2014; 77(3): 443-53.
[http://dx.doi.org/10.1111/tpj.12361] [PMID: 24147867]

[73]   Qing LS, Shan XQ, Xu XM, *et al.* Rapid probe and isolation of bioactive compounds from *Dioscorea panthaica* using human serum albumin functionalized magnetic nano-particles (HSA-MNPs)-based ligand fishing coupled with electrospray ionization mass spectrometry. Rapid Commun Mass Spectrom 2010; 24(22): 3335-9.
[http://dx.doi.org/10.1002/rcm.4777] [PMID: 20973009]

[74]   Bonifácio BV, Silva PB, Ramos MA, Negri KM, Bauab TM, Chorilli M. Nanotechnology-based drug delivery systems and herbal medicines: A review. Int J Nanomedicine 2014; 9: 1-15.
[http://dx.doi.org/10.2147/IJN.S52634] [PMID: 24363556]

[75]   Silva Garcia Praca F, Silva Garcia Medina W, Petrilli R, Vitoria Lopes Badra Bentley M. Vitoria Lopes Badra Bentley M Liquid crystal nanodispersions enable the cutaneous delivery of photosensitizer for topical PDT: Fluorescence microscopy study of skin penetration. Curr Nanosci 2012; 8(4): 535-40.
[http://dx.doi.org/10.2174/157341312801784203]

[76]   Khuda-Bukhsh AR, Bhattacharyya SS, Paul S, Boujedaini N. Polymeric nanoparticle encapsulation of a naturally occurring plant scopoletin and its effects on human melanoma cell A375. J Chin Integr Med 2010; 8(9): 853-62.
[http://dx.doi.org/10.3736/jcim20100909] [PMID: 20836976]

[77]   Kumari A, Yadav SK, Yadav SC. Biodegradable polymeric nanoparticles based drug delivery systems. Colloids Surf B Biointerfaces 2010; 75(1): 1-18.
[http://dx.doi.org/10.1016/j.colsurfb.2009.09.001] [PMID: 19782542]

[78]   Das S, Das J, Samadder A, Boujedaini N, Khuda-Bukhsh AR. Apigenin-induced apoptosis in A375 and A549 cells through selective action and dysfunction of mitochondria. Exp Biol Med 2012; 237(12): 1433-48.
[http://dx.doi.org/10.1258/ebm.2012.012148] [PMID: 23354402]

[79]   Rajendran R, Radhai R, Kotresh TM, Csiszar E. Development of antimicrobial cotton fabrics using herb loaded nanoparticles. Carbohydr Polym 2013; 91(2): 613-7.
[http://dx.doi.org/10.1016/j.carbpol.2012.08.064]

[80]   Zheng X, Kan B, Gou M, *et al.* Preparation of MPEG–PLA nanoparticle for honokiol delivery *in vitro.* Int J Pharm 2010; 386(1-2): 262-7.
[http://dx.doi.org/10.1016/j.ijpharm.2009.11.014] [PMID: 19932160]

[81]   Paul S, Bhattacharyya SS, Samaddar A, Boujedaini N, Khuda-Bukhsh AR. Anticancer potentials of root extract of Polygala senega against benzo[a]pyrene-induced lung cancer in mice. J Chin Integr Med 2011; 9(3): 320-7.
[http://dx.doi.org/10.3736/jcim20110314] [PMID: 21419086]

[82]   Souto EB, Severino P, Santana MHA, Pinho SC. Nanopartículas de lipídios sólidos: Métodos clássicos de produção laboratorial. Quim Nova 2011; 34: 1762-9.
[http://dx.doi.org/10.1590/S0100-40422011001000009]

[83]   Kolenyak dos Santos F, Helena Oyafuso M, Priscila Kiill C, Palmira Daflon-Gremiao M, Chorilli M. Nanotechnology-based drug delivery systems for treatment of hyperproliferative skin diseases-a review. Curr Nanosci 2013; 9: 159-67.
[http://dx.doi.org/10.2174/1573413711309010027]

[84]   Chen-yu G, Chun-fen Y, Qi-lu L, *et al.* Development of a Quercetin-loaded nanostructured lipid carrier formulation for topical delivery. Int J Pharm 2012; 430(1-2): 292-8.
[http://dx.doi.org/10.1016/j.ijpharm.2012.03.042] [PMID: 22486962]

[85]   Kakkar V, Singh S, Singla D, Kaur IP. Exploring solid lipid nanoparticles to enhance the oral

bioavailability of curcumin. Mol Nutr Food Res 2011; 55(3): 495-503.
[http://dx.doi.org/10.1002/mnfr.201000310] [PMID: 20938993]

[86]   Rossetti FC, Fantini MCA, Carollo ARH, Tedesco AC, Lopes Badra Bentley MV. Analysis of liquid crystalline nanoparticles by small angle X-ray diffraction: Evaluation of drug and pharmaceutical additives influence on the internal structure. J Pharm Sci 2011; 100(7): 2849-57.
[http://dx.doi.org/10.1002/jps.22522] [PMID: 21337546]

[87]   Bernardi DS, Pereira TA, Maciel NR, *et al.* Formation and stability of oil-in-water nanoemulsions containing rice bran oil: *In vitro* and *in vivo* assessments. J Nanobiotechnol 2011; 9: 1-9.
[http://dx.doi.org/10.1186/1477-3155-9-44] [PMID: 219521076]

[88]   Desenvolvimento de nanoemulsões contendo extratos dos frutos de Syagrus romanzoffiana (Cham) Glassman e estudo fitoquímico destes extratos Brazil: Faculdade de Ciências Farmacêuticas. . Portuguese: UFOP 2010.

[89]   Odeh F, Al-Jaber H, Khater D. Nanoflora—how nanotechnology enhanced the use of active phytochemicals App Nanotechnol. Drug Deliv 2014; 343-68.
[http://dx.doi.org/10.5772/58704]

[90]   Thapa RK, Khan GM, Parajuli-Baral K, Thapa P. Herbal Medicine Incorporated nanoparticles: Advancements in herbal treatment. Asian J Biomed Pharm Sci 2013; 3: 7-14.

[91]   Zhang C, Gu C, Peng F, *et al.* Preparation and optimization of triptolide-loaded solid lipid nanoparticles for oral delivery with reduced gastric irritation Mol Molecules 2013; 18(11): 13340-56.
[http://dx.doi.org/10.3390/molecules181113340] [PMID: 24172242]

[92]   Hu L, Xing Q, Meng J, Shang C. Preparation and enhanced oral bioavailability of cryptotanshinone-loaded solid lipid nanoparticles. AAPS PharmSciTech 2010; 11(2): 582-7.
[http://dx.doi.org/10.1208/s12249-010-9410-3] [PMID: 20352534]

[93]   Lima AM, Pizzol CD, Monteiro FBF, *et al.* Hypericin encapsulated in solid lipid nanoparticles: Phototoxicity and photodynamic efficiency. J Photochem Photobiol B 2013; 125: 146-54.
[http://dx.doi.org/10.1016/j.jphotobiol.2013.05.010] [PMID: 23816959]

[94]   Odeh F, Ismail SI, Abu-Dahab R, Mahmoud IS, Al Bawab A. Thymoquinone in liposomes: A study of loading efficiency and biological activity towards breast cancer. Drug Deliv 2012; 19(8): 371-7.
[http://dx.doi.org/10.3109/10717544.2012.727500] [PMID: 23043626]

[95]   Ravindran J, Nair HB, Sung B, Prasad S, Tekmal RR, Aggarwal BB. RETRACTED: Thymoquinone poly (lactide-co-glycolide) nanoparticles exhibit enhanced anti-proliferative, anti-inflammatory, and chemosensitization potential. Biochem Pharmacol 2010; 79(11): 1640-7.
[http://dx.doi.org/10.1016/j.bcp.2010.01.023] [PMID: 20105430]

[96]   Solanki SS, Sarkar B, Dhanwani RK. Microemulsion drug delivery system: For bioavailability enhancement of ampelopsin. ISRN Pharm 2012; 2012: 1-4.
[http://dx.doi.org/10.5402/2012/108164] [PMID: 22830055]

[97]   Russi S, Verma HK, Laurino S, *et al.* Adapting and surviving: Intra and extra-cellular remodeling in drug-resistant gastric cancer cells. Int J Mol Sci 2019; 20(15): 3736.
[http://dx.doi.org/10.3390/ijms20153736] [PMID: 31370155]

[98]   Singh A, Dilnawaz F, Mewar S, Sharma U, Jagannathan NR, Sahoo SK. Composite polymeric magnetic nanoparticles for co-delivery of hydrophobic and hydrophilic anticancer drugs and MRI imaging for cancer therapy. ACS Appl Mater Interfaces 2011; 3(3): 842-56.
[http://dx.doi.org/10.1021/am101196v] [PMID: 21370886]

CHAPTER 14

Metallic Nanoparticles Synthesized Through Medicinal Plants: Therapeutic Improvement

Bushra Hafeez Kiani[1,*]

[1] *Department of Biological Sciences (Female Campus), International Islamic University, Islamabad, 44000, Pakistan*

Abstract: The field of nanotechnology has developed new medicinal nanoparticles that have various uses in pharmaceutics and healthcare. Distinct macroscopic and microscopic entities including plants, fungi, microalgae, bacilli, and seaweed have been used to biosynthesize nanoparticles. Naturally-occurring chemicals like flavonoids, tannins, alkaloids, steroids, and saponins are abundantly present in plants. A potentially unharmful method to produce nanoparticles can be through extracts of different plants. As plant extracts carry many specialized metabolites, they can act as stabilizers and reducers in bioreduction reactions that take place in metallic nanoparticle production. The production of metallic nanoparticles by biological techniques is an easier, cheaper, and more environmentally sound option in comparison with other physical and chemical techniques that are extremely toxic and unsafe for biological use. Greener nanoparticles like Co, Cu, Ag, Pd, Au, ZnO, Pt, and Fe_3O_4 have been biosynthesized using medicinal plants. These nanoparticles have various uses in pharmaceutics ranging from gene delivery, drug delivery, pathogen detection, tissue engineering, and protein detection. Not only that but, metallic nanoparticles can also potentially be remedies to different acute diseases including hepatitis, human immunodeficiency virus, malaria, and even cancer. Improvements in drug delivery and tissue engineering have been made possible by nanotechnology and this has greatly facilitated translational level studies that relate to pharmaceutics. In this chapter, green syntheses of metallic nanoparticles through medicinal plants along with their uses in therapeutic improvements are described.

Keywords: Medicinal plant, Metallic nanoparticles, Nano-packaging, Nano-degradation, Phytotherapy, Reducing agent.

INTRODUCTION

Phytotherapy employs plant extracts as medicines to cure diseases. The majority of phytochemical formulations have been in use long before contemporary medicines, since the Stone Age. Improvements in the phytotherapy domain by

---
* **Corresponding author Bushra Hafeez Kiani:**Department of Biological Sciences (Female Campus), International Islamic University, Islamabad, 44000, Pakistan; E-mail: bushra.hafeez@iiu.edu.pk

Zulqurnain Khan, Azra Yasmin & Naila Safdar (Eds.)
All rights reserved-© 2023 Bentham Science Publishers

different techniques and procedures have enabled the engineering of new drugs through conventional formulations. Conventional healthcare approaches include several concepts which are focused on the balance between the atmosphere and body functions [1]. Medicinal plants are being used for centuries; however, their existence has been imperiled recently thanks to contemporary approaches to alternative medicine. Approximately 85 percent of the global populace utilizes phytotherapy and plant-derived chemicals and medicines for treating a variety of illnesses [2]. A majority of phytoconstituents have been utilized for manufacturing medicinal goods since the chemicals produced in plants have a variety of recognized advantages [3]. These chemicals, also called specialized metabolites, are not direct contributors to the developing stages of a plant but rather are involved in the defensive mechanism of plants.

Secondary metabolites assist in plant-surrounding interactions thus giving plants a greater survival rate. The development of these metabolites has enabled intimate interactions of plants with a variety of prokaryotes [4]. Many works of research conducted on plants suggest that these specialized metabolites could be used as raw materials in artificial or chemically synthesized biologically active pharmaceutical compounds. Secondary metabolites have been discovered to be structurally comparable to chemical messengers or hormones. This offers a diverse spectrum of plant-based therapeutically important compounds. Furthermore, such metabolites are prevalent in combinations, this increases the absorption and effectiveness of the compounds while minimizing crop resistance [3]. Species possessing powerful therapeutic qualities include *Curcuma longa* (reduces inflammation and risk of cancer), *Withania somnifera* (increases bile pigment concentrations), *Aloe barbadensis* (reduces inflammation), *Ocimum tenuiflorum* (reduces toxicity), and *Zingiber officinale* (aids in digestion) [5].

Garcinia indica, Magnolia champaca, creeping rootstalk of *Curcuma aromatica, Vernonia anthelmintica* seedlings, and *Nyctanthes arbourtristis* foliage, are well-known plant species utilized in fighting different infections [6]. *Ruta graveolens* and *Ambrosia arborescens* foliage has been used as excellent blood purifiers. The bloom and foliage of *Borago officinalis* are used for treating sore throats, coughs, and other related problems. When compared, most components of *Matricaria recutita L.* are reported for alleviating abdominal pain as well as similar gastrointestinal problems [7]. Specialized metabolites include alkaloids, flavonoids, isoprenoids, and phenols hold great therapeutic potential. Isoprenoids are compounds present in nature that act as transmitters among symbionts, multiple plant bodies, or plant species with pollinators [8].

Isoprenoid-rich plant species have several therapeutic characteristics, including chemoprevention, bactericidal, fungicidal, virucidal, and inflammation-reducing activities [9]. Peppermint camphor, cannabinoid, and qinghaosu are examples of isoprenoids that have been widely utilized in pharmaceutical medicines. Paclitaxel, a well-known chemotherapeutic drug recognized for its mitosis-inhibiting properties in cancerous tissues, is an isoprenoid derived from *Taxus brevifolia* [10]. Alkaloids are nitrogenous plant compounds often produced as wound responses. Therefore, alkaloids have been widely used in medicine development in contemporary healthcare. Researchers have reported bactericidal, inflammation-reducing, and antioxidant properties in alkaloids derived from diverse plants like *Rutaceae* or *Solanaceae*. Cocaine, galantamine, caffeine, calotropin, quinine, and nicotine and phenols are the most common type of specialized metabolite present in a broad variety of plant species [11].

Secondary metabolites in addition to their role in providing pigment and flavor are extensively researched in regard to their antioxidative, inflammation-reducing, antibacterial, immune-modulating, and disinfecting characteristics. Phenols derived from plants are well-recognized for their ability to induce oxidative stress and potentially interfere with intracellular processes. Plant phenols like ellagic acid and glucose ester are present in berry plants like strawberries and have been shown to have chemotherapeutic effects on prostate, mouth, and colorectal carcinoma. This further insinuates the biological importance of plant-based phenolic compounds [12]. Flavonoids are plant polyphenolic compounds essential for the coloring of flowers. Flavonoids are frequently utilized for medicinal use because of their virucidal, anti-allergy, cancer-fighting, oxidation-inhibiting, and inflammation-reducing abilities [8]. Quercetin, a pharmaceutically important flavonoid, possesses oxidation-inhibiting and inflammation-reducing properties that could be utilized for preventing tissue injuries [13].

Limonin and naringin are citrus-based flavonoids that show chemotherapeutic effects and can scavenge free radicals [12]. Specialized metabolite classes such as aetheroleums, glycosides, tannins, and saponins have been in use in conventional medicine. Other metabolites like flavonoids and alkaloids have been reported to act as reducers and stabilizers to sustainably produce nanoparticles. Following nanoparticles have increased capability of reducing metal salts plants that contain a high ratio of secondary metabolites. Microscopy, Zeta potential, Dynamic Light Scattering, and Ultraviolet-Visible spectroscopy are some of the techniques that arc commonly used to study nanoparticles. Nanoparticles that are created are coated with different required medicinal metabolites that have various benefits. They are easily used in food, remediation, medicinal, agricultural, environmental, and pesticide industries due to increased surface area and small size, and high reactivity ratio. Synthesis of both plant-based nanoparticles and green carbon dots that could be applied as nanotherapeutics agents have been made possible due to advancements in the field of green nanotechnology. Insufficient data on toxic levels of nanoparticles and its impact on the environment has affected the expansion and large-scale use of green nanoparticles [14].

Distribution of Medicinal Plants

Approximately 7.2 percent of plant species that are known are used as herbal medicines due to their medicinal properties [15]. According to the statistics reports, these plants are not grown evenly throughout the world. Rural areas of Africa, some areas of Asia, Central America and South America greatly depend on herbal medicines. According to studies conducted in 2017, 90 percent of the people living in Germany depend on herbal medicines. India, China, and European countries are the places where the highest ratio of herbal plants is seen due to conservation methodologies *e.g., in-situ* and *ex-situ.*

As medicinal plants are now used in pharmaceutics, different techniques like *in-situ* and *ex-situ* conservation, synthetic seed technology, and plant tissue culture are applied to conserve these plants. The growth of annual plants and plant industry-based economy is also enhanced by these techniques [16]. According to the work done by Schippmann (2002), some species of *Apocynaceae, Menispermaceae, Asclepiadaceae, Guttiferae, Apiaceae,* and *Canellaceae* are greatly important for medicinal plants. Countries including China, the USA, Nepal, the Philippines, Malaysia, Thailand, Pakistan, India, Vietnam, Indonesia, and Thailand are using approximately 52,885 species of plants that are recognized for medicinal importance [17].

There is increased awareness of side effects and toxic levels of synthetic medicines. Countries like Western Europe, South- East Asia, the USA, and Japan are extensively working on medicines derived from plants. 60-90 percent of herbal plants are cultivated, the ratio of plants that are harvested from the wild is low, and cultivated species are in the ratio of 3-6 percent, expressing that some species are dominating in the trade of medicinal plants [17]. *Phyllanthus emblica, Zingiber officinale, Glycyrrhiza glabra, Berberis aristata, Senna alexandrina, Withania somnifera, Strychnos vomica, Terminalia bellirica,* and *Eclipta prostrata* are major plants on which Indian pharmaceuticals are working [18]. Different parts of plants are used as infusions, oral sprays, tonics, syrups, and decoction to cure infertility and blood pressure. *Cinnamomum camphora, Aloe vera, Mentha piperita, Momordica charantia,* and *Gentiana lutea* are also used for this purpose [19].

According to the research by Schippmann (2002), the leading countries that import medicinal plants are Hong Kong, Germany, Japan, and the USA and they spend almost 1 billion USD on imports. Export of approximately 600 million USD is done by China, Germany, the USA, and India [17]. In 2013, there was a

10.7 percent increase in trade market value and that is about 6.2 billion USD. Forty-five percent of total imports are contributed by the USA, Japan, Hong Kong, and China and they were the top importers. While the top exporters were China, India, Hong Kong, the USA, and Germany, and contributed about 52.7 percent of export [20]. The total revenue of the entire world made by medicinal herbs or their output in 2010 was 60,000 million US dollars and this value is likely to increase to 5,000 billion US dollars by 2050 [21].

Metal Nanoparticles and Medicinal Plants

Green synthesis of biomolecules involves the use of molecules that are part of bacteria, plants, and yeast as reducing agents, and hence the process is made efficient and environment friendly. Particles that are smaller than 100 nanometers in diameter have unique antimicrobial, anticancer, anti-larval, and catalytic properties. The isolation process of plant extracts is easy and molecules like proteins, polyphenols, alkaloids, terpenoids, phenolic acid, and sugars work as reducing agents. Products produced by these reducing agents function for stabilizing particles that are formed.

For the production of nanoparticles, plant extracts are mixed with the required metal ions in the solution. When metal nanoparticles are formed from plant extracts, there is a change in color during incubation. A reduction in metal ions results in color change as it induces effects in Surface Plasmon excitation. Depending on the conditions in which these nanoparticles are formed, they may have varying colors, shapes, and sizes. Different shapes that have been reported under different conditions are rod-shaped, triangular, spherical, and polyhedral [22].

Nanoparticles usually consist of three layers; the core consists of a reduced metal surface that is covered by a shell and surface layer. The surface layer is responsible for the reactions and functions of particles, moreover, it is the area from where capping molecules are fixed and work for particle stability. The shape of a nanoparticle affects its properties to a great extent as it controls the level/amount of content between the surface layer and surroundings [23, 24]. The use of *Magnolia kobus* leaves for synthesizing copper nanoparticles using copper sulfate pentahydrate as the metal ion was studied. The presence of copper nanoparticles of size 45-110 nanometers covered with molecules from the extract was confirmed by techniques like UV-Visible Spectroscopy, High-Resolution Transmission Electron Microscopy, Energy Dispersive X-Ray Spectroscopy, and Scanning Electron Microscopy [25].

Different concentrations of nanoparticles that could have anti-larval properties against *Anopheles stephensi, Aedes aegypti,* and *Culex quinquefasciatus* were prepared. Depending upon the stages of development, the death rate was between 89-100 percent and statistical tests proved that nanoparticles have anti larval and adulticidal effects [26]. The antimicrobial efficacy of the copper nanoparticles toward *Escherichia coli* ranged between 70 to 99 percent, confirming their utility in clinical application as powerful antibacterial drugs [25]. Researchers found that copper nanoparticles produced with extracts of *Lawsonia inermis* leaves and a mixture of copper ions seemed to have the necessary dimension as well as morphological characteristics to function as conductors in an aqueous medium, this contains a wide range of uses in physical and material science techniques [27]. The antimicrobial activity of ZnO nanoparticles produced using *Hibiscus subdariffa* foliage and a zinc acetate solution was effective towards *Subtilis aureus* and *Escherichia coli*. The hypoglycemic effects of these nanoparticles were also studied using *Mus musculus*. Despite extremely minimal levels of nanoparticles, this research indicated that zinc oxide nanoparticles reduced bloodstream glucose concentrations inside the examined *Mus musculus* [27, 28]. Iron-polyphenol complex nanoparticles obtained from extracts of *Salvia officinalis* foliage can be used to remove flocculants due to their absorption capability. It was confirmed that nanoparticles have a positive charge on their surface and dye removal activity was effective for ethyl violet [29] (Fig. **1**).

Major applications for which plant nanoparticles are used are enhanced levels of plant growth. Increased level of seed growth and seed germination in the wheat seed are seen by using zinc oxide nanoparticles obtained from *Aloe barbadensis* [30]. In recent agriculture, an increase in seed properties and the rate of plant growth by using nanoparticles holds much importance.

DIFFERENT PLANT PARTS USED FOR PRODUCING METALLIC NANOPARTICLES

The physicochemical properties of plant-based nanomaterials have got more attention due to their extensive applications. Metallic nanoparticles like platinum, gold, titanium oxide, silver, copper, nickel, zinc, and magnetite are obtained from natural resources and are studied in detail. Different parts of plants are used for preparing metallic nanoparticles in different shapes and sizes by using different biological techniques. Large changes in metal concentration and quantity of plant extract in a medium cause alteration of biosynthetic reactions and hence changes in the shape and size of nanoparticles [31] (Table **1**).

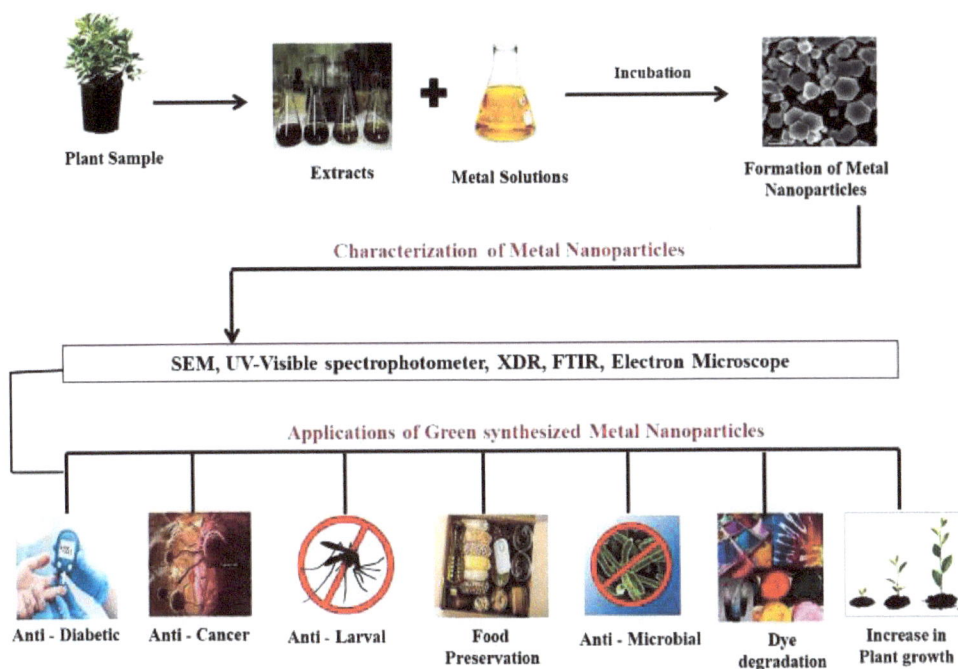

Fig. (1). Diagrammatic representation of isolation, characterization and applications of nanoparticles synthesized from Plants.

Table 1. Well-known medicinal plants used for the synthesis of metal nanoparticles.

S. No.	Plant Species	Metals Used	Applications	References
1	*Glycyrrhiza glabra*	Silver	Anti-ulcer properties	[41]
2	*Strychonos vomica*	Silver	Antimicrobial properties	[42]
3	*Bacopa monnieri*	Platinum	Antioxidant activity and Neuro-protective ability	[43]
4	*Eclipta prostrate*	Gold	Antibacterial properties	[44]
5	*Terminalia bellirica*	Gold	Lethality against Brine shrimps	[45]
6	*Withania somnifera*	Silver	Antifungal, bactericidal effect	[46]
7	*Zingiber officinale*	Silver	Antibacterial activity	[47]
8	*Hibiscus subdariffa*	Zinc Oxide	Antibacterial activity	[48]
9	*Camellia sinensis*	Iron	Degrade common dye contaminants	[49]
10	*Salvia officinalis*	Iron	Good adsorption capability	[50]

Stem as a Source for Nanoparticle Synthesis

Stem methanolic extract of *Coleus aromaticus* is used for the synthesis of silver nanoparticles to create Ag and *Coleus aromaticus* complex. Plants extract consists of an aldehyde group that reduces silver ions into metallic silver nanoparticles. Different functional groups like AC,O, C, and N indicate amide I, polypeptides that are responsible for capping ionic substances into metallic nanoparticles [32]. The biosynthesis of silver crystals at a molecular level is being studied. But some models on the mechanism of nanoparticles are being proposed in pathogenic organisms.

Fruits-mediated Synthesis of Metallic Nanoparticles

Plant fruit body extract with silver nitrate solution of varying concentrations of *Tribulus terrestris* was used [48] to get environment-friendly silver nanoparticles having specified morphology. The extract consists of active phytochemical compounds that could be used for single-step reduction reactions [33]. Palladium nanoparticles can be achieved by using polyphenols from grapes and they also work efficiently against bacterial diseases [34]. Optimum physiochemical parameters to create nanomaterial can be effectively used as a threat to different endemic diseases.

Seeds as Source

Extract of fenugreek seed consists of high flavonoid levels and different bioactive compounds like saponin, lignin, and vitamins. By using powerful reducing agents to reduce chloroauric acid, fenugreek seed extract can work as a better surfactant. Seed extract contains COO_, C,N, and C,C functional groups. Functional groups present in metabolites work as surfactants of gold nanoparticles and flavonoids stabilize the electrostatic stabilization of gold nanoparticles [35].

Leaves-mediated Synthesis of Nanoparticles

The use of plant extracts as mediators for obtaining nanoparticles has been declared. Different plant leaves like *Murraya koenigii, Centella Asiatica and Alternanthera sessilis* have been observed. *Piper nigrum* leaves contain a biologically active compound that is involved in the synthesis of nanoparticles by an environmentally friendly approach. Longumine and Piper longminine are extracts of *Piper nigrum* that have the ability to function as a capping agent for the biosynthesis of silver nanoparticles to increase the cytotoxic effect of cancerous cells [36]. Synthesis of nanoparticles using leaves extracted from plants like Artemisia nilagirica has been reported [37].

Flowers as a Source for NPs Production

Environment-friendly synthesis of gold nanoparticles by rose petals was studied. The maximum concentration of sugars and proteins is present in the medium extract. Due to these functional groups, we can reduce tetrachloroaurate salt into bulk gold nanoparticles (GNPs) [38]. Similarly, groups of plants having a wide variety like *Catharanthus roseus* and *Clitoria ternatea* are used to obtain metallic nanoparticles of desired shape and size. *Nyctanthes arbortristis* and *Mirabilis jalapa* flower extracts are used as reducing agents for the synthesis of gold nanoparticles [39, 40].

APPLICATIONS OF PLANT-BASED METAL NANOPARTICLES

Antimicrobial Agent

Due to the antimicrobial characteristics of silver, it is used to source metallic nanoparticles. Increased surface area and capped surface by active substances of plant extracts with silver nanoparticles result in increased medicinal properties. Silver nanoparticles work for maximum microorganisms like monoderms and diderms. They work by creating pores in the bacterial cell wall or denaturation of proteins [51]. Antimicrobial effects against monoderms like *Staphylococcus* sp. and diderms like *Pantoea agglomerans*, *Klebsiella* sp., and *Rahnella* sp. in low concentrations are shown by silver nanoparticles of size 10-20nm obtained by leave extracts of *Impatiens balsamina and Lantana camara* [24].

Antimicrobial properties of silver nanoparticles (10-70 nm) obtained from the fruit extract of *Emblica Officinalis* against pathogens of human-like *Escherichia coli, Klebsiella pneumonia, Staphylococcus aureus,* and *Bacillus subtilis* were reported in the literature. It is confirmed by Fourier Transform-Infrared Spectroscopy that phytoconstituents of plant extracts can work as capping agents and are there on particle surfaces [52]. Silver nanoparticles of size 14.6 nanometers are obtained from extracts of *Ocimum tenuiflorum* foliage that express antibacterial properties against diderms. Plant extract of the same plant produces silver nanoparticles of size 25 nanometers which shows antibacterial properties and the zone of inhibition is shown at 14mm for fungi like *Candida albicans* and monoderms like *Escherichia coli* exhibiting a zone of inhibition of 13 mm [53]. The antibacterial property of nanoparticles obtained from extracts of *Azadirachta indica* foliage is increased 2.8 times if silver nanoparticles are incorporated into the GIC without having any effect on any other medical quality of cement [54]. Silver nanoparticles are very important for covering wounds that stop the chances of bacterial infection and express good signs of healing. Particles were mostly present on the surface meaning that they did not penetrate the cell; moreover, cell death was also not reported [55]. Improved bactericidal properties and an increase in fibroblast migration resulting in a good healing process were seen by using nanoparticles of size 20 nanometers and obtained from the extracts of *Melia azedarach* foliage [56].

Nanoparticles can also be used in tissue engineering in which they are used on prosthetics to reduce rejection chances. Nanoparticles of titanium are found to be great for the bone healing process. Other fields in medicines where they can be applied include an increase in the drug delivery process, in which specified receptors are involved to deliver drugs correctly [57]. Polymethyl methacrylate, Glass polyalkenoate (ionomer) cement, and calcium phosphate cement are some of the bone cements used in bone surgeries to bind prosthetics and bone and help in proper binding and shock absorption [58]. Sometimes bone cement can cause complications after surgery due to increased bacterial infection because of biofilm formation and hence resulting in transplant rejection. Silver nanoparticles obtained from *Cucurma aromatica* rhizomes have antibiofilm properties (94% at 50 micrograms per milliliter) that are effective in coating Polymethyl methacrylate [59]

Anti-larval Activity

Natural resources, especially plants have greatly affected the formation of insecticides against vectors, mainly mosquitos. Currently, the main focus is on the synthesis of gold or silver nanoparticles from plants to control mosquitoes and their effect on mosquito larvae [60].

Nanoparticles obtained from different natural sources can thus be used to get many different insecticides and insect repellents without using any toxic compound. Silver nanoparticles were obtained from both dry and fresh foliage and fruit extracts of *Solanum nigrum L.* to study the anti-larval property of nanoparticles against *Anopheles stephensi* and *Culex quinquefasciatus*. It was confirmed that silver nanoparticles have both anti-larval and antibacterial properties when they were tested on *Staphylococcus aureus*, *Bacillus mycoides*, *Bacillus subtilis*, *Pseudomonas putida*, *Escherichia coli*, *Aeromonas salmonicida*, *Pseudomonas aeruginosa*, *Bacillus licheniformis*, and *Pseudomonas fluorescens* and hence they can be used in the field of medicine against different diseases that are due to pathogens [61].

According to recent studies, green nanoparticles obtained from plants hold an important place in the fields of biomedical and environmental studies. It was studied by Shelar that silver nanoparticles separated from the fruit peel extract of *Momordica charantia* have helminthicidal and larvicidal activity against *P. posthuma* and *A. albopictus, A. aegypti* respectively. Even 20 parts per million destroyed 85 percent of *A. albopictus* and *A. aegypti* larvae and very low concentrations were not toxic [62].

Antiviral Effects of Metallic Nanoparticles

Plants-based nanoparticles are an alternate way of getting drugs that are used for addressing and monitoring the growth of different pathogens that have viral properties. Entry of a virus in a host cell can be very dangerous as it starts the translational process at a faster speed so that it can increase in number. The biosynthesis of silver nanoparticles can work as a powerful broad-spectrum antiviral agent that can stop virus-cell function. Bio-silver nanoparticles that have the persuasive anti-HIV ability at early stages of reverse transcription mechanism were studied [63]. Metallic nanoparticles obtained from natural sources have various binding sites that can bind with the gp120 of the virus membrane and help to stop viral function. Bio-based nanoparticles are working effectively for cell-free viruses and cell-associated viruses. Moreover, post-entry stages of the HIV-1 life cycle are being stopped by silver and gold nanoparticles. Hence, metallic nanoparticles can work effectively as antiviral drugs against retroviruses [64].

Anti-inflammatory Action of Nanoparticles

One of the important wound healing mechanisms is anti-inflammatory. Anti-inflammation is a series of steps that produce immune-responsive molecules like cytokines and interleukins that can be produced from keratinocytes such as B lymphocytes and macrophages. Endocrine systems secrete many inflammatory mediators like antibodies and enzymes. Primary immune organs are responsible for the secretion of other anti-inflammatory agents like Interleukin-1, Interleukin-2 and these mediators speed up the healing process. Moreover, inflammatory mediators are used in biochemical pathways and to control the spread of diseases.

In positive wound repair mechanisms and tissue regeneration during inflammatory functions, gold nanoparticles obtained from plants can be utilized. Gold and platinum nanoparticles obtained from natural sources can be used as a good alternative to treat inflammation in a natural way [46, 45].

Anti-diabetic Agent

Enzymes like pancreatic α-amylase and intestinal α-glucosidase release glucose in the blood after the reduction of oligosaccharides and disaccharides, to treat diabetes that is not depending upon insulin concentration. This glucose level can be reduced by using nanoparticles obtained from plants to inhibit the function of these enzymes [65]. Copper nanoparticles synthesized from *Plumbago zeylanica* and *Gnidia glauca* have the best inhibitory effect against porcine α-amylase

(50.99 ± 4.27 percent) and α-glucosidase (88.60 ± 0.78 percent) as compared to standard Acarbose [66]. Antidiabetic effects were observed using zinc oxide nanoparticles obtained from extracts of *Nasturtium officinale* foliage in alloxan-diabetic Wister rats. Metallic nanoparticles that are capped with metabolites obtained from extracts of *Nasturtium officinale* foliage decrease glucose levels in the blood and increase insulin levels whereas zinc oxide nanoparticles without extract express poor results, showing how important plant extracts are for the anti-diabetic effect of nanoparticles [67].

According to Rajkumar (2018) nanoparticles obtained from the leaf extract of *Andrographis paniculata* when mixed with α- amylase enzyme and incubated along with starch and dinitro salicylic acid reagent express anti-diabetic activity, showing that it can be used as a powerful anti-diabetic molecule [68]. *Tephrosia tinctoria* can be used to synthesize silver nanoparticles to evaluate their anti-diabetic and anti-oxidative properties. Nanoparticles obtained from plants have effective anti-diabetic properties as they inhibit carbohydrate digestion [69]. Similarly, gold nanoparticles were synthesized by reducing auric chloride with *Cassia auriculata*. The study shows that gold nanoparticles obtained from plant sources can be effective as anti-diabetic drugs [70]. Anti-diabetic properties were also observed in streptozocin-induced diabetic mice after using zinc oxide nanoparticles. Nanoparticle exposure caused a decrease in blood sugar levels in mice (59.58% and 48.27%) for the plant extract sintered at 60° (PZN60) and 100° (PZN100), respectively [27].

Anti-cancer Agent

The unchecked division of cells is known as cancer which results in tumor formation as cells lose their ability to apoptosis. Nanoparticles start apoptosis in cells by ROS generation resulting in the damage of organelle. Nanoparticles obtained from plants initiate the activity of apoptotic proteins and stop the working of anti-apoptotic proteins by starting extrinsic and intrinsic pathways of apoptosis. These pathways kill cells and decrease tumor progression and growth. Silver nanoparticles obtained from *Nepeta deflersiana* extracts can be employed as anticancer agents against HeLa cell lines. Cell shrinkage, increase in apoptosis, loss of adhesion, induction in mitochondrial membrane potential (121 percent at 10 micrograms per milliliter), decrease in glutathione level (69 percent at 25 micrograms per milliliter), increase in ROS (207 percent at 10 micrograms per milliliter) and increase in lipid peroxidation levels (65 percent at 25 micrograms per milliliter) are some of those events that are observed in a cell that is treated with metallic nanoparticles for one day [71]. According to these results, nanoparticles obtained from plants have different properties like inducing oxidative stress that causes the death of neoplastic cells.

Silver nanoparticles obtained from the extract of *Rheum ribes* were observed for anti-cancer properties on breast carcinoma cell lines (MDA-MB-231). After incubation of 48 hours, the observed rate for cell survival was 50 percent, and on contact with metallic nanoparticles cancer cells undergo apoptosis and the IC50 value comes out to be 98.96 micrograms per milliliter [72]. Apoptotic properties against renal carcinoma cell lines A498 and Sw-156 were studied using gold nanoparticles obtained from *Curcuma wenyujin* extracts. Mitochondrial and nuclear damage was seen immediately after contact of the cell with nanoparticles and was treated instantly by apoptosis. CC50 value came out to be 25 micrograms per milliliter and 40 micrograms per milliliter, respectively [73]. Silver nanoparticles obtained from *Withania coagulans* extracts experimented against SiHa cell lines. Nanoparticles are cytotoxic and cells start apoptosis within 2 days of incubation. IC50 value came out to be 13.74 micrograms per milliliter and the cells showed shrinkage and cell adhesion [74]. Gold and silver nanoparticles can be obtained by treating *Butea monosperma* with $HAuCl_4$ and $AgNO_3$ consecutively and were then examined for their biocompatibility with normal endothelial cells and cancer cell lines (B16F10, MCF-7, HNGC2 & A549). Synthesized nanoparticles were then used to develop the required setup for drug delivery using FDA-approved doxorubicin and were then studied for targeted drug delivery and also seen to stop cell proliferation [75].

ADDITIONAL APPLICATION OF METAL NANOPARTICLES

Food Preservation

Nanoparticle-based polymeric layers are extensively used in food preservation. Research on this technique using cellulose or sodium alginate has expressed good results having less food spoiling bacteria. But toxic levels of using nanoparticles in food preservation are still not completed [76]. Silver nanoparticles obtained from extracts of *Fatsia japonica* foliage are applied as a preservative agent for fruit belonging to the citrus family. Nanoparticles have significantly reduced rotting caused due to *Penicillium italicum* [77].

Nanocoating on packed or stored foodstuff reduces spoilage as it acts as a barrier when food is exposed to certain environments like moisture, air, water, sunlight, and humidity. Hence, it keeps the original color, taste, and texture of foodstuff, which is extremely important for packed food or export purposes. Nanocoatings may be different types like nanofibers, nanotubes, or nano layering. The coating process has increased the total life of food materials [78]. Antioxidant properties seen in silver nanoparticles obtained from *Aesculus hippocastanum* foliage extract were more as compared to silver nanoparticles obtained from plant extracts [79]. Silver nanoparticles of size 17 nanometers obtained from *Zingiber officinale* extract expressed increased antibacterial and antioxidant properties, showing the importance of silver nanoparticles in preserving foodstuff [80]. Metallic nanoparticles have been frequently utilized in the food preserving and hygiene domain. Food preservation includes the concepts of encapsulating food, wastewater purification, and moisture control, while food safety includes disease resistance, preventing allergies, biofilm production, as well as the reduction of metallic ion concentrations in foodstuff. Improving the nutritional values and changing the morphology of foods are included in food functionality [81]. Moderately-sized oily triglycerides may be generated by refining specific organic materials like palm and coconut. They could be employed as a nano-encapsulating substance against consumable goods. Antimicrobial properties of *Coffea arabica* foliage-assisted biosynthesized silver nanoparticles were tested against *Escherichia coli* and Staphylococcus aureus which are common contaminants of poultry items, meats, milk, and other *dairy items*. As a result, prior to packing, enclosing food products using nanoparticles can extend their storability [82].

Dye Degradation

Many industrial sectors, such as the textiles, dyeing, and suede industries, dump dyes into aquatic ecosystems. Approximately 20 percent of the dyes employed are wasted due to the poor uptake rate of the colored material. Some dyeing pigments, like azo dyes, build up in the aquatic environment but do not break down in nature. These could be cancerous, mutation-inducing, or act as an allergen. Life forms may indeed be damaged by these colored pigments as well as prevent plants inside the aquatic environment from absorbing light from the sun [83].

Nanoparticles sourced through plants in the presence of sun or ultraviolet light exhibit photocatalytic properties against dyes. Metallic nanoparticles, on the other hand, can catalyze processes driven by chemical reducers such as sodium borohydride. Plant-based metallic nanoparticles have increased active sites for the degrading process as they have increased surfaces. The electrons are excited by the light from the sun which strikes the particles, forming reactive species which degrade the dyes. Due to the substantial discrepancies in redox potentials, degrading reactions driven by sodium borohydride is not beneficial kinetic models. The nanoparticles, on the contrary, serve as catalysts inside the electron-transferring process, therefore, enabling the process [83].

Zinc nanoparticles of 29 nanometers, produced from extracts of *Ziziphus* fruits, have been found to break down popular dyes like Methylene blue and Eriochrome Black T approximately to 85 percent in a first-order process under the sun and can be reprocessed 4 times [84]. The dye-degrading characteristics of silver nanoparticles produced by the extract of *Alpinia nigra* fruits were investigated against popular dyes including Methyl Orange, Orange G, and Rhodamine B. In the presence of the sun, the degrading rate was reported to be 85.9, 83.4, and 79.9 percent, sequentially. The nanoparticles had been repurposed thrice with no problems [85]. The dye degrading of Nile blue and reactive yellow160 was studied with copper oxide nanoparticles made from the extract of *Psidium guajav* foliage. After 2 hours of sun exposure, the deterioration levels reached 93 percent and 81 percent, correspondingly [86].

Camellia sinensis-sourced iron nanoparticles were utilized for breaking down Malachite green under varied temperatures, dye concentrations, and pH . This assay was carried out at 3, 6, and 9 pH. The efficiency of removal increased with the pH, as pH 9 showed about 100 percent effectiveness. The effectiveness of the dye decreased from 97 percent to 84 percent as the quantity rose from 30 to 200 milligrams per liter. The heat seemed to have an influence as well, as the removal effectiveness improved after the temperature was raised from 288 to 308 Kelvin. According to the hypothesis, the dye's breakdown mechanism is attributable to its surface adhesion on the nanoparticle and subsequently oxidizing the iron present in the NPs [87].

Plant Growth Enhancers

Nano-agriculture involves nanoparticles produced by various metals of several forms and dimensions to boost the synthesis of specialized metabolites and plant development. Nanoparticles affect plants as changes in genetic expression cause changes in the expression of several biological processes that affect development. When nanoparticles were given to plants, they were reported to enhance reactive oxidative species molecules by generating a burst of oxidative stress; this raises the level of antioxidants in the crops that might impact the synthesis of specialized metabolites. This could probably be due to the nanoparticles' antibacterial characteristics, which could influence the topsoil flora, and increase the bioavailability or absorption of nutrients. Metallic nanoparticles have been reported as hazardous if found in excess quantities because they break the cellular walls of the plants. However, small quantities of nanoparticles could penetrate the walls of plants *via* plasmodesmata and affect development [88].

An optimal number of nanoparticles is determined by the plant's pore size and the metal. Sulfur nanoparticles produced utilizing 20-nanometer extracts of *Melia azedarach* have been investigated for the enhancement of plant development in *Cucurbita pepo*. The height of stems, length of roots, branch volume, and thickness of stems, all grew dramatically after the nanoparticles were administered. In comparison with the control group, the total height of the plant's roots and shoots increased thrice. They hypothesized that sulfur nanoparticles reduced the number of harmful topsoil microbes, hence lowering illness rates in crops while boosting topsoil sulfur bioavailability [89].

Gloriosa superba was given gold nanoparticles made using the extract of *Terminalia arjuna* to evaluate the improvement in the rate of germinating seeds. *Gloriosa superba* showed an increase in the rate of germinating seeds to 39.67 percent and the rate of development was almost twice as fast after being exposed to a 1,000 micromolar dosage. In comparison to the control group, the leaf count, secondary roots, and stem thickness all increased. The enhanced rate of growth appeared to be caused by the improved oxygen and water permeability in the seedlings after treatment using gold nanoparticles, as well as interference with plant hormone concentrations, for example, elevated concentrations of gibberellic acid [90].

The efficiency of silver nanoparticles made from *Thuja occidentalis* plant extract as a growth stimulator in *Phaseolus vulgaris* was investigated. When these nanoparticles were applied to the topsoil, they gathered and raised the porosity and pH of the topsoil. The bioavailability of carbon and nitrogen in the topsoil was raised as well, probably as a result of the interactions among silver nanoparticles and plant root cells. The number of leaves, the rate of reactions catalyzed by enzymes, and chloroplast composition improved as a consequence [91].

The developmental effects of zinc oxide nanoparticles produced from extracts of *Aloe barbadensis Mill* were studied on wheat seedlings. The results were evaluated against chemically produced zinc oxide nanoparticles. A 62 milligrams per liter dosage of chemical-based nanoparticles given to wheat plants exhibited 105 percent shoot elongation and 47 percent root growth. The exterior of nanoparticles has an accumulation of amino acids and bioflavonoids from botanical extracts, which is assumed to be responsible for the increased growth percentage of these wheat plants [92].

RECENT ADVANCES

There has been a major improvement in the methods and applicability of green nanotechnology since the beginning of the 21st century. Bioremediating techniques, consumable nanoparticles extracted from plant parts, green eco-friendly carbon dots, magnetic hyperthermia for generating nanoparticles, and quorum sensing inhibition are among the most significant breakthroughs. Consumable plant-based nanoparticles are found in nature and have been utilized in delivering therapeutics. These edible nanoparticles are rich in phosphatides and glycolipids, and may also have active plant chemicals that give them their therapeutic characteristics. The drug delivery system based on lipid nanoparticles has many restrictions, along with being lowly biocompatible and harmful to the body [93].

Nanoparticles derived by plants are utilized for encapsulating therapeutics or small interfering RNA and are also consumable since these particles are harmless and have high biocompatibility. Centrifuging consumable plant parts can create consumable nanoparticles. Plant-based nanoparticles have been reported to have varied locations for internalizing depending on how they enter the body. Nanoparticles derived from *Zingiber officinale* have been employed for treating IBD in humans [94]. These nanoparticles are primarily concentrated in the epithelium of intestines, where they stimulate the secretion of inflammation-reducing cytokines including Interleukin-10 and Interleukin 22. It also inhibits the release of inflammation-inducing cytokines like tumor necrosis factor-alpha [94]. It was discovered that juices of *Citrus limon* contained chemotherapeutic nanoparticles that cause apoptosis induction through the TNF-related apoptosis-inducing ligand mechanism [95].

A newer type of nanoparticles known as carbon dots was created to replace quantum dots. The majority of the restrictions, including poor water solubility and severe toxicity of metallic compounds in quantum dots, were overcome by carbon dots while maintaining the bright fluorescence. These carbon dots, on the other hand, were made using extremely pure compounds, which makes the procedure expensive. As a result, a sustainable production technique for carbon dots has been recently established, in which different methods including devolatilization and microwave irradiation have been employed on different plant portions to manufacture eco-friendly carbon dots. The carbon dots produced using ginger has cancer-fighting properties against cancerous cells in tissues of the liver, lungs, cervix, and breasts while being harmless on healthy cell lines. It was demonstrated that carbon dots made using roses can be utilized as a tetracycline sensor [96].

The production of chemical compounds in biofilms after reaching specified levels of population is a characteristic seen in a few bacterial species and is called quorum sensing. Quorum sensing plays an important role in bacterial pathogenicity and has lately become a potential drug treatment. The silver nanoparticles produced utilizing extract of *Carum coticum* exhibited inhibition properties towards quorum sensing in diderms. Researchers have recently reported the generation of nanoparticles by living macroalgae without requiring the removal of specialized metabolites *via* sunlight-mediated synthesis. This seaweed had Ag nanoparticles bonded onto their surfaces as a result of the biosynthesis. Therefore, seaweed *Ulva Armoricana* could be utilized for bioremediation due to its ability to collect gold contaminants in water and synthesize nanoparticles in sunlight [97].

DRAWBACKS OF GREEN METAL NANOPARTICLES

Due to its small size and properties, nanoparticles have gained significant attention. As they can easily cross blood barriers, enter smaller capillaries, and are taken up by all cells, so are used in the field of medicine and for drug delivery. Metal nanoparticles that are made by following conventional techniques could have adverse effects as they may undergo chemical reduction during the use of harmful chemicals. Therefore, green nanoparticles got this fame as the origin of capping metal is plant and they are nontoxic. One of the major drawbacks that green nanotechnology cannot overcome is the changes that are undergone in nanoparticles when they get in contact with different inorganic and organic environmental factors. Particles may lose their stability or may agglomerate when they are oxidized by different natural factors in the environment and metal ions would be released, hence becoming the cause of toxicity in plants, animals, and humans or loss in properties that depends upon size. Some of the properties of nanoparticles may also change when they come in contact with the low pH environment of the stomach or the high pH of the intestine [98].

Insufficient knowledge about the morphology and properties of nanoparticles is also one of the drawbacks of using green nanotechnology. The properties and stability of particles produced from different plants greatly depend on respective capping agents. Different research papers that are being published on this topic fail to conclude the compound that is responsible for the creation of these particles. Insufficient data is available on the way these particles are formed. Secondary metabolites like flavonoids and phenolics [99] are involved in the synthesis of nanoparticles and this is also reported in some of the papers but the mechanism is still unknown.

The reproducibility of results is affected by insufficient knowledge. Nanoparticles formed from plants are usually of different sizes and shapes. This difference is because large amounts of plant metabolites are present within the extract. Particles obtained from various plants have different shapes and sizes like spherical, triangular, and even some are in the shape of a flower. The quantity of plant extract added into the metal solution, temperature, and pH also greatly affects the morphology of particles. Some of the limitations of nanoparticles are insufficient knowledge of efficiency measurements and standardization of the synthesis techniques [100].

FUTURE PROSPECTS

As green nanotechnology is a manageable and eco-friendly process, it has great scope in the near future in the field of research. Recent advancement has been made in the field of the use of nanotechnologies like carbon dots for bio-sensing, targeted drug delivery, and many other products that are of commercial importance. Less solubility in an aqueous medium is one of the disadvantages of this technique [100]. Hence more research should be made on the synthesis of carbon dots and enclosing drugs in nanoparticles made up of plants.

The synthesis of homogeneously distributed nanoparticles is affected due to less information available on the formation of nanoparticles using plant extracts which makes the reproducibility of results complex. As extracts consist of a vast variety of compounds, the study of these mechanisms becomes difficult. Studies on the use of purified extracts will increase the productivity of the process and allow scalability [100]. Insufficient information is also one of the drawbacks of not using green nanoparticles on an industrial scale [99]. Further research on production, functionalization, and the toxicity of transformation, would make possible the use of green nanotechnology at the commercial level.

Properties of metal nanoparticles to be used as plant growth enhancers against larval agents can be used in nanobio-fertilizers and bio Nano bio-pesticides in the near future. As small concentrations of nanoparticles can produce great results due to increased surface area, small concentrations of chemicals will be introduced into the soil. But more research should be done on environmental factors and the impact of nanoparticle transformation on plants. Toxicology analysis should be done before using these compounds at a practical level. If proper control measures and accurate information is gathered, the phenomena of green nanotechnology can be applied to decrease losses faced in the agricultural industry and to increase yield.

CONCLUSION

As nanoparticles produced by green synthesis are more useful in comparison with the commercially available ones, it's an emerging field for study. Extracts obtained from therapeutic plants work by reducing metal salts into nanoparticles and also act as capping and stabilizing agents. Fourier Transform Infrared Spectroscopy and X-ray diffraction techniques have shown that on the surface of nanoparticles, all those medicinal properties are present that are exhibited by secondary metabolites. Such nanoparticles can be used as antimicrobial, antioxidant, or anticancer agents. These properties can also be used to resolve industrial problems like water pollution, and bacterial infections in medicinal devices and food packaging. Analysis of nanoparticles obtained from plants or any other source has shown that those obtained from plants exhibit better applications and administer the best solution for different problems as shown in this article. Therefore, thoroughly studying the mechanisms and control properties of nanoparticles provides excellent solutions for commercially related problems.

REFERENCES

[1]   Wachtel-Galor S, Benzie IF. Herbal medicine. Biomolecular and Clinical Aspects. 2nd., CRC Press/Taylor & Francis 2011.

[2]   Ferreira TS, Moreira CZ, Cária NZ, Victoriano G, Silva WF Jr, Magalhães JC. Phytotherapy: An introduction to its history, use and application. Rev Bras Plantas Med 2014; 16(2): 290-8.
      [http://dx.doi.org/10.1590/S1516-05722014000200019]

[3]   Wink M. Introduction: Biochemistry, physiology and ecological functions of secondary metabolites. Annual plant reviews : Biochemistry of plant secondary metabolism. 2010; 40: pp. 1-9.
      [http://dx.doi.org/10.1002/9781444320503.ch1]

[4]   Kennedy DO, Wightman EL. Herbal extracts and phytochemicals: plant secondary metabolites and the enhancement of human brain function. Adv Nutr 2011; 2(1): 32-50.
      [http://dx.doi.org/10.3945/an.110.000117] [PMID: 22211188]

[5]   Kumar S, Dobos GJ, Rampp T. The significance of ayurvedic medicinal plants. J Evid Based Complementary Altern Med 2017; 22(3): 494-501.
      [http://dx.doi.org/10.1177/2156587216671392] [PMID: 27707902]

[6]   Bhagwat MK, Datar AG. Antifungal activity of herbal extracts against plant pathogenic fungi. Arch Phytopathol Pflanzenschutz 2014; 47(8): 959-65.
      [http://dx.doi.org/10.1080/03235408.2013.826857]

[7]   Morales F, Padilla S, Falconí F. Medicinal plants used in traditional herbal medicine in the province of Chimborazo, Ecuador. Afr J Tradit Complement Altern Med 2016; 14(1): 10-5.
      [http://dx.doi.org/10.21010/ajtcam.v14i1.2] [PMID: 28331911]

[8]   Kabera JN, Semana E, Mussa AR, He X. Plant secondary metabolites: biosynthesis, classification, function and pharmacological properties. J Pharm Pharmacol 2014; 2: 377-92.

[9]   Joshi J, Nepal S, Zhang Q, Zhang LJ. Water in the hindu kush himalaya. ChamSpringer 2019; pp. 333-59.

[10] Bergman ME, Davis B, Phillips MA. Medically useful plant terpenoids: Biosynthesis, occurrence, and mechanism of action. Molecules 2019; 24(21): 3961.
      [http://dx.doi.org/10.3390/molecules24213961] [PMID: 31683764]

[11] Amirkia V, Heinrich M. Alkaloids as drug leads : A predictive structural and biodiversity-based analysis. Phytochem Lett 2014; 10: 40-8.
      [http://dx.doi.org/10.1016/j.phytol.2014.06.015]

[12] Zhang Y, Seeram NP, Lee R, Feng L, Heber D. Isolation and identification of strawberry phenolics with antioxidant and human cancer cell antiproliferative properties. J Agric Food Chem 2008; 56(3): 670-5.
      [http://dx.doi.org/10.1021/jf071989c] [PMID: 18211028]

[13] Parasuraman S, Anand David AV, Arulmoli R. Overviews of biological importance of quercetin: A bioactive flavonoid. Pharmacogn Rev 2016; 10(20): 84-9.
      [http://dx.doi.org/10.4103/0973-7847.194044] [PMID: 28082789]

[14] Allkin B. t Least 28,187 Plant Species are Currently Recorded as Being of Medicinal Use. Willis KJ. State of the World's Plants. London (UK)Royal Botanic Gardens, Kew 2017.

[15] Dalar A, Mukemre M, Unal M, Ozgokce F. Traditional medicinal plants of Ağrı Province, Turkey. J

Ethnopharmacol 2018; 226: 56-72.
[http://dx.doi.org/10.1016/j.jep.2018.08.004] [PMID: 30099095]

[16]   Chen SL, Yu H, Luo HM, Wu Q, Li CF, Steinmetz A. Conservation and sustainable use of medicinal plants: Problems, progress, and prospects. Chin Med 2016; 11(1): 37.
[http://dx.doi.org/10.1186/s13020-016-0108-7] [PMID: 27478496]

[17]   Schippmann U, Leaman DJ, Cunningham AB. Impact of cultivation and gathering of medicinal plants on biodiversity: Global trends and issues. Biodiversity and the ecosystem approach in agriculture, forestry and fisheries. FAO 2002; pp. 142-67.

[18]   Mazid M, Khan TA, Mohammad F. Medicinal plants of rural India: A review of use by Indian folks. Indo Global Journal of Pharmaceutical Sciences 2012; 2(3): 286-304.
[http://dx.doi.org/10.35652/IGJPS.2012.35]

[19]   Ugbogu OC, Akukwe AR. The antimicrobial effect of oils from Pentaclethra macrophylla Bent, Chrysophyllum albidum G. Don and Persea gratissima Gaerth F on some local clinical bacteria isolates. Afr J Biotechnol 2009; 8: 285-7.

[20]   Tripathi HI, Suresh R, Kumar SA, Khan FE. International trade in medicinal and aromatics plants: A case study of past 18 years. Curr Res Med Aromat Plants 2017; 39: 1-7.

[21]   Nirmal SA, Pal SC, Otimenyin SO, et al. Contribution of herbal products in global market. Pharma Rev 2013; pp. 95-104.

[22]   Rout A, Jena PK, Sahoo D, Bindhani BK. Green synthesis of silver nanoparticles of different shapes and its antibacterial activity against Escherichia coli. Int J Curr Microbiol Appl Sci 2014; 3: 374-83.

[23]   Sankar R, Karthik A, Prabu A, Karthik S, Shivashangari KS, Ravikumar V. Origanum vulgare mediated biosynthesis of silver nanoparticles for its antibacterial and anticancer activity. Colloids Surf B Biointerfaces 2013; 108: 80-4.
[http://dx.doi.org/10.1016/j.colsurfb.2013.02.033] [PMID: 23537829]

[24]   Aritonang HF, Koleangan H, Wuntu AD. Synthesis of silver nanoparticles using aqueous extract of medicinal plants'(Impatiens balsamina and Lantana camara) fresh leaves and analysis of antimicrobial activity. Int J Microbiol 2019; 2019: 1-8.
[http://dx.doi.org/10.1155/2019/8642303] [PMID: 31354833]

[25]   Lee HJ, Lee G, Jang NR, Yun JH, Song JY, Kim BS. Biological synthesis of copper nanoparticles using plant extract. Nanotechnology 2011; 1: 371-4.

[26]   Mondal NK, Hajra A. Synthesis of copper nanoparticles (CuNPs) from petal extracts of marigold (Tagetes sp.) and sunflower (Helianthus sp.) and their effective use as a control tool against mosquito vectors. Int J Mosq Res 2016; 6: 1-15.

[27]   Cheirmadurai K, Biswas S, Murali R, Thanikaivelan P. Green synthesis of copper nanoparticles and conducting nanobiocomposites using plant and animal sources. RSC Advances 2014; 4(37): 19507-11.
[http://dx.doi.org/10.1039/c4ra01414f]

[28]   Katheresan V, Kansedo J, Lau SY. Efficiency of various recent wastewater dye removal methods: A review. J Environ Chem Eng 2018; 6(4): 4676-97.
[http://dx.doi.org/10.1016/j.jece.2018.06.060]

[29]   Wang Z, Fang C, Mallavarapu M. Characterization of iron–polyphenol complex nanoparticles synthesized by Sage (Salvia officinalis) leaves. Environmental Technology & Innovation 2015; 4: 92-7.
[http://dx.doi.org/10.1016/j.eti.2015.05.004]

[30]   Muhd Julkapli N, Bagheri S, Bee Abd Hamid S. Recent advances in heterogeneous photocatalytic decolorization of synthetic dyes. ScientificWorldJournal 2014; 2014: 1-25.
[http://dx.doi.org/10.1155/2014/692307] [PMID: 25054183]

[31]   Chandran SP, Chaudhary M, Pasricha R, Ahmad A, Sastry M. Synthesis of gold nanotriangles and

silver nanoparticles using Aloe vera plant extract. Biotechnol Prog 2006; 22(2): 577-83.
[http://dx.doi.org/10.1021/bp0501423] [PMID: 16599579]

[32]   Vanaja M, Rajeshkumar S, Paulkumar K, Gnanajobitha G, Malarkodi C, Annadurai G. Phytosynthesis and characterization of silver nanoparticles using stem extract of Coleus aromaticus. Int J Mater Biomater Appl 2013; 3: 1-4.

[33]   Gopinath V, MubarakAli D, Priyadarshini S, Priyadharsshini NM, Thajuddin N, Velusamy P. Biosynthesis of silver nanoparticles from Tribulus terrestris and its antimicrobial activity: A novel biological approach. Colloids Surf B Biointerfaces 2012; 96: 69-74.
[http://dx.doi.org/10.1016/j.colsurfb.2012.03.023] [PMID: 22521683]

[34]   Amarnath K, Kumar J, Reddy T, Mahesh V, Ayyappan SR, Nellore J. RETRACTED: Synthesis and characterization of chitosan and grape polyphenols stabilized palladium nanoparticles and their antibacterial activity. Colloids Surf B Biointerfaces 2012; 92: 254-61.
[http://dx.doi.org/10.1016/j.colsurfb.2011.11.049] [PMID: 22225943]

[35]   Mittal AK, Chisti Y, Banerjee UC. Synthesis of metallic nanoparticles using plant extracts. Biotechnol Adv 2013; 31(2): 346-56.
[http://dx.doi.org/10.1016/j.biotechadv.2013.01.003] [PMID: 23318667]

[36]   Justin Packia Jacob S, Finub JS, Narayanan A. Synthesis of silver nanoparticles using piper longum leaf extracts and its cytotoxic activity against Hep-2 cell line. Colloids Surf B Biointerfaces 2012; 91: 212-4.
[http://dx.doi.org/10.1016/j.colsurfb.2011.11.001] [PMID: 22119564]

[37]   Vijayakumar M, Priya K, Nancy FT, Noorlidah A, Ahmed ABA. Biosynthesis, characterisation and anti-bacterial effect of plant-mediated silver nanoparticles using Artemisia nilagirica. Ind Crops Prod 2013; 41: 235-40.
[http://dx.doi.org/10.1016/j.indcrop.2012.04.017]

[38]   Noruzi M, Zare D, Khoshnevisan K, Davoodi D. Rapid green synthesis of gold nanoparticles using rosa hybrida petal extract at room temperature Spectrochim Acta A Mol Biomol Spectrosc 2011; 79(5): 1461-5.
[http://dx.doi.org/10.1016/j.saa.2011.05.001]

[39]   Das RK, Gogoi N, Bora U. Green synthesis of gold nanoparticles using nyctanthes arbortristis flower extract. Bioprocess Biosyst Eng 2011; 34(5): 615-9.
[http://dx.doi.org/10.1007/s00449-010-0510-y] [PMID: 21229266]

[40]   Vankar PS, Bajpai D. Preparation of gold nanoparticles from Mirabilis jalapa flowers Biophys 2011; 47: 157-60.

[41]   Sreelakshmy V, Deepa MK, Mridula P. Green synthesis of silver nanoparticles from Glycyrrhiza glabra root extract for the treatment of gastric ulcer. J Dev Drugs 2016; 5: 1-5.

[42]   Subbaiah KV, Savithramma N. Validation and characterization of silver nanoparticles from Strychnos nux-vomica–an important ethnomedicinal plant of Kurnool district. Andhra Pradesh, India Int J Pharma Biol Sci 2014; 4: 45-53.

[43]   Nellore J, Pauline C, Amarnath K. Bacopa monnieri phytochemicals mediated synthesis of platinum nanoparticles and its neurorescue effect on 1-methyl 4-phenyl 1, 2, 3, 6 tetrahydropyridine-induced experimental parkinsonism in zebrafish. J Neurodegener Dis 2013; 2013: 1-8.
[http://dx.doi.org/10.1155/2013/972391] [PMID: 26317003]

[44]   Swaminadham V, Acharyulu NP, Diwakar BS, Sastry YN. Synthesis and anti-bacterial studies of gold nanoparticles from eclipta prostrata. Int J Recent Innov Trends Comput Commun 2015; 3: 45-8.

[45]   Kesarla MK, Mandal BK, Bandapalli PR. Gold nanoparticles by *Terminalia bellirica* aqueous extract – a rapid green method. J Exp Nanosci 2014; 9(8): 825-30.
[http://dx.doi.org/10.1080/17458080.2012.725257]

[46]   George JA, Sundar S, Paari KA. A review on metal nanoparticles from medicinal plants: Synthesis,

characterization and applications. Nanosci Nanotechnol Asia 2021; 11(5): e201020187020.
[http://dx.doi.org/10.2174/2210681210999201020142942]

[47] Yang N, Li F, Jian T, *et al.* Biogenic synthesis of silver nanoparticles using ginger (Zingiber officinale) extract and their antibacterial properties against aquatic pathogens. Acta Oceanol Sin 2017; 36(12): 95-100.
[http://dx.doi.org/10.1007/s13131-017-1099-7]

[48] Bala N, Saha S, Chakraborty M, *et al.* Green synthesis of zinc oxide nanoparticles using Hibiscus subdariffa leaf extract: Effect of temperature on synthesis, anti-bacterial activity and anti-diabetic activity. RSC Advances 2015; 5(7): 4993-5003.
[http://dx.doi.org/10.1039/C4RA12784F]

[49] Weng X, Huang L, Chen Z, Megharaj M, Naidu R. Synthesis of iron-based nanoparticles by green tea extract and their degradation of malachite. Ind Crops Prod 2013; 51: 342-7.
[http://dx.doi.org/10.1016/j.indcrop.2013.09.024]

[50] Nayak PS, Pradhan S, Arakha M, *et al.* Silver nanoparticles fabricated using medicinal plant extracts show enhanced antimicrobial and selective cytotoxic propensities. IET Nanobiotechnol 2019; 13(2): 193-201.
[http://dx.doi.org/10.1049/iet-nbt.2018.5025] [PMID: 31051451]

[51] Ankamwar B, Damle C, Ahmad A, Sastry M. Biosynthesis of gold and silver nanoparticles using Emblica Officinalis fruit extract, their phase transfer and transmetallation in an organic solution. J Nanosci Nanotechnol 2005; 5(10): 1665-71.
[http://dx.doi.org/10.1166/jnn.2005.184] [PMID: 16245525]

[52] Aazam ES, Zaheer Z. Growth of Ag-nanoparticles in an aqueous solution and their antimicrobial activities against Gram positive, Gram negative bacterial strains and Candida fungus. Bioprocess Biosyst Eng 2016; 39(4): 575-84.
[http://dx.doi.org/10.1007/s00449-016-1539-3] [PMID: 26796584]

[53] Laschke MW, Augustin VA, Sahin F, *et al.* Surface modification by plasma etching impairs early vascularization and tissue incorporation of porous polyethylene (M edpor®) implants. J Biomed Mater Res B Appl Biomater 2016; 104: 1738-48.
[http://dx.doi.org/10.1002/jbm.b.33528] [PMID: 26355709]

[54] Lawrence J, Nho R. The role of the mammalian target of rapamycin (mTOR) in pulmonary fibrosis. Int J Mol Sci 2018; 19(3): 778.
[http://dx.doi.org/10.3390/ijms19030778] [PMID: 29518028]

[55] Chinnasamy G, Chandrasekharan S, Bhatnagar S. Biosynthesis of silver nanoparticles from Melia azedarach: Enhancement of antibacterial, wound healing, antidiabetic and antioxidant activities. Int J Nanomedicine 2019; 14: 9823-36.
[http://dx.doi.org/10.2147/IJN.S231340] [PMID: 31849471]

[56] Vaishya R, Chauhan M, Vaish A. Bone cement. J Clin Orthop Trauma 2013; 4(4): 157-63.
[http://dx.doi.org/10.1016/j.jcot.2013.11.005] [PMID: 26403875]

[57] Arora M, Chan EK, Gupta S, Diwan AD. Polymethylmethacrylate bone cements and additives: A review of the literature. World J Orthop 2013; 4(2): 67-74.
[http://dx.doi.org/10.5312/wjo.v4.i2.67] [PMID: 23610754]

[58] Thomas R, Snigdha S, Bhavitha KB, Babu S, Ajith A, Radhakrishnan EK. Biofabricated silver nanoparticles incorporated polymethyl methacrylate as a dental adhesive material with antibacterial and antibiofilm activity against *Streptococcus mutans*. 3 Biotech 2018; 8(9): 404.

[59] Adhikari U, Ghosh A, Chandra G. Nano particles of herbal origin: A recent eco-friend trend in mosquito control. Asian Pac J Trop Dis 2013; 3(2): 167-8.
[http://dx.doi.org/10.1016/S2222-1808(13)60065-1]

[60] Rawani A, Ghosh A, Chandra G. Mosquito larvicidal and antimicrobial activity of synthesized nano-

crystalline silver particles using leaves and green berry extract of Solanum nigrum L. (Solanaceae: Solanales). Acta Trop 2013; 128(3): 613-22.
[http://dx.doi.org/10.1016/j.actatropica.2013.09.007] [PMID: 24055718]

[61] Shelar A, Sangshetti J, Chakraborti S, Singh AV, Patil R, Gosavi S. Helminthicidal and larvicidal potentials of biogenic silver nanoparticles synthesized from medicinal plant *Momordica charantia*. Med Chem 2019; 15(7): 781-9.
[http://dx.doi.org/10.2174/1573406415666190430142637] [PMID: 31208313]

[62] Suriyakalaa U, Antony JJ, Suganya S, *et al.* Hepatocurative activity of biosynthesized silver nanoparticles fabricated using Andrographis paniculata. Colloids Surf B Biointerfaces 2013; 102: 189-94.
[http://dx.doi.org/10.1016/j.colsurfb.2012.06.039] [PMID: 23018020]

[63] Chue-Gonçalves M, Pereira GN, Faccin-Galhardi LC, Kobayashi RKT, Nakazato G. Metal nanoparticles against viruses: possibilities to fight SARS-CoV-2. Nanomaterials 2021; 11(11): 3118.
[http://dx.doi.org/10.3390/nano11113118] [PMID: 34835882]

[64] Patra JK, Das G, Shin HS. Facile green biosynthesis of silver nanoparticles using *Pisum sativum* L. outer peel aqueous extract and its antidiabetic, cytotoxicity, antioxidant, and antibacterial activity. Int J Nanomedicine 2019; 14: 6679-90.
[http://dx.doi.org/10.2147/IJN.S212614] [PMID: 31695363]

[65] Jamdade DA, Rajpali D, Joshi KA, *et al.* Gnidia glauca-and Plumbago zeylanica-mediated synthesis of novel copper nanoparticles as promising antidiabetic agents. Adv Pharmacol Sci 2019; 2019: 1-11.
[http://dx.doi.org/10.1155/2019/9080279] [PMID: 30886631]

[66] Bayrami A, Ghorbani E, Rahim Pouran S, Habibi-Yangjeh A, Khataee A, Bayrami M. Enriched zinc oxide nanoparticles by *Nasturtium officinale* leaf extract: Joint ultrasound-microwave-facilitated synthesis, characterization, and implementation for diabetes control and bacterial inhibition. Ultrason Sonochem 2019; 58: 104613.
[http://dx.doi.org/10.1016/j.ultsonch.2019.104613] [PMID: 31450359]

[67] Rajakumar G, Thiruvengadam M, Mydhili G, Gomathi T, Chung IM. Green approach for synthesis of zinc oxide nanoparticles from *Andrographis paniculata* leaf extract and evaluation of their antioxidant, anti-diabetic, and anti-inflammatory activities. Bioprocess Biosyst Eng 2018; 41(1): 21-30.
[http://dx.doi.org/10.1007/s00449-017-1840-9] [PMID: 28916855]

[68] Rajaram K, Aiswarya DC, Sureshkumar P. Green synthesis of silver nanoparticle using tephrosia tinctoria and its antidiabetic activity. Mater Lett 2015; 138: 251-4.
[http://dx.doi.org/10.1016/j.matlet.2014.10.017]

[69] Ganesh Kumar V, Dinesh Gokavarapu S, Rajeswari A, *et al.* Facile green synthesis of gold nanoparticles using leaf extract of antidiabetic potent *Cassia auriculata*. Colloids Surf B Biointerfaces 2011; 87(1): 159-63.
[http://dx.doi.org/10.1016/j.colsurfb.2011.05.016] [PMID: 21640563]

[70] Al-Sheddi ES, Farshori NN, Al-Oqail MM, *et al.* Anticancer potential of green synthesized silver nanoparticles using extract of Nepeta deflersiana against human cervical cancer cells (HeLA). Bioinorg Chem Appl 2018; 2018: 1-12.
[http://dx.doi.org/10.1155/2018/9390784] [PMID: 30515193]

[71] Aygün A, Gülbağça F, Nas MS, *et al.* Biological synthesis of silver nanoparticles using *Rheum ribes* and evaluation of their anticarcinogenic and antimicrobial potential: A novel approach in phytonanotechnology. J Pharm Biomed Anal 2020; 179: 113012.
[http://dx.doi.org/10.1016/j.jpba.2019.113012] [PMID: 31791838]

[72] Liu R, Pei Q, Shou T, Zhang W, Hu J, Li W. Apoptotic effect of green synthesized gold nanoparticles from *Curcuma wenyujin* extract against human renal cell carcinoma A498 cells. Int J Nanomedicine 2019; 14: 4091-103.
[http://dx.doi.org/10.2147/IJN.S203222] [PMID: 31239669]

[73]   Tripathi D, Modi A, Narayan G, Rai SP. Green and cost effective synthesis of silver nanoparticles from endangered medicinal plant withania coagulans and their potential biomedical properties. Mater Sci Eng C 2019; 100: 152-64.
[http://dx.doi.org/10.1016/j.msec.2019.02.113] [PMID: 30948049]

[74]   Patra S, Mukherjee S, Barui AK, Ganguly A, Sreedhar B, Patra CR. Green synthesis, characterization of gold and silver nanoparticles and their potential application for cancer therapeutics. Mater Sci Eng C 2015; 53: 298-309.
[http://dx.doi.org/10.1016/j.msec.2015.04.048] [PMID: 26042718]

[75]   Souza VGL, Fernando AL. Nanoparticles in food packaging: Biodegradability and potential migration to food—A review. Food Packag Shelf Life 2016; 8: 63-70.
[http://dx.doi.org/10.1016/j.fpsl.2016.04.001]

[76]   Zhang J, Si G, Zou J, Fan R, Guo A, Wei X. Antimicrobial effects of silver nanoparticles synthesized by fatsia japonica leaf extracts for preservation of citrus fruits. J Food Sci 2017; 82(8): 1861-6.
[http://dx.doi.org/10.1111/1750-3841.13811] [PMID: 28727146]

[77]   Sing T, Shukla S, Kumar P, Wahla V, Bajpai VK, Rather IA. Corrigendum: Application of nanotechnology in food science: Perception and overview. Front Microbiol 2017; 8: 2517.
[http://dx.doi.org/10.3389/fmicb.2017.02517] [PMID: 29255457]

[78]   Küp FÖ, Çoşkunçay S, Duman F. Biosynthesis of silver nanoparticles using leaf extract of Aesculus hippocastanum (horse chestnut): Evaluation of their antibacterial, antioxidant and drug release system activities. Mater Sci Eng C 2020; 107: 110207.
[http://dx.doi.org/10.1016/j.msec.2019.110207] [PMID: 31761206]

[79]   Kalantari K, Afifi AM, Moniri M, Moghaddam AB, Kalantari A, Izadiyan Z. Autoclave assisted synthesis of AgNPs in *Z. officinale* extract and assessment of their cytotoxicity, antibacterial and antioxidant activities. IET Nanobiotechnol 2019; 13(3): 262-8.
[http://dx.doi.org/10.1049/iet-nbt.2018.5066] [PMID: 31053688]

[80]   Bajpai VK, Kamle M, Shukla S, *et al.* Prospects of using nanotechnology for food preservation, safety, and security. J Food Drug Anal 2018; 26(4): 1201-14.
[PMID: 30249319]

[81]   Dhand V, Soumya L, Bharadwaj S, Chakra S, Bhatt D, Sreedhar B. Green synthesis of silver nanoparticles using coffea arabica seed extract and its antibacterial activity. Mater Sci Eng C 2016; 58: 36-43.
[http://dx.doi.org/10.1016/j.msec.2015.08.018] [PMID: 26478284]

[82]   Gudelj I, Hrenović J, Dragičević T, Delaš F, Šoljan V, Gudelj H. [Azo dyes, their environmental effects, and defining a strategy for their biodegradation and detoxification]. Arh Hig Rada Toksikol 2011; 62(1): 91-101.
[http://dx.doi.org/10.2478/10004-1254-62-2011-2063] [PMID: 21421537]

[83]   Golmohammadi M, Honarmand M, Ghanbari S. A green approach to synthesis of ZnO nanoparticles using jujube fruit extract and their application in photocatalytic degradation of organic dyes. Spectrochim Acta A Mol Biomol Spectrosc 2020; 229: 117961.
[http://dx.doi.org/10.1016/j.saa.2019.117961]

[84]   Baruah D, Yadav RNS, Yadav A, Das AM. Alpinia nigra fruits mediated synthesis of silver nanoparticles and their antimicrobial and photocatalytic activities. J Photochem Photobiol B 2019; 201: 111649.
[http://dx.doi.org/10.1016/j.jphotobiol.2019.111649] [PMID: 31710925]

[85]   Singh J, Kumar V, Kim KH, Rawat M. inventors. Environ Res 2019; 177: 1-12.

[86]   Weng X, Huang L, Chen Z, Megharaj M, Naidu R. Synthesis of iron-based nanoparticles by green tea extract and their degradation of malachite. Ind Crops Prod 2013; 51: 342-7.
[http://dx.doi.org/10.1016/j.indcrop.2013.09.024]

[87]   Rastogi A, Zivcak M, Sytar O, *et al.* Impact of metal and metal oxide nanoparticles on plant: A critical review. Front Chem 2017; 5: 78.
[http://dx.doi.org/10.3389/fchem.2017.00078] [PMID: 29075626]

[88]   Salem NM, Albanna LS, Awwad AM, Ibrahim QM, Abdeen AO. Green synthesis of nano-sized sulfur and its effect on plant growth. J Agric Sci 2016; 8: 188-94.

[89]   Gopinath K, Gowri S, Karthika V, Arumugam A. Green synthesis of gold nanoparticles from fruit extract of terminalia arjuna, for the enhanced seed germination activity of gloriosa superba. J Nanostructure Chem 2014; 4(3): 115.
[http://dx.doi.org/10.1007/s40097-014-0115-0]

[90]   Das P, Barua S, Sarkar S, *et al.* Plant extract–mediated green silver nanoparticles: Efficacy as soil conditioner and plant growth promoter. J Hazard Mater 2018; 346: 62-72.
[http://dx.doi.org/10.1016/j.jhazmat.2017.12.020] [PMID: 29247955]

[91]   Singh J, Kumar S, Alok A, *et al.* The potential of green synthesized zinc oxide nanoparticles as nutrient source for plant growth. J Clean Prod 2019; 214: 1061-70.
[http://dx.doi.org/10.1016/j.jclepro.2019.01.018]

[92]   Yang C, Zhang M, Merlin D. Advances in plant-derived edible nanoparticle-based lipid nano-drug delivery systems as therapeutic nanomedicines. J Mater Chem B Mater Biol Med 2018; 6(9): 1312-21.
[http://dx.doi.org/10.1039/C7TB03207B] [PMID: 30034807]

[93]   Zhang M, Viennois E, Prasad M, *et al.* Edible ginger-derived nanoparticles: A novel therapeutic approach for the prevention and treatment of inflammatory bowel disease and colitis-associated cancer. Biomaterials 2016; 101: 321-40.
[http://dx.doi.org/10.1016/j.biomaterials.2016.06.018] [PMID: 27318094]

[94]   Raimondo S, Naselli F, Fontana S, *et al. Citrus limon* -derived nanovesicles inhibit cancer cell proliferation and suppress CML xenograft growth by inducing trail-mediated cell death. Oncotarget 2015; 6(23): 19514-27.
[http://dx.doi.org/10.18632/oncotarget.4004] [PMID: 26098775]

[95]   Tejwan N, Saha SK, Das J. Multifaceted applications of green carbon dots synthesized from renewable sources. Adv Colloid Interface Sci 2020; 275: 102046.
[http://dx.doi.org/10.1016/j.cis.2019.102046] [PMID: 31757388]

[96]   Qais FA, Shafiq A, Ahmad I, Husain FM, Khan RA, Hassan I. Green synthesis of silver nanoparticles using Carum copticum: Assessment of its quorum sensing and biofilm inhibitory potential against gram negative bacterial pathogens. Microb Pathog 2020; 144: 104172.
[http://dx.doi.org/10.1016/j.micpath.2020.104172] [PMID: 32224208]

[97]   Jorge de Souza TA, Rosa Souza LR, Franchi LP. Silver nanoparticles: An integrated view of green synthesis methods, transformation in the environment, and toxicity. Ecotoxicol Environ Saf 2019; 171: 691-700.
[http://dx.doi.org/10.1016/j.ecoenv.2018.12.095] [PMID: 30658305]

[98]   Terenteva EA, Apyari VV, Dmitrienko SG, Zolotov YA. Formation of plasmonic silver nanoparticles by flavonoid reduction: A comparative study and application for determination of these substances. Spectrochim Acta A Mol Biomol Spectrosc 2015; 151: 89-95.
[http://dx.doi.org/10.1016/j.saa.2015.06.049] [PMID: 26125987]

[99]   Gopinath K, Gowri S, Karthika V, Arumugam A. Green synthesis of gold nanoparticles from fruit extract of *Terminalia arjuna* , for the enhanced seed germination activity of Gloriosa superba. J Nanostructure Chem 2014; 4(3): 115.
[http://dx.doi.org/10.1007/s40097-014-0115-0]

[100]   Aslani F, Bagheri S, Muhd Julkapli N, Juraimi AS, Hashemi FSG, Baghdadi A. Effects of engineered nanomaterials on plants growth: An overview. Sci World J 2014; 2014: 1-28.
[http://dx.doi.org/10.1155/2014/641759]

CHAPTER 15

Carbon Nanostructures and Medicinal Plants

Kalakotla Shanker[1,*], Sushil Y. Raut[2], Tamatam Sunilkumar Reddy[1], Divya Pa[2], S.P. Dhanabal[1] and Kristina Apryatina[3]

[1] *Department of Pharmacognosy & Phyto-Pharmacy, JSS College of Pharmacy, JSS Academy of Higher Education & Research, Ooty, Nilgiris, Tamil Nadu, India*

[2] *Department of Pharmaceutics, Dr. D.Y. Patil Institute of Pharmaceutical Sciences and Research, Pimpri, Pune-411018, India*

[3] *Department of High Molecular Compounds and Colloid Chemistry, Faculty of Chemistry, Lobachevsky State University of Nizhny Novgorod (UNN), Moscow, Russia*

Abstract: It has been a decade since the widespread usage of carbon nanostructures (CNSs) in biomedical research. A few examples are the use of CNSs in medication, for protein administration and in instruments to provide nucleic acids to treat cancer and other chronic diseases. The near-infrared optical characteristics of CNSs allowed them to be used in diagnostics and in non-invasive and very sensitive imaging equipment. In recent years, the scientific and industrial sectors have paid increasing attention to the physical and chemical properties of various nanomaterials. Structure, electronics, water, and more may all be derived from them. This chapter will focus on carbon nanomaterials and related nanostructures, which are designed to give the most up-t--date research results. There is a broad acceptance of traditional medicine in many societies, with over 60 percent of the world's population and over 80 percent of the population in developing countries depending on medicinal plants for medical reasons. Among the many reasons for this are the ease of use, affordability, and low cost. It is believed that nanotechnology will play a significant role in medicinal plant research and drug delivery in the near future. These nano-drug delivery devices may boost the activity of medicinal plants, but also solve some of their limitations. Nanocarriers aiding in the treatment of cancer, diabetes, and other life-threatening illnesses by delivering herbal chemicals will also be discussed in this chapter.

Keywords: Biological cargoes, Carbon nanostructures (CNSs), Laser ablative therapy, Multiwalled carbon nanotubes (MWCNT), Nanocarriers.

* **Corresponding author Kalakotla Shanker:** Department of Pharmacognosy & Phyto-Pharmacy, JSS College of Pharmacy, JSS Academy of Higher Education & Research, Ooty, Nilgiris, Tamil Nadu, India; E mail: drshanker@jssuni.edu.in

Zulqurnain Khan, Azra Yasmin & Naila Safdar (Eds.)

INTRODUCTION

Buckyballs' discovery in 1985 sparked a fast expansion of the carbon family. Fullerenes, carbon nanotubes, graphene, and nanodiamonds all have different shapes and forms (Fig. **1**) owing to the occurrence of three hybridization types of carbon: sp^3 (the most common), sp^2 (the second most common), and sp^1 (the third most common). Physical and chemical qualities have attracted growing attention from the scientific and industrial worlds in recent years. Structure, electronics, energy, water, and more are all possible uses of them. Carbon nanomaterials and related nanostructures will be covered in this special issue, which is intended to provide the most recent research findings in this field. In addition to CNTs and graphene, the articles that have been approved span a wide range of carbon-based materials including amorphous carbon films, carbon nanocomposite, graphite-like layered materials, carbon nanofibers, and so on. They cover a broad variety of subjects relevant to the production and uses of these nanocarbon materials in the disciplines of mechanics and tribology; thermal management; power generation through photovoltaics; energy storage *via* lithium-ion batteries; and medication delivery *via* photocatalysis. Carbon nanomaterials' electrical, optical, and surface adsorption characteristics have been calculated theoretically using density functional theory in two studies as well [1].

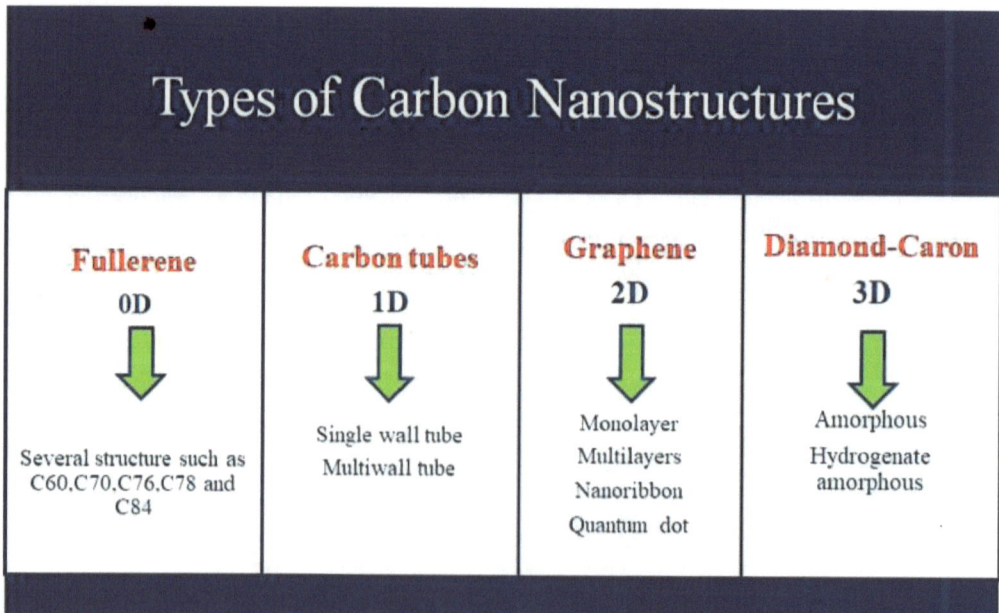

Fig. (1). Various carbon nanostructures.

Carbon Nanotubes

A Carbon Nanotube is a tube-shaped carbon-based material with a nanometer-wide width. To put it in perspective, that's one billionth of a millimetre, or a thousand times smaller than the width and length of an average human hair [2].

For example, carbon may be used to build structures with diverse characteristics by combining in different ways. Carbon's sp^2 hybridization creates a multilayer structure with weak van der Waals bonding and strong in plane limits. Around an average hollow, a few to several dozen MWCNT cylinders with regular periodic interlayer spacing are arranged in concentric cylinders. For multiwall nanotube pictures analysed in real space, we found a wide variety of interlayer spacings (0.34 to 0.39 nm) [3].

There is a continuous hexagonal lattice with carbon molecules at the corners of the hexagons in graphite's layer.

The structures of carbon nanotubes may be used to classify them:

• Nanotubes with one wall (SWNT)
• Nanotubes with several walls (MWNT)

CARBON NANOTUBES SYNTHESIS

Carbon nanotubes are typically synthesised using one of three techniques: arc discharge, laser ablation, or chemical vapour deposition (CVD). Each approach may be tailored to the researcher's individual study objective.

1. Arc Discharge
2. Deposition of Chemical Vapors
3. Laser ablative therapy

Arc Discharge

MWNTs and SWNTs were first synthesised using an arc discharge. The anode and cathode of an arc discharge are typically made of two high-purity graphite electrodes [4]. In an environment of 400 mbar of Helium, a DC current (about 100 A°) was sent through two high-purity graphite separated (by about 1–2 mm) electrodes, causing them to melt. A carbon rod is formed at the cathode of an experimental arc discharge equipment after a length of time of arc discharge. MWNTs may be made using this process, but SWNTs can also be made by adding a metal catalyst, such as Fe, Co, Ni, Y or Mo, to either the anode or the cathode of the reactor [5].

Deposition of Chemical Vapours

Chemical vapour deposition (CVD) was initially described in 1993 by Endo and his research team to build MWNTs. It was three years later when Dai in Smalley's group was able to modify CO-based CVD to produce SWNTs. You can use CVD to transfer energy from an energy source like a plasma or a resistively heated coil to carbon molecules in the gaseous state. Carbon dioxide, methane, carbon monoxide, and acetylene are all hydrocarbons that are used in the CVD process. A quartz tube filled with hydrocarbons is heated to 720 degrees Celsius in an oven. The figure is showing the vapour deposition of chemical compounds to produce pure carbon molecules at high temperatures, hydrocarbons are broken down to their hydrogen carbon bonds. When the carbon is heated and coated with a catalyst (often a first-row transition metal like Ni, Fe, or Co), it will diffuse toward the substrate and bond. For example, low power consumption, a wide temperature range, and the ability to scale up the process were some of CVD's benefits. This process may create both MWNTs and SWNTs depending on the temperature, with SWNT production occurring at a higher temperature than MWNT production [6, 7].

Laser Ablative Therapy

Laser ablation was used by Smalley and his colleagues to create carbon nanotubes in 1995. High-power lasers were used in the laser ablation method to evaporate carbon from a graphite target that was heated to high temperatures. This method may be used to make MWNTs and SWNTs. For the arc discharge process to work, the graphite targets must have metal particles added as catalysts. The number and type of catalysts, laser power and wavelength, temperature, pressure, kind of inert gas, and fluid dynamics around the carbon target all affect the amount and quality of carbon nanotubes generated. Carbon targets with 1.2 percent cobalt/nickel and 98.8 percent graphite composite are focused on by the laser in an argon environment (500 Torr) at 1200°C in a quartz tube furnace. In 1996, Smalley's group accomplished these requirements for the manufacturing of SWNTs. Vapours from the high temperature chamber are carried by argon gas, which is then collected by a cooler collector located downstream. This approach yields SWNTs with a diameter range of 1.0 to 1.6 nm. Laser ablation generated carbon nanotubes that were up to 90% pure compared to those produced by the arc discharge process, and the widths of the nanotubes were very tapered [8, 9].

GREEN SYNTHESIS OF CARBON NANOTUBES

Carbon nanotubes are widely exploited as breakthrough materials in a variety of sectors due to their diverse features. Carbon precursors are put on a substrate (*e.g.*, silicon, silica, quartz, zeolite, calcium carbonate, *etc.*) and heated to high temperatures, which results in the creation of nanotubes. Hydrocarbons (*e.g.*, methane, ethane, carbon monoxide) are the most often employed carbon precursors [10]. However, they are connected with a variety of disadvantages, such as the fact that the majority of these compounds are toxic or create ecologically dangerous by-products or waste. These processes are carried out under a variety of severe experimental circumstances, such as high temperatures (*e.g.*, 1700°C), caustic solvents, metallic catalysts, high currents (*e.g.*, 100-400 A), and so on, prompting significant protective measures and additional purification phases. Apart from these limitations, other concerns include inadequate yield, high manufacturing costs, and the integration of structural imperfections in CNTs. To overcome these issues, scientists have come up with a variety approach or modified existing ones that use green chemistry principles and, to a great extent, avoid the employment of these extreme reaction conditions [11]. The development of green chemistry, defined as the use of natural resources as a raw material to generate a specific chemical substance or compounds, has become a prominent area of study in various nations, notably in Asia, which has plentiful natural ingredient sources [12].

Green synthesis promotes the use of natural sources (such as plant extracts, animal wastes, microbes, and viruses), avoiding the use of harmful hydrocarbons as carbon sources; second, the product is free of metal catalyst impurities, requiring no further purification procedures [13, 14]. As carbon precursors for CNT synthesis, a variety of natural materials such as camphor powder, chicken feather waste [15], and oils (such as turpentine, eucalyptus, coconut, and neem) may be employed. Green synthesis is cost-effective because natural catalysts and raw materials are abundant, and some expensive equipment, such as sputtering and dip coating, is not required. furthermore, Green resources are also more desirable since they are non-toxic. Green synthesis enables the decreased generation of by-products and waste, the direct use of safer chemicals, solvents, and energy-efficient procedures, and avoiding of harsh temperature/pressure or vacuum. This reduces power consumption and conserves energy resources, hence supporting economical CNT production [10].

A thorough review of the advantages of green synthesis of CNTs over conventional processes is provided in Table **1**. Many researchers have reported in the literature on green synthesis approaches that entail particular processes including supercritical drying, plant extract aided synthesis, water-assisted synthesis, and interfacial polymerization, which are described below and has been summarized in Table **2**.

Table 1. Benefits of green synthesis *vs* conventional synthesis of CNTs production [10].

Property	Green Synthesis	Conventional Synthesis
Raw materials	Natural resources (as carbon precursors or catalysts) *e.g.*, eucalyptus oil, coconut oil, neem oil, feather waste or chicken fat *etc.*	Toxic compounds as carbon sources and catalysts, such as hydrocarbons such as methane, ethane, benzene, toluene, xylene
By products	Not toxic	Toxic gaseous by-products
Operating conditions	Less power consumption (lower temperatures)	High energy consumption (Extremely high temperatures, for instance 1700°C, usage of high currents)
Equipment	Simple and low-cost	Costly equipment for sputtering, dip coating *etc.*
Synthesis time	Short	Prolonged
Final product	Does not contain toxic substances	Require additional purification processes due to presence of toxic metallic particles.
Economical aspects	Highly economical	High costs

Table 2. Examples of CNTs developed by various workers using green synthesis.

S. No.	Green Carbon Precursor	Green Synthesis Type and Process Conditions	Result	References
1	Coconut shell	Process: one step water assisted synthesis Temperature: 700°C Dried at 60°C	SWCNTs Diameter: 123 nm	[12]
2	Coconut oil	Process and temperature: CVD in one furnace at 350°C and pyrolysis in the second furnace at 1175°C. Gas: Nitrogen Substrate: Quartz Catalyst: Ferrocene	MWCNTs Diameter: 1.7 – 2.1 nm Length: 50 μm	[17]
3	Walnut extract	Process: CVD Substrate: silicon wafer Temperature: 575 °C Gas: Argon	MWCNTs Diameter: 8-15 nm Length: ultra-long 3600 μm	[13]
4	Neem oil	Process: spray pyrolysis + CVD Substrate: Quartz Temperature: 750-850 °C Gas: Argon Catalyst: Ferrocene	MWCNTs Diameter: 10-30 nm Length: 20-40 μm	[18]

(Table 2) cont.....

S. No.	Green Carbon Precursor	Green Synthesis Type and Process Conditions	Result	References
5	Coconut oil and Olive oil	Process: Pyrolysis Substrate: Silicon wafer Temperature: 900 °C Gas: Argon Catalyst: $NiCl_2$	SWCNTs Diameter: 27-31 nm	[16]
6	Chicken fat	Process: CVD Substrate: Silicon wafer Temperature: 750°C Gas: Argon Catalyst: Ferrocene	Mixture of single and multi-walled CNTs Diameter: 18-78 nm Length: 35 μm	[22]

Plant Extract Assisted Synthesis

Hydrocarbon is degraded into carbon and hydrogen during CNT production. A supersaturation condition is obtained when the hydrogen (after producing hydrogen gas) is ejected from the metal. The carbon then crystallises into long, cylindrical tubes adhering to the substrate as a result of this. Tripathi *et al.* employed a green technique to synthesise multi-walled carbon nanotubes (MWCNTs). Plants including *Cynodon dactylon*, *Azadirachta indica*, *Juglans regia* and *Thevetia peruviana* were used to extract the green leaves from which CNTs were produced using the CVD process. The leaves were thoroughly cleaned and dried before being crushed and extracted using Soxhlet equipment. A drop of extract (previously treated with methanol) was placed on a silicon wafer and dried under argon gas to prepare a sample for CVD. In a CVD chamber, the sample was heated to 575°C while still under argon gas, then acetylene gas was added for 10 minutes after 575°C was reached. This led to the formation of MWCNTs [13].

Pyrolysis of Natural Oils

In some other cases, the natural carbon source may also be burned to produce graphite and then treated using a well-known procedure. Hamid *et al.* developed a green synthesis of CNTs by pyrolysis employing natural oils (such as olive oil and coconut oil) as precursors [16]. Pyrolysis typically includes the decomposition of organic substances at high temperatures in the absence of oxygen. The precursor solution is pumped, often with the assistance of argon gas, into a high-temperature (*e.g.*, 900°C) hot furnace holding a silicon wafer. During this procedure, carbon is deposited onto the silicon wafer inside the furnace, which is then allowed to cool while the argon gas is treated. The approach is useful since it makes use of readily accessible and affordable natural carbon sources.

Paul *et al.* used a similar approach to develop CNTs using coconut oil as a precursor. However, instead of argon gas and silicon wafers, the workers employed nitrogen gas and quartz in this case. Initially, the CVD process was carried out in one furnace at 350°C, with the evaporated oil vapours collected in a second furnace and subjected to pyrolysis at 1175°C. The samples were brought to room temperature instantly and dried in an oven at 400°C for 2 hours. This allowed the formation of CNTs [17].

Kumar *et al.* used the spray pyrolysis method in conjunction with CVD to create CNTs from neem oil. As a catalyst, iron (in the form of ferrocene) was used, and argon was used as an inert gas. For around 10 minutes, a combination of neem oil and ferrocene, aided by argon gas, was allowed to travel through a quartz tube kept at roughly 800°C. As a result, CNTs were formed and deposited on the walls of the quartz tube [18].

One Step Water Assisted Synthesis

One-step water assisted synthesis is another method for the eco-friendly production of CNTs [19]. In this process, graphite taken from a natural source is swiftly heated to a high temperature and then rapidly cooled by immersing it in cold water. The graphite's abrupt drop in temperature is thought to have provided enough energy for the graphite to crimp and form a tube shape. Hakim *et al.* demonstrated that CNTs may be synthesised using coconut shells in a one-step water-assisted synthesis. As a carbon source for CNTs, graphite or other compounds with a similar structure to graphite were created during the burning of coconut shell. The graphite flakes formed after the combustion of coconut shells were heated in a furnace until they reached a temperature of 700°C using this approach. The red-hot graphite was removed immediately, submerged in cold water, and kept for many hours. After that, the end product was taken out and dried in an oven at 60°C to eliminate any remaining moisture. SEM, TEM, and FTIR were used to confirm the formation of nanotubes [12].

Green Synthesis Using Animal Waste

Many researchers have developed carbon nano products from animal wastes, such as carbon dots from chicken eggs [20], graphene from animal bones, and insects [21], as well as CNTs made from chicken fat. Suriani *et al.* proved that chicken fat can be used to synthesize CNTs. They initially heated the chicken fat into oil at 200°C, eventually purified and processed for CNT synthesis employing ferrocene as a catalyst and silicon as a substrate using CVD. The key advantages of this synthesis are that the chicken fat (which is otherwise wasted and pollutes the environment) is efficaciously used. Furthermore, CNTs are synthesised in an environmentally benign way [22]

Supercritical Carbon Dioxide Drying

The synthesis of conjugated CNT products using supercritical carbon dioxide has also been reported in the literature. Zhuyin *et al.* synthesized CNT-graphene aerogels using this approach. Graphene oxide and CNT dispersions were combined in the presence of HCl as a catalyst while vitamin C was added. This resulted in the formation of a graphene-CNT hydrogel, which was then dried using supercritical CO2 to form a CNT-graphene aerogel [23].

Interfacial Polymerization

In addition, the notion of green chemistry has been used for the synthesis of nanocomposites of CNT formed by direct conjugation of CNTs with other ligands, such as the nanocomposites of CNTs and polyaniline (PANI) synthesised by Nguyen *et al.* through interfacial polymerization. The synthesis of CNTs/PANI nanocomposites was aided by a biphasic system composed of non-toxic (or green) liquids, water (containing dispersed CNTs) and an ionic liquid, BMIMBF4 (containing dissolved aniline). Initial mixing of the two phases led to the creation of PANI at the interface, which then migrated into the aqueous area containing CNTs to produce CNT/PANI nanocomposites [24].

APPLICATIONS OF CARBON NANOTUBES IN DRUG DELIVERY

Tissue Engineering

For cell tracing, microenvironment sensing, and scaffolding for incorporation into the host's body, carbon nanotubes may be employed as a tissue engineering material of choice. Tissue formation assessment may be improved by using them as optical and radiotracer contrast agents. Cell development might be guided by CNTs with new features like electrical conductivity. So, CNTs are employed in tissue engineering as a result [25].

Design & Development of Pharmaceuticals

Carbon nanotubes (CNTs) are employed as medication delivery and diagnostic carriers. They enable the covalent and non-covalent insertion of several components and the generation of new therapeutic candidates. CNTs may be functionalized in a variety of ways to carry a variety of moieties for targeting, imaging, and treatment. One of them has a fluorescein probe in addition to the antifungal medicine amphotericin B and the anticancer substance methotrexate. The biological activity of the medicine is maintained, but CNTs are capable of reducing the undesirable effects associated with drug delivery alone [26].

As a result of the growing interest in nanobiotechnology, nanoparticle-based drug discovery and delivery form a significant field of nanomedicine, which is also pushed by the pharmaceutical industry's hunt for novel medications [27].

Applications in Biomedicine

Carbon nanotubes possess practical utility in biological and biomedical fields. A novel class of bioactive carbon nanotubes is coupled with protein, carbohydrates, or nucleic acids; it is an illustration of the bio nanotechnology "bottom up" production concept. They pave the way for an entirely new and interesting research approach in the field of chemical biology, with the goal of identifying and altering the behavior of cells at the molecular level. There is a bio modification of carbon nanotubes for the purpose of delivering genetic elements to cells in a targeted manner, which might be employed in innovative therapeutic techniques [28].

System of Targeted Drug Delivery

Nanoparticles' unique features enable them to interact with intricate biological operations. This rapidly increasing area necessitates multidisciplinary study and offers prospects for the creation and development of multifunctional strategies capable of targeting, diagnosing, and treating catastrophic illnesses such as cancer [28].

Nanomedicine

Carbon nanotubes are currently being employed in diagnostic and therapeutic applications. The majority of these applications involve injecting or implanting carbon nanotubes and their matrixes into patients. The toxicological and pharmacological properties of these carbon nanotube-based nanomedicines are under intensive investigation. Carbon nanotubes are being developed pharmaceutically in order to become practical and effective nanomedicines [29].

System for the Delivery of Anticancer Drugs

Chemotherapeutic chemicals are traditionally administered due to their lack of selectivity, which results in a deadly impact on healthy tissues. Due to the fact that therapeutic and diagnostic substances may functionalize carbon nanotubes (CNTs), optimizing CNTs as drug carriers would open the way for their employment as nanovectors in cells. Various articles have discussed the embedding of cisplatin (Cis-Diammine Dichloro Platinum-CDDP – a platinum-based chemotherapeutic medication) in single-wall carbon nanotubes (SWCNTs). *In-vitro* study was done once the anticancer medication was discharged with certainty. The capacity of tubes containing cisplatin to suppress the viability of prostate cancer cells (PC3 and DU145) demonstrated the delivery system's efficacy.

Transportation and Distribution of Biological Cargoes

SWNTs are internalized for transport and delivery into cells through endocytosis, which seems to have no negative impact on the delivered cargo or the breached cell. The development of SWNT as a novel class of cellular transporters has several fascinating potential applications for SWNT-based systems for drug administration, protein delivery, gene therapy, and cancer treatment [30.]

Transdermal Drug Delivery System with Electro-Sensitive Delivery

To regulate medication release, an electrospinning approach was used to create an electro sensitive transdermal drug delivery device. The matrix was created using polyethylene oxide and pentaerythritol triacrylate polymers. To boost the electrical sensitivity, multi-walled carbon nanotubes were utilized. The release experiment was conducted at various voltage levels. SEM (Scanning Electron Microscopy) and TEM (Transmission Electron Microscopy) were used to detect carbon nanotubes in the centre of electrospun fibers. The amount of medication released was raised by increasing the applied electric voltage.

These findings were attributed to the carbon additive's superior electrical conductivity. The benefits of the electrosensitive transdermal medication delivery system were increased by the inclusion of carbon nanotubes [31].

Transporters of Intracellular Protein

Numerous proteins spontaneously associate with the sidewalls of single-walled carbon nanotubes. This straightforward nonspecific binding method enables the formation of noncovalent protein nanotube conjugates. The proteins are easily delivered through mammalian cells *via* the endocytosis route, which is facilitated by nanotubes. Once released from the endosomes, internalized protein nanotube conjugates may reach the cytoplasm of cells and conduct biological activities, as indicated by cytochrome c-mediated apoptosis induction. Carbon nanotubes constitute a novel type of molecular transporter that may be valuable in the future for protein delivery *in vitro* and *in vivo* [32].

Systematic Delivery of Soft Drugs

Polymers and lipids are combined with surfactants to create soft delivery systems. Micelles, liquid crystal phases, liposomes and polymer gels, as well as more unique structures like carbon nanotubes, polyelectrolyte multilayer capsules, and liquid crystal particles, are all examples of soft drug delivery systems [33].

CARBONS NANOTUBES: RECENT ADVANCES

Carbon nanotubes have received a lot of attention in the scientific community. In 1991, Iijima disclosed them. Numerous researchers have observed remarkable physical and chemical features. Diamond has lower electrical conductivity and electronic characteristics. As CNTs have mechanical qualities that are superior to those of any currently available material, they hold great promise for the creation of fundamentally new material systems in the twenty-first century. SWNTs have become one of the most widely investigated nanostructures due to their unique features. Metal nanoparticles benefit greatly from their intrinsic physical features, which make them perfect supports. Electrodeposition may be used to alter SWNTs for use in catalytic applications. Many mild or oxidative pretreatments are required to prepare raw SWNT material for electrochemical investigations. Customized carbon nanotubes have been made using a variety of therapeutically relevant compounds. The delivery of proteins, nucleic acids, medicines, antibodies, and other therapies using functionalized carbon nanotubes has been confirmed [34, 35].

CARBON NANOSTRUCTURES AND MEDICINAL PLANTS

Most people in underdeveloped countries and more than 60% of the global population still use medicinal plants directly for medical reasons, making conventional medicine the favoured basic healthcare approach in many communities [36].

Among the many reasons for this is the ease of use, affordability, and low costs [37]. Plants have a long history of being used to treat a variety of human ailments. Various plant components, such as the leaf, stem, bark, and root, are utilised to prevent, alleviate, or convert disorders to normal. Due to the fact that the practice of "herbal remedies" does not strictly correspond to scientific facts, orthodox medicine regards "herbal medications" as alternative medicine. However, the majority of pharmaceutical medicines presently prescribed by physicians originated as herbal cures, including opium, aspirin, digitalis, and quinine. Today's medicine relies on active chemicals derived from higher plants, and around 80% of these active components demonstrate a good association between their current therapeutic applications and their historic use [38].

There has been a rise in recent years in both the search for and usage of plant-based medications and drug delivery through nanotechnology. In order to find phytochemicals and hints that may be turned into medications to cure various ailments, scientists such as pharmacologists, microbiologists, botanists, and phytochemists search the Earth.

The application of nanotechnology in herbal medicine, and more especially in medication delivery, is expected to grow quickly in this approach. Nano herbal medication delivery systems offer the potential to improve the activity of medicinal plants while also addressing some of their drawbacks. Nanocarriers assist in curing deadly diseases including cancer, diabetes, and others with herbal compounds [39].

The lack of scientific justification and processing difficulties meant that herbal medicines were not considered for the development of novel formulations for a long time; however, today's modern phytopharmaceutical research can address the pharmacokinetics, mechanism of action, site of action, and dose accuracy requirements of herbal medicines in novel drug delivery systems such as nanoparticles, microemulsions, matrix systems, and solid dispersions to solve these scientific needs. By combining natural pharmaceuticals into contemporary dosage forms, they can be used more effectively and in a better manner. Designing new medicine delivery methods for herbal ingredients is one way to accomplish that goal [40].

THE NEED FOR NANO-SIZED HERBAL REMEDIES DELIVERY SYSTEMS

The nano-sized herbal delivery method was chosen to solve the disadvantages of traditional herbal medicine delivery systems [41].

- The use of nanoparticles to focus herbal medicines to specific organs enhances selectivity, drug delivery, efficacy, and safety, allowing for dosage reduction and increased patient compliance.
- Nanoparticles can be used to boost the solubility of herbal drugs and aid in their localization, resulting in increased effectiveness [42].
- Due to their unusual size and large loading capacity, they appear to be capable of delivering significant quantities of medications to disease locations.
- By delivering the medicine in microscopic particles, the total surface area of the drug is increased, resulting in a more rapid breakdown in circulation.
- Demonstrates an EPR (enhanced permeation and retention) effect, *i.e.*, improved penetration across barriers owing to the tiny size and retention due to inadequate lymphatic outflow, as seen in tumour [43].
- Without the inclusion of any specific ligand moiety, it exhibits passive targeting to the disease site of action.
- Reduces adverse effects.

NANOPARTICLE-BASED DRUG DELIVERY SYSTEM FOR HERBAL MEDICINES

NDDS is the most comprehensive and ongoing strategy for enhancing performance in a consistent manner. This is accomplished by the calculation of NDDSs for herbal extract components [44].

The nanocarriers should take several key aspects into account. The medication should be provided on a need-to-know basis. Nano-sized herbal medicine medications have a great deal of promise for the decorative movement, as they have the potential to solve several obstacles associated with plant medicine. The delivery of herbal medicines using nanocarriers will deliver the most beneficial component of the drug to the targeted site, where it will overcome the full barrier of photocatalytic acid and pH characteristics [45].

Thus, the incorporation of herbal remedies into the real practice of medicine is a distinct method aimed at treating a number of sequential diseases [46].

Nanotechnology enables a novel form of drug release because of the medication's characteristic tininess and restricted discharge. Thus, the incorporation of "herbal medicine" into nanocarriers expands its potential for treating a number of chronic conditions. Due to its potential for expansion from the micro to the molecular level, this branch of medical technology has become the most chosen. The importance of information of health benefits and medication formulation has grown exponentially as the trends in medicine release systems have changed [47].

In recent decades, an herbal medication delivered by nanocarriers has garnered considerable interest due to its future potential and unique features that make these materials useful in a variety of human activities. Thus, nano herbal systems have a potential future for increasing activity and resolving the difficulties associated with plant medicine.

The primary reason for using nanocarriers for herbal medications is that before they enter the bloodstream, their action is destroyed in the stomach's very acidic pH or is metabolised by the liver [48, 49]. Due to the lack of evidence for the drug's therapeutic effect in this suboptimal amount on the affected region, the drug has been formulated using carriers to increase the bioavailability and therapeutic activity of the herbal drug molecule in the affected region and to protect the drug from the acidic environment.

Secondary metabolites, or biologically active molecules generated during secondary metabolism, are essential for medicinal plants phytotherapeutic effects and have been adopted by mankind since the dawn of time. This is mostly owing to the therapeutic properties of their secondary metabolites (some of which are also taken as medications in contemporary medicine) or their structures, which have served as a platform for the creation of new effective synthetic pharmaceuticals [50]. Using nanoparticles that function as elicitors is one technique to boost the synthesis of these secondary metabolites. Nanoparticles, on the other hand, may have a variety of different advantages on medicinal plants, depending on particle size, content, concentration, and application method (*e.g.*, increased plant growth, improved photosynthesis, and overall performance). On either hand, they have the ability to harm plants mechanically, adversely change morphological and biochemical features of plants, and have cytotoxic and genotoxic effects, especially at high doses. The effects of different nanoparticles on the synthesis of beneficial secondary metabolites in these plants grown in hydroponic systems, soil, hairy root, or *in vitro* cultures are given special consideration [51, 52].

CARBON NANOTUBES –PLANT INTERACTION

Due to the exponential development of nanotechnology and artificial nanomaterials, it is important to understand how nanoparticles interact with living species for biosafety. In the case of nanomaterial-plant interactions, this information is vital for ecological risk assessment and for developing nanotechnological applications in agriculture to boost crop yields and minimize pesticide use. According to Nowack and Bucheli, nanomaterial phytotoxicology didn't exist until 2007 [53]. The potential toxicity of nanoparticles on plants has not been well studied, and the current findings are descriptive and inconsistent with minimal information on the underlying mechanisms of action. Gene expression, DNA damage, and ROS production may be altered by nanomaterial interactions. Different plant species, ontogenetic phases and cultivars have very variable responses [54].

Impact of Carbon Nanotubes on Plant Growth and Development

In comparison to fullerenes, there is a lot of research on plant interactions with different forms of carbon nanotubes, including impacts on seed germination, early plant development, cell culture, gene expression, and many physiological functions. SWCNTs seem to be more toxic than MWCNTs owing to their smaller size, and toxicity is enhanced further by the functionalization of nanotubes [55]. Because CNTs have high tensile strength, piercing impacts may cause mechanical injury to tissues.

SWCNTs have mostly been studied in early seedlings grown in aqueous suspensions or in different culture mediums supplemented with SWCNTs. A variety of plant species, including fig plants (*Ficus carica*) [54], maize (*Zea mays*)

[56], and tomato (*Solanum Lycopersicum*) seedlings [57], have been shown to have stimulatory effects on early seedling growth. SWCNTs had a dose-dependent influence on seedling biomass formation in salvia (*Salvia macrosiphon*), pepper (*Capsicum annuum*), and tall fescue (*Festuca arundinacea*), where 10-30 mg L1 of SWCNTs boosted seedling biomass formation, while 40 mg L1 of SWCNTs had a detrimental effect on seedling growth. Blackberry (*Rubus adenotrichos*) produced *in vitro* in a culture media supplemented with functionalized carboxy-SWCNTs (SWCNTs-COOH) showed a similar response [58]. Short-term applications (24 and 48 hours) of SWCNTs functionalized with poly-3-aminobenzenesulfonic acid and non-functionalized SWCNTs in six important crops [cabbage (*brassica oleracea)*, carrot (*Daucus carota*), cucumber (*Crocus sativus*), lettuce (*Lactuca sativa*), onion (*Allium cepa*), and tomato (*Solanum Lycopersicum*)] had contradictory effects. Tomato (*S. Lycopersicum*) root elongation was inhibited by non-functionalized nanotubes, whereas cucumber (*C. sativus*) and onion (*Allium cepa*) root elongation were stimulated (*A. cepa*). Functionalized nanotubes inhibited root elongation in lettuce (*L. sativa*) but had no effect on cabbage (*B. oleracea*) or carrot (*D. carota*) [55]. Small-seeded species including lettuce (*L. sativa*), onion (*A. cepa*), and tomato (*S. Lycopersicum*) seemed to be more sensitive than large-seeded species with a lower surface to volume ratio, providing a smaller surface area for interactions with SWCNTs, according to the authors.

A limited number of studies have connected the impact of SWCNT exposure on plant morphological features with molecular changes. Increased development of seminal roots was connected with increased expression of the relevant genes (SLR1, RTCS) in maize (*Z. mays*) cultivated on Murashige and Skoog medium modified 20 mg L1 SWCNTs, but the suppression of root hair growth was represented by downregulation of root hair-related genes (RTH1, RTH3). Furthermore, SWCNTs, like plants under stress, may increase histone deacetylation, most likely as a response to SWCNT accumulation in the root cortex [56].

The vast majority of studies looking at the impact of MWCNTs on plant growth and development have utilized hydroponics or agar media as growth substrates, with soil culture being used only in a few cases. MWCNT effects have also been explored in the reproductive stage of plant development, in addition to germination and early growth investigations. In tomato (*Solanum Lycopersicum*) cultivated in soil with MWCNT additions, Khodakovskaya *et al.* found a doubling of flower setting and yield, which was not apparent in control soils just treated with activated charcoal [57]. De La Torre-Roche *et al.* on the other hand, found no evidence of MWCNT soil amendments having an impact on zucchini (*Cucurbita pepo*) and tomato plants (*S. Lycopersicum*) [59]. This is most likely owing to the low likelihood of CNT interaction with plant tissues and the restricted mobility of MWCNTs in soil [60].

Changes in gene expression, as well as damage to DNA and chromatin structures, are often linked to morphological features in plants treated with MWCNTs. In response to MWCNT treatments, Ghosh *et al.* found DNA damage, micronucleus production, and chromosomal aberration in onion roots (*A. cepa*) [61]. Application of MWCNTs to the growth medium induced overexpression of various biotic stress-related genes, such as subtilisin-like endoprotease, meloidogyne-induced giant cell protein, and threonine deaminase, in tomato roots (*S. Lycopersicum*), and this was also observed after SWCNT treatments in maize (*Z. mays*) seedlings. Increased expression of aquaporins (water channel proteins) in tomato (*S. Lycopersicum*) seedlings, soybean (*Glycine max*), maize (*Z. mays*), and barley (*Hordeum vulgare*) seedlings treated with MWCNTs seed treatments may represent a similar stress response [54]. Highly scattered MWCNTs with distinct functional groups attached to the surface-induced up-regulation of aquaporin gene expression in tomato seedlings (*S. Lycopersicum*), but MWCNTs in the form of massive aggregates were ineffective [62]. The observations imply that CNTs operate as stress factors, capable of inducing plant defense responses and hormesis effects, as well as toxicities, depending on the degree of the stimulus.

Impact on Secondary Metabolite Production

MWCNTs increased enzyme and antioxidant activity and stimulated the production of several secondary metabolites and antioxidants, with a maximum dosage of 250 g/mL [63]. *In vitro* treatment of *Satureja khuzistanica Jamzad* plants with 250 mg/L MWCNTs resulted in a RA content of 140.49 mg/g (d.w.), but *in vivo* exposure to *S. khuzistanica* to MWCNTs for 24 h resulted in a RA content of 7.13 mg RA/g (d.w.). Ghorbanpour and Hadian [64] previously demonstrated that MWCNTs stimulated callus induction, secondary metabolite production, and increased antioxidant activity in the medicinal plant *S. khuzestanica* cultivated *in vitro*.

Spraying two-month-old *Salvia verticillata* L leaves with MWCNTs (0–1000mg/L) produced absorption through the epidermal cells layer into the parenchymal cells of the exposed leaves; the treatment impaired assimilation pigments and generated dose-dependent oxidative stress. The activity and gene expression patterns of RA synthase is linked with RA accumulation at 50 and 1000 mg/L MWCNTs. Lower ROS levels at lower MWCNT doses promoted secondary metabolite synthesis, but larger amounts enhanced oxidative stress [54]. *Salvia nemorosa* callus irradiated at 70 Gy produced a high-yielding cell line that produced 18.53, 5.21, 1.9, and 7.59 mg/g d.w. of RA, salvianolic acid B, ferulic acid, and cinnamic acid, respectively. Irradiated callus cell suspension culture stimulated with 100 mg/L MWCNT-COOH (20–30 nm) produced 268.47 g/L fresh and 22.17 g/L dried biomass. Both -irradiation and MWCNT-COOH increased antioxidant activity. These results revealed that increased secondary metabolite synthesis in *S. nemorosa* by combining -irradiation with MWCNT-COOH might be used for large-scale phenolic compound production [65].

MWCNTs boosted biomass production in *Catharanthus roseus* callus culture maintained in the dark, and encouraged the biosynthesis of total generated alkaloids, resulting in a considerable improvement in the production of vinblastine and vincristine alkaloids [66].

CYTOTOXICITY AND GENOTOXICITY OF CARBON NANOTUBES

MWCNT's uptake in *Allium cepa* root cells altered cellular morphology, harmed membrane integrity and mitochondrial function, and resulted in significant DNA damage, micronucleus formation, chromosome aberration, and the formation of inter nucleosomal fragments, all of which are signs of apoptotic cell death. The cyto-genotoxic action of MWCNT was confirmed by the accumulation of cells in the sub-G0 phase of the cell cycle, as well as significant increases in CpG methylation and 5-methyl-deoxycytidine levels [61].

For obtaining these bioactive acids on a larger scale, the use of MWCNT-COOH as an elicitor of rosmarinic acid and salvianolic acid B in a cell suspension culture prepared from -irradiated (70- Gy) *Salvia nemorosa* L callus, which allows for >13- and 14-fold higher concentrations of these secondary metabolites than in the control, can be recommended [67].

CONCLUSION

Carbon nanostructures have been used in biomedical research since decades distinctively in relation to medication and protein administration for the treatment of cancer and other chronic diseases. Beside this, they are also used in diagnostics and imaging equipments because of their near infrared optical characteristics. Current chapter shed light on different methods used for the green synthesis of CNS as well as their use in therapeutics for delivering herbal chemicals and overcome the short comings associated with the use of phyto-drugs.

REFERENCES

[1]   Jiang J, Yang J, Lin J, Huang Z, Wang SC. Carbon nanomaterials and related nanostructures: Synthesis, characterization and application, J. Nanomat 2014; pp. 1-0.

[2]   Mintmire JW, Dunlap BI, White CT. Are fullerene tubules metallic? Phys Rev Lett 1992; 68(5): 631-

4.
[http://dx.doi.org/10.1103/PhysRevLett.68.631] [PMID: 10045950]

[3]   Ajayan PM, Ebbesen TW. Nanometre-size tubes of carbon. Rep Prog Phys 1997; 60(10): 1025-62.
[http://dx.doi.org/10.1088/0034-4885/60/10/001]

[4]   Iijima S, Ichihashi T. Single-shell carbon nanotubes of 1-nm diameter. Nature 1993; 363(6430): 603-5.
[http://dx.doi.org/10.1038/363603a0]

[5]   Bethune DS, Kiang CH, de Vries MS, *et al.* Cobalt-catalysed growth of carbon nanotubes with single-atomic-layer walls. Nature 1993; 363(6430): 605-7.
[http://dx.doi.org/10.1038/363605a0]

[6]   Endo M, Takeuchi K, Igarashi S, Kobori K, Shiraishi M, Kroto HW. The production and structure of pyrolytic carbon nanotubes (PCNTs). J Phys Chem Solids 1993; 54(12): 1841-8.
[http://dx.doi.org/10.1016/0022-3697(93)90297-5]

[7]   Dai H, Rinzler AG, Nikolaev P, Thess A, Colbert DT, Smalley RE. Single-wall nanotubes produced by metal-catalyzed disproportionation of carbon monoxide. Chem Phys Lett 1996; 260(3-4): 471-5.
[http://dx.doi.org/10.1016/0009-2614(96)00862-7]

[8]   Guo T, Nikolaev P, Thess A, Colbert DT, Smalley RE. Catalytic growth of single-walled manotubes by laser vaporization. Chem Phys Lett 1995; 243(1-2): 49-54.
[http://dx.doi.org/10.1016/0009-2614(95)00825-O]

[9]   Thess A, Lee R, Nikolaev P, *et al.* Crystalline ropes of metallic carbon nanotubes. Science 1996; 273(5274): 483-7.
[http://dx.doi.org/10.1126/science.273.5274.483] [PMID: 8662534]

[10]   Wani TU, Mohi-ud-din R, Wani TA, *et al.* Green synthesis, spectroscopic characterization and biomedical applications of carbon nanotubes. Curr Pharm Biotechnol 2021; 22(6): 793-807.
[http://dx.doi.org/10.2174/1389201021999201110205615] [PMID: 33176640]

[11]   Tang SLY, Smith RL, Poliakoff M. Principles of green chemistry: Productively. Green Chem 2005; 7(11): 761-2.
[http://dx.doi.org/10.1039/b513020b]

[12]   Hakim YZ, Yulizar Y, Nurcahyo A, Surya M. Green synthesis of carbon nanotubes from coconut shell waste for the adsorption of pb(ii) ions. Acta Chimica Asiana 2018; 1(1): 6-10.
[http://dx.doi.org/10.29303/aca.v1i1.2]

[13]   Tripathi N, Pavelyev V, Islam SS. Synthesis of carbon nanotubes using green plant extract as catalyst: Unconventional concept and its realization. Appl Nanosci 2017; 7(8): 557-66.
[http://dx.doi.org/10.1007/s13204-017-0598-3]

[14]   Alam MS, Garg A, Pottoo FH, *et al.* Gum ghatti mediated, one pot green synthesis of optimized gold nanoparticles: Investigation of process-variables impact using Box-Behnken based statistical design. Int J Biol Macromol 2017; 104(Pt A): 758-67.
[http://dx.doi.org/10.1016/j.ijbiomac.2017.05.129] [PMID: 28601649]

[15]   Gao L, Li R, Sui X, Li R, Chen C, Chen Q. Conversion of chicken feather waste to N-doped carbon nanotubes for the catalytic reduction of 4-nitrophenol. Environ Sci Technol 2014; 48(17): 10191-7.
[http://dx.doi.org/10.1021/es5021839] [PMID: 25089346]

[16]   Hamid ZA, Azim AA, Mouez FA, Rehim SSA. Challenges on synthesis of carbon nanotubes from environmentally friendly green oil using pyrolysis technique. J Anal Appl Pyrolysis 2017; 126: 218-29.
[http://dx.doi.org/10.1016/j.jaap.2017.06.005]

[17]   Paul S, Samdarshi SK. Carbon microtubes produced from coconut oil. N Carbon Mater 2010; 25(5): 321-4.
[http://dx.doi.org/10.1016/S1872-5805(09)60036-6]

[18]   Kumar R, Tiwari RS, Srivastava ON. Scalable synthesis of aligned carbon nanotubes bundles using green natural precursor: Neem oil. Nanoscale Res Lett 2011; 6(1): 92.
[http://dx.doi.org/10.1186/1556-276X-6-92] [PMID: 21711585]

[19]   Kang Z, Wang E, Gao L, *et al.* One-step water-assisted synthesis of high-quality carbon nanotubes directly from graphite. J Am Chem Soc 2003; 125(45): 13652-3.
[http://dx.doi.org/10.1021/ja037399m] [PMID: 14599190]

[20]   Wang J, Wang CF, Chen S. Amphiphilic egg-derived carbon dots: Rapid plasma fabrication, pyrolysis process, and multicolor printing patterns. Angew Chem Int Ed 2012; 51(37): 9297-301.
[http://dx.doi.org/10.1002/anie.201204381] [PMID: 22907831]

[21]   Akhavan O, Bijanzad K, Mirsepah A. Synthesis of graphene from natural and industrial carbonaceous wastes. RSC Advances 2014; 4(39): 20441-8.
[http://dx.doi.org/10.1039/c4ra01550a]

[22]   Suriani AB, Dalila AR, Mohamed A, *et al.* Vertically aligned carbon nanotubes synthesized from waste chicken fat. Mater Lett 2013; 101: 61-4.
[http://dx.doi.org/10.1016/j.matlet.2013.03.075]

[23]   Sui Z, Meng Q, Zhang X, Ma R, Cao B. Green synthesis of carbon nanotube–graphene hybrid aerogels and their use as versatile agents for water purification. J Mater Chem 2012; 22(18): 8767-87771.
[http://dx.doi.org/10.1039/c2jm00055e]

[24]   Nguyen VH, Shim JJ. Green synthesis and characterization of carbon nanotubes/polyaniline nanocomposites. J Spectro 2015; pp. 1-9.

[25]   Harrison BS, Atala A. Carbon nanotube applications for tissue engineering. Biomaterials 2007; 28(2): 344-53.
[http://dx.doi.org/10.1016/j.biomaterials.2006.07.044] [PMID: 16934866]

[26]   Prato M, Kostarelos K, Bianco A. Functionalized carbon nanotubes in drug design and discovery. Acc Chem Res 2008; 41(1): 60-8.
[http://dx.doi.org/10.1021/ar700089b] [PMID: 17867649]

[27]   Jain KK. Advances in use of functionalized carbon nanotubes for drug design and discovery. Expert Opin Drug Discov 2012; 7(11): 1029-37.
[http://dx.doi.org/10.1517/17460441.2012.722078] [PMID: 22946637]

[28]   Yang W, Thordarson P, Gooding JJ, Ringer SP, Braet F. Carbon nanotubes for biological and biomedical applications. Nanotechnology 2007; 18(41): 412001.
[http://dx.doi.org/10.1088/0957-4484/18/41/412001]

[29]   Foldvari M, Bagonluri M. Carbon nanotubes as functional excipients for nanomedicines: II. Drug delivery and biocompatibility issues. Nanomedicine 2008; 4(3): 183-200.
[http://dx.doi.org/10.1016/j.nano.2008.04.003] [PMID: 18550450]

[30]   Wong Shi Kam N, Dai H. Single walled carbon nanotubes for transport and delivery of biological cargos. Phys Status Solidi, B Basic Res 2006; 243(13): 3561-6.
[http://dx.doi.org/10.1002/pssb.200669226]

[31]   Im JS, Bai BC, Lee YS. The effect of carbon nanotubes on drug delivery in an electro-sensitive transdermal drug delivery system. Biomaterials 2010; 31(6): 1414-9.
[http://dx.doi.org/10.1016/j.biomaterials.2009.11.004] [PMID: 19931904]

[32]   Kam NWS, Dai H. Carbon nanotubes as intracellular protein transporters: Generality and biological functionality. J Am Chem Soc 2005; 127(16): 6021-6.
[http://dx.doi.org/10.1021/ja050062v] [PMID: 15839702]

[33]   Malmsten M. Soft drug delivery systems. Soft Matter 2006; 2(9): 760-9.
[http://dx.doi.org/10.1039/b608348j] [PMID: 32680216]

[34]   Vairavapandian D, Vichchulada P, Lay MD. Preparation and modification of carbon nanotubes:

Review of recent advances and applications in catalysis and sensing. Anal Chim Acta 2008; 626(2): 119-29.
[http://dx.doi.org/10.1016/j.aca.2008.07.052] [PMID: 18790113]

[35]   Prakash S, Kulamarva A. Recent advances in drug delivery: Potential and limitations of carbon nanotubes. Recent Pat Drug Deliv Formul 2007; 1(3): 214-21.
[http://dx.doi.org/10.2174/187221107782331601] [PMID: 19075888]

[36]   Shrestha PM, Dhillion SS. Medicinal plant diversity and use in the highlands of dolakha district, nepal. J Ethnopharmacol 2003; 86(1): 81-96.
[http://dx.doi.org/10.1016/S0378-8741(03)00051-5] [PMID: 12686446]

[37]   Asase A, Kokubun T, Grayer RJ, *et al.* Chemical constituents and antimicrobial activity of medicinal plants from ghana: *Cassia sieberiana, haematostaphis barteri, mitragyna inermis* and *pseudocedrela kotschyi.* Phytother Res 2008; 22(8): 1013-6.
[http://dx.doi.org/10.1002/ptr.2392] [PMID: 18618525]

[38]   Sarkar S, Zaidi S, Chaturvedi AK, Srivastava R, Dwivedi PK, Shukla R. Search for a herbal medicine: Antiasthmatic activity of methanolic extract of *Curcuma longa.* J Pharmacogn Phytochem 2015; 3: 59-72.

[39]   Goyal A, Kumar S, Nagpal M, Singh I, Arora S. Potential of novel drug delivery systems for herbal drugs. Indian J Pharm Educ Res 2011; 45: 225-35.

[40]   Bonifácio BV, Silva PB, Ramos MA, Negri KM, Bauab TM, Chorilli M. Nanotechnology-based drug delivery systems and herbal medicines: A review. Int J Nanomedicine 2014; 9: 1-15.
[PMID: 24363556]

[41]   Ansari SH, Sameem M, Islam F. Influence of nanotechnology on herbal drugs: A Review. J Adv Pharm Technol Res 2012; 3(3): 142-6.
[http://dx.doi.org/10.4103/2231-4040.101006] [PMID: 23057000]

[42]   Sharma M. Applications of nanotechnology based dosage forms for delivery of herbal drugs. Res rev j pharm nanotechnol 2014; 2: 23-30.

[43]   Chidambaram M, Manavalan R, Kathiresan K. Nanotherapeutics to overcome conventional cancer chemotherapy limitations. J Pharm Pharm Sci 2011; 14(1): 67-77.
[http://dx.doi.org/10.18433/J30C7D] [PMID: 21501554]

[44]   Yadav D, Suri S, Choudhary AA, Sikender M. Novel approach: Herbal remedies and natural products in pharmaceutical science as nano drug delivery systems. Int J Pharm Tech 2011; 3: 3092-116.

[45]   Singh RP, Singh SG, Naik H, Jain D, Bisla S. Herbal excipients in novel drug delivery system. Int J Comp Pharm 2011; 2: 1-7.

[46]   Bairwa N, Sethiya N, Mishra SH. Protective effect of stem bark of *ceiba pentandra* linn. against paracetamol-induced hepatotoxicity in rats. Pharmacognosy Res 2010; 2(1): 26-30.
[http://dx.doi.org/10.4103/0974-8490.60584] [PMID: 21808535]

[47]   Wu XY, Lee PI. Preparation and characterization of thermal- and pH-sensitive nanospheres. Pharm Res 1993; 10(10): 1544-7.
[http://dx.doi.org/10.1023/A:1018900114881] [PMID: 8272420]

[48]   Swamy MK, Sinniah UR. Patchouli (*pogostemon cablin* benth.): Botany, agrotechnology and biotechnological aspects. Ind Crops Prod 2016; 87: 161-76.
[http://dx.doi.org/10.1016/j.indcrop.2016.04.032]

[49]   Siddiqui AA, Iram F, Siddiqui S, Sahu K. Role of natural products in drug discovery process. Int J drug Dev Res 2014; 6: 172-204.

[50]   Kralova K, Jampilek J. Responses of medicinal and aromatic plants to engineered nanoparticles. Appl Sci 2021; 11(4): 1813.
[http://dx.doi.org/10.3390/app11041813]

[51]   Sharon M, Ed. History of nanotechnology: From prehistoric to modern times Advances in nanotechnology & applications. Hoboken, NJ, USA: Wiley 2019.
[http://dx.doi.org/10.1002/9781119460534]

[52]   Shukla AK, Iravani S, Eds. Green synthesis, characterization and applications of nanoparticles. Micro & Nano Technologies Series. Amsterdam: Elsevier 2019; p. 523.

[53]   Nowack B, Bucheli TD. Occurrence, behaviour and effects of nanoparticles in the environment. Environ Pollut Barking Essex 2007; 150: 5-22.

[54]   Zaytseva O, Neumann G. Carbon nanomaterials: Production, impact on plant development, agricultural and environmental applications. Chem Biol Technol Agric 2016; 3(1): 17.
[http://dx.doi.org/10.1186/s40538-016-0070-8]

[55]   Cañas JE, Long M, Nations S, *et al.* Effects of functionalized and nonfunctionalized single-walled carbon nanotubes on root elongation of select crop species. Environ Toxicol Chem 2008; 27(9): 1922-31.
[http://dx.doi.org/10.1897/08-117.1] [PMID: 19086209]

[56]   Yan S, Zhao L, Li H, *et al.* Single-walled carbon nanotubes selectively influence maize root tissue development accompanied by the change in the related gene expression. J Hazard Mater 2013; 246-247: 110-8.
[http://dx.doi.org/10.1016/j.jhazmat.2012.12.013] [PMID: 23291336]

[57]   Khodakovskaya MV, de Silva K, Nedosekin DA, *et al.* Complex genetic, photothermal, and photoacoustic analysis of nanoparticle-plant interactions. Proc Natl Acad Sci 2011; 108(3): 1028-33.
[http://dx.doi.org/10.1073/pnas.1008856108] [PMID: 21189303]

[58]   Flores D, Chacón R, Alvarado L, Schmidt A, Alvarado C, Chaves J. Effect of using two different types of carbon nanotubes for blackberry *(rubus adenotrichos) in vitro* plant rooting, growth and histology. Am J Plant Sci 2014; 5(24): 3510-8.
[http://dx.doi.org/10.4236/ajps.2014.524367]

[59]   De La Torre-Roche R, Hawthorne J, Deng Y, *et al.* Multiwalled carbon nanotubes and c60 fullerenes differentially impact the accumulation of weathered pesticides in four agricultural plants. Environ Sci Technol 2013; 47(21): 12539-47.
[http://dx.doi.org/10.1021/es4034809] [PMID: 24079803]

[60]   Kasel D, Bradford SA, Šimůnek J, Pütz T, Vereecken H, Klumpp E. Limited transport of functionalized multi-walled carbon nanotubes in two natural soils. Environ Pollut 2013; 180(180): 152-8.
[http://dx.doi.org/10.1016/j.envpol.2013.05.031] [PMID: 23770315]

[61]   Ghosh M, Bhadra S, Adegoke A, Bandyopadhyay M, Mukherjee A. MWCNT uptake in *Allium cepa* root cells induces cytotoxic and genotoxic responses and results in DNA hyper-methylation. Mutat Res 2015; 774: 49-58.
[http://dx.doi.org/10.1016/j.mrfmmm.2015.03.004] [PMID: 25829105]

[62]   Villagarcia H, Dervishi E, de Silva K, Biris AS, Khodakovskaya MV. Surface chemistry of carbon nanotubes impacts the growth and expression of water channel protein in tomato plants. Small 2012; 8(15): 2328-34.
[http://dx.doi.org/10.1002/smll.201102661] [PMID: 22514121]

[63]   Mondal A, Basu R, Das S, Nandy P. Beneficial role of carbon nanotubes on mustard plant growth: an agricultural prospect. J Nanopart Res 2011; 13(10): 4519-28.
[http://dx.doi.org/10.1007/s11051-011-0406-z]

[64]   Lin D, Xing B. Phytotoxicity of nanoparticles: inhibition of seed germination and root growth. Environ Pollut Barking Essex 2007; 150: 243-50.
[http://dx.doi.org/10.1016/j.envpol.2007.01.016]

[65]   Wang X, Han H, Liu X, Gu X, Chen K, Lu D. Multi-walled carbon nanotubes can enhance root

elongation of wheat *(Triticum aestivum)* plants. J Nanopart Res 2012; 14(6): 841.
[http://dx.doi.org/10.1007/s11051-012-0841-5]

[66]   McGehee DL, Alimohammadi M, Khodakovskaya MV. Carbon-based nanomaterials as stimulators of production of pharmaceutically active alkaloids in cell culture of *catharanthus roseus*. Nanotechnology 2019; 30(27): 275102.
[http://dx.doi.org/10.1088/1361-6528/ab1286] [PMID: 30901766]

[67]   Heydari HR, Chamani E, Esmaielpour B. Cell line selection through gamma irradiation combined with multi-walled carbon nanotubes elicitation enhanced phenolic compounds accumulation in *salvia nemorosa* cell culture. Plant Cell Tiss Org Cult 2020; 142(2): 353-67.
[http://dx.doi.org/10.1007/s11240-020-01867-6]

CHAPTER 16

Nano Elicitors and Bioactive Plant Metabolites

Yamin Bibi[1,*], **Sobia Nisa**[2], **Kulsoom Zahara**[1] and **Abdul Qayyum**[3]

[1] *Department of Botany, PMAS-Arid Agriculture University Rawalpindi, Rawalpindi-46300, Pakistan*

[2] *Department of Microbiology, The University of Haripur, Haripur, 22620, Pakistan*

[3] *Department of Agronomy, The University of Haripur, Haripur, 22620, Pakistan*

Abstract: Nature has given plants the ability to produce a wide variety of secondary metabolites including alkaloids, phenolics, terpenoids and saponins. These metabolites provide them a defense mechanism against biological and non-biological stress factors. On the other hand, the same metabolites have proved to be effective against different dreadful human diseases. The efficacy of such metabolites ranges from antimicrobial to anticancerous effects. Bioactivity-guided characterization is one of the useful strategies that have been employed to identify, purify and characterize active components. These bioactive components have proved useful in future drug discovery. Elicitors are defined as signaling metabolites with the ability to induce biochemical and physiological processes in plants resulting in the activation of plants defense mechanisms. Elicitation is a useful tool as it leads to the generation of stress conditions and hence the accumulation of bioactive secondary metabolites in plants. Various strategies have been adopted to enhance the production of bioactive secondary metabolites including plant cell and tissue culture and use of signaling metabolites. Nowadays, nano-elicitors have emerged as an effective tool to enhance the production of pharmacologically important compounds. Various classes of nanoparticles (NPs) have been reported to be utilized as nano-elicitors like metallic NPs, metallic oxide NPs and carbon nanotubes with positive effects on phytochemical profile. The possible mechanism of nanomaterials as elicitors is the interaction with plant genomes by increasing the expression level of genes involved in the biosynthesis of active metabolites. Despite triggering biosynthetic potential of plants, certain negative effects have been observed in plants' primary metabolism like lower chlorophyll content, a decrease in cell viability, a decline in sugar content and suppressed seed germination. Thus, there is a need to develop biocompatible nanoparticles for use as nanoelicitors in plants to avoid the negative impacts of the used entities.

Keywords: Bioactive metabolites, biocompatibility, carbon nanotubes (CNT), nano-elicitors, nano-tubes.

---

* **Corresponding author Yamin Bibi**: Department of Botany, PMAS-Arid Agriculture University Rawalpindi, Rawalpindi-46300, Pakistan; E-mail: dryaminbibi@uaar.edu.pk

Zulqurnain Khan, Azra Yasmin & Naila Safdar (Eds.)

INTRODUCTION

Phytochemistry is the foundation of medicinal practices based on herbal products [1]. A good grasp of plant biochemistry can lead to a better comprehension of its potential medicinal value. Primary metabolites' role in essential life functions, such as growth, reproduction, cell division, respiration, and storage, has been described by modern chemistry [2]. By-products of activities like the Krebs or citric acid cycle, glycolysis, photosynthesis, and related pathways produce these metabolites.

Phytochemicals are a group of biochemical molecules produced by the plant cell due to the primary metabolic pathways. Albrecht Kossel, the Nobel Laureate in Physiology or Medicine in 1910, coined the term "secondary metabolite."Czapek, on the other hand, classified them as end-products after thirty years. Secondary metabolites, according to Czapek, are formed as a result of secondary nitrogen metabolism changes. Progress in analytical techniques, such as chromatography, enabled the isolation of an increasing number of these compounds in the mid-twentieth century, laying the groundwork for establishing phytochemistry as a study.

Phytochemicals with antiviral, antibacterial, and antifungal properties can protect plants from infections. They also produce UV-absorbing chemicals, which protect the leaves from the harmful effects of direct sunlight. Some forage grasses, such as clover or alfalfa, have been found to have estrogenic characteristics and interact with the fertility of grazing animals. These metabolites are natural sources of biologically active compounds that are widely utilized in commercial and healthcare industries. Plants have attributed a significant role in both Western and conventional medicine systems. For thousands of years, plant-derived therapeutic agents have been a part of the evolution of human healthcare [3].

Classes of Bioactive Metabolites

Plant metabolites are classified into various groups based on their chemical structures. These classes include:

Phenolics

Compounds with hydroxylated aromatic rings, with the hydroxy group connected directly to the phenyl, substituted phenyl, or other aryl groups, are phenolic compounds. With over 8000 known structures, phenolic compounds are extensively dispersed in plants and are among the most abundant secondary metabolites [4]. Plant phenolics are primarily responsible for defense against predators, pathogens, parasites, and ultraviolet radiation. They are universally found in all plant parts and are an integral part of the human diet [5].

Phenolics are well-known components of plant meals (cereals, legumes, fruits, vegetables, olive oil, chocolate, *etc.* and beverages (beer, tea, wine, coffee, and so on), and are partially responsible for the organoleptic characteristics of these foods. For instance, they add to the astringency and bitterness of fruits. Flavonoids, phenolic acids, tannins, and the less well-known stilbenes and lignans, are all examples of plant phenolics [6]. There are two types of phenolic acids: benzoic acid derivatives like gallic acid and cinnamic acid derivatives like coumaric, caffeic, and ferulic acid. Caffeic acid is the most abundant phenolic acid in many fruits and vegetables, and it is frequently esterified with quinic acid, as in chlorogenic acid, the most abundant phenolic compound in coffee. Ferulic acid, which is found in cereals and is esterified to hemicelluloses in the cell wall, is another frequent phenolic acid [7].

Flavonoids

Flavonoids are a type of polyphenol that is abundant in human diets. The flavan nucleus is its basic structure, consisting of 15 carbon atoms grouped in three rings (C6-C3-C6). Flavones, flavonols, flavonols, flavanones, isoflavones, and anthocyanins are the six subgroups of flavonoids [8]. Because of the degree of methoxylation, glycosylation, prenylation, and hydroxylation in each subgroup, structural diversity in each subgroup is relatively high. Quercetin, a flavonol abundantly found in broccoli, onion, and apple; catechin, an anthocyanin found in tea; cyanidin-glycoside, an anthocyanin found in berries; naringenin, a flavanone abundantly found in grapefruit; and isoflavones, glycitein, and daidzein in soybean.

Tannins

Tannins are another important class of polyphenols involved in human diets and can be categorized into two classes. Gallotannins and ellagitannins are two types of hydrolyzable tannins with a core of glucose or another polyol esterified with gallic acid or hexahydroxydiphenic acid. The various options in oxidative linking are primarily responsible for their structural diversity. Many oligomeric compounds with molecular weights of 2,000 to 5,000 Daltons result from intermolecular oxidation among molecules.

Flavan-3-ol oligomers or polymers are linked *via* an interflavan carbon bond in condensed tannins. They are sometimes referred to as proanthocyanidins because, when heated in a low pH alcoholic solution, they break down into anthocyanidins *via* an acid-catalyzed oxidation reaction [9].

Alkaloids

Alkaloids are a large group of chemical compounds with a heterocyclic ring, including nitrogen atom(s). Their definition is challenging since these chemicals do not, from any perspective, characterize a homogeneous group of compounds [10]. Based on chemical structure, alkaloids can be divided into different types, *e.g.*, phenyl isoquinolines, phenylethylamines, piperidines, indoles, acridones, aromatics, quinolines, quinozolines, isoquinolines, carbolines, ephedras, ergots, bisindoles, pyrrolidines, pyrroloindoles, pyridines, purines, pyrrolizidines, tropanes, simple tetrahydroisoquinolines, terpenoids and steroids [11].

Though humans have been using alkaloids as teas, potions, and medicine for at least 3000 years, their identification was not made until the nineteenth century. Nowadays, over 3000 alkaloids are recognized in over different 4000 plant species. Alkaloids exhibit various pharmacological properties, including analgestics, local anesthetics, cardiotonics, respiratory relaxants, vasoconstriction, muscle relaxants, and antineoplastic [12].

Examples of some alkaloids:

Morphine

Morphine is a well-known alkaloid and is commonly utilized in medicine. It is a powerful narcotic used to treat pain, but its usefulness is limited due to its addictive characteristics. Codeine, a Methyl ether derivative produced in the opium poppy, has analgesic properties and is non-addictive [10]

Nicotine

Nicotine is isolated from tobacco plants *viz. Nicotiana tabacum*. Nicotine has tranquilizing properties and addictive properties. This compound is tremendously toxic, triggering respiratory paralysis at high dosages. It is a multimodal ganglion cholinergic-receptor agonist with actions mediated by binding to receptors in the adrenal medulla, autonomic ganglia, neuromuscular junction, and brain [13].

Caffeine

Caffeine occurs in numerous plant species like tea (Camellia sinensis), guarana (Paullinia cupana), coffee (Coffea spp.), kola (Cola acuminata), and mate (Ilex paraguariensis). In raw coffee beans, it is bounded to chlorogenic acid and released in the roasting process. Therapeutically, it pertains to diuretic properties and acts as a stimulant of the central nervous system, cardiovascular and respiratory systems.

Vinblastine

Vinblastine has been used to treat diabetes and high blood pressure since identified from *Catharanthus roseus* G. Along with the other vinca alkaloids, vinorelbine, vincristine, and vindesine, this molecule is crucial for cancer prevention.

Saponins

Saponins are bioactive molecules containing a polycyclic aglycone moiety connected to a monosaccharide or oligosaccharide chain and either a steroid or triterpenoid. Saponins have been found in over 500 plants belonging to at least 90 families. They have been found in all plant parts, including leaves, fruits, stems, flowers, roots, and bulbs, but are accumulated in the roots of most plant species, including *Eleutherococcus senticosus, Gentiana lutea, Glycyrrhiza spp., Digitalis purpurea, Dioscorea villosa,* and *Panax ginseng*, among others [14].

Saponins have been discovered to have a variety of pharmacological effects. Sedative, anticancer, molluscicidal, piscicidal, spermicidal, expectorant, and analgesic effects, to name a few. The expectorant and antitussive effects of glycyrrhizin extracted from glycyrrhiza radix are well established. It is also utilized for treating chronic hepatitis. Anti-inflammatory activities of saponins extracted from *Bupleurum falcatum* (Apiaceae) and *Phytolacca americana* roots are said to have anti-inflammatory qualities in Korean medicine [15].

Terpenes

Terpenes are the most diverse and extensive secondary metabolites found in plants. The word "terpene" is derived from the word "turpentine," which means "resin." Terpenes are composed of 5-carbon isoprene units that can be arranged in various ways. The number of isoprene units in a molecule affects its classification [16].

Hemiterpenes

Hemiterpenes constitute only one isoprene unit. Hemiterpenoids include the isoprene unit and its oxygen-containing derivatives, such as angelic acid (*Angelica archangelica*) and isovaleric acid (*Vaccinium myrtillus*).

Monoterpenes

Monoterpenes constitute two isoprene molecules with a molecular formula $C_{10}H_{16}$. They are the chief constituents of plant essential oils. They occur in a variety of plant families, *i.e.*, Apiaceae, Lamiaceae, Rutaceae, and Pinaceae

Monoterpenes are classified into aldehydes (citronellal), ketones (Carvone), unsaturated hydrocarbons (limonene), alcohol esters (linalyl acetate), and alcohols (linalool). These compounds have varied therapeutic potential. Compounds like menthol and camphor are utilized as anti-itching and analgesic agents.

Sesquiterpenes

Sesquiterpenes contain three isoprene units and have the chemical formula $C_{15}H_{24}$. Sesquiterpenes have been classified into over 200 structural categories so far. These molecules can be categorized as: acyclic (farnesol), monocyclic (bisabolol), and bicyclic (bisabolol) (caryophyllene). Antiprotozoal, antibacterial, and antifungal properties can exist in several Sesquiterpene lactones. *Vernonia colorata* sesquiterpenes inhibit Entamoeba histolytica, Helenalin from *Arnica montana* flowers has cardiotonic qualities, and Atractylodis rhizoma from *Atractylodis macrocephala*, is used as a painkiller, diuretic, and anti-inflammatory in therapeutic practice.

Diterpenes

Diterpenes have four isoprene units with the chemical formula $C_{20}H_{32}$. Acyclic and macrocyclic compounds are the two types of chemicals. The ring systems present in macrocyclic diterpenes are also categorized as 6-membered ringed structures, as well as fused 5- and 7-membered ringed structures. Vitamin K1 and Vitamin A are diterpenes. Like all other terpene groups, Diterpenes offer a wide range of pharmacological activities, such as antibacterial, antineoplastic, antifungal, analgesic, antiprotozoal activities, and anti-inflammatory.

Sesterterpenes

Terpenes have 25 carbon atoms and five isoprene units from seed oils of *Camellia spp.* Geranyl farnesol exhibits cytotoxic activity.

Triterpenes

Triterpenes have a chemical formula of $C_{30}H_{48}$ and are composed of six isoprene units with more than 4000 triterpenoids identified as triterpenes that make up a substantial portion of the lipid constituents of all plants. They act as precursors of steroids in both plants and animals. There are around 40 primary forms of triterpenes and steroids.

BIOACTIVITIES OF PLANT-BASED NATURAL PRODUCTS

Natural chemicals derived from plants and their derivatives have long been known to be rich sources of therapeutic substances. They provide unlimited possibilities for discovering novel drugs, typically due to a plethora of diverse chemicals. Phytochemicals' practical biological actions have been firmly stated by scientific groups worldwide. According to the researchers, plant secondary metabolites have antioxidant, anticancer, analgesic, antibacterial, antidiarrheal, and wound-healing effects [17].

Antioxidant Activity

Antioxidants are compounds that prevent oxidative damage by quenching reactive oxygen species. All living organisms have a defense mechanism against free radicals that stabilize free radicals, but when the production level of these free radicals surpasses, oxidative stress is generated [18]. Many antioxidant-based drugs have been used to treat diseases caused by this oxidative stress. However, synthetic antioxidants are restricted due to their carcinogenic properties; therefore, interest in natural antioxidants has been increasing. Almost all of the plant species examined have high antioxidant potential, and two-thirds of them are regarded to have medicinal potential. The discovery and subsequent isolation of ascorbic acid from plant sources prompted researchers to pay attention to exogenous plant antioxidants. Since then, the antioxidant activities of plants have attracted a great deal of consideration. Exacerbated oxidative stress has long been recognized as a significant contributor to the progression of various life-threatening diseases [19].

There are now 19 *in vitro* and 10 *in vivo* methods for determining antioxidant activity that is frequently used to evaluate sample antioxidant activity [20]. Up to now, thousands of secondary metabolites have been recognized in plants. Structurally these metabolites are either nitrogen-deficient (terpenoids and phenolics) or nitrogen-containing (alkaloids). Alkaloids are found in around 20% of plant species, with tropane, terpenoid, purine, and indole alkaloids being the most common. However, the radical scavenging action of alkaloids appears to be moderate in in-conditions. Monoterpenes, sesquiterpenes, and diterpenes have been recognised as having significant antioxidant properties among terpenoids [21]. In both *in vivo* and *in vitro* experiments, carotenoids and tetraterpenes were found to exhibit potent antioxidant action; however, a few highly prized carotenoids, such as beta-carotene, demonstrated prooxidant effects at high concentrations. When compared to all secondary metabolites, phenolic antioxidants appear to be the most important, as they have been shown to have a beneficial antioxidant effect in both *in vivo* and *in vitro* tests [22].

Production of Reactive Oxygen Species and Defensive System in Plants

Mitochondria and Chloroplasts are the two main places of reactive oxygen species production within plant cells. They are also in charge of maintaining a delicate balance between energy production and control of ROS generation. Peroxisomes are a third significant site of reactive oxygen species production, *i.e.*, superoxide ($O_2\bullet$ -), nitric oxide (NO$\bullet$), and hydrogen peroxide (H_2O_2) (Parcheta *et al.*, 2021).

In plants, reactive oxygen species are generated at photosystems I and II of the chloroplasts membrane, complex I, complex III, and ubiquinone of the mitochondrial electron transport chain and matrix of the peroxisome. The general procedure of free radical production is shown in Fig. (1).

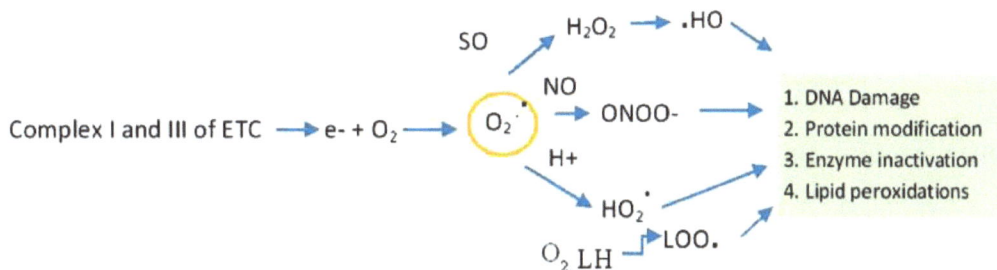

Fig. (1). Outline of free radical production.

To avoid the harmful effects of free radicals, plants have excellent enzymatic and non-enzymatic defensive systems. Enzymatic systems comprise, glutathione peroxidase (GPx), catalase (CAT), SOD, and glutathione reductase (GR), whereas non-enzymatic systems involve the production of antioxidants *i.e.*, ascorbic acid, phenolic acids, proline, carotenoids, glutathione, flavonoids, *etc* [19].

Methods to Analyze Plants Antioxidant Potential

Antioxidant activity in plants is measured using a variety of *in vitro* techniques. (Table **1**); though, every assay has its limits concerning applicability. Numerous approaches usually have been used to evaluate antioxidant potential. Antioxidant tests are divided into two categories based on the inactivation mechanism: reaction-based electron transfer (ET) methods and hydrogen atom transfer (HAT). HAT-based approaches assess an antioxidant's ability to generate stable compounds by donating hydrogen. SET-based procedures evaluate the capacity to transfer one electron to decrease any molecule, despite these methodologies being more relevant to radical chain-breaking antioxidant capacity [21].

Table 1. Well-known *in-vitro* tests used for determining the antioxidant activity of plants.

S. No	Assay	Mechanism
1.	Beta carotene assay	Hydrogen Atom Transfer
2.	Oxygen Radical Absorbance capacity	Hydrogen Atom Transfer

(Table 1) cont.....

S. No	Assay	Mechanism
3.	Lipid peroxidation inhibition capacity	Hydrogen Atom Transfer
4.	Total Radical trapping antioxidant parameter	Hydrogen Atom Transfer
5.	Copper reduction Assay	Single electron transfer
6.	Ferric-reducing antioxidant assay	Single electron transfer
7.	Total Phenolic content assay	Single electron transfer
8.	2, 2'-Azino-Bis-3-Ethylbenzothiazoline-6-Sulfonic Acid assay	Both Hydrogen Atom and Single electron transfer
9.	2,2-Diphenyl-1-picrylhydrazyl assay	Both Hydrogen Atom and Single electron transfer

Antimicrobial Activity

An antimicrobial is a substance that destroys microorganisms or halts their growth. Plants are recognized to have antimicrobial properties because of their bioactive compounds. These days, microbial infections and medication resistance have been significant defies that threaten human health. Worldwide these microbes are accountable for billions of deaths annually. In 2013, 9.2 million deaths were reported due to these infections, *i.e.*, approximately 17% of total deaths. The manifestation of the advancement of resistance has affected the present antibacterial medicines to become less effective [23]. In current years, numerous approaches are recommended to overcome antibiotic resistance. One of the suggested approaches to attaining this aim involves the blend of other molecules with the existing antibiotics, which seemingly reinstates the required antimicrobial activity. Regarding this situation, phytochemicals have shown potent activities, whereas many investigators have utilized natural products to perform against bacterial resistance. When used singly or in combination, these phytochemicals can boost the activity against a wide range of bacteria [24].

Mechanisms of Antimicrobial Activity

The antibacterial agent's activity is primarily attributable to two mechanisms: chemical interference with the production or function of necessary components of microbes or evading the traditional methods of antimicrobial resistance. There are several targets for the antimicrobial agents that involve protein biosynthesis, cell membrane destruction; DNA replication and repair; cell-wall biosynthesis, and inhibition of a metabolic pathway [25].

PLANT-DERIVED CHEMICALS

Even though artificial antimicrobial agents are being widely used in various countries, the use of plant-based natural medicines has attracted the interest of various researchers. These compounds have shown favorable results in overcoming the development of antibiotic resistance in microbial pathogens [26]. Among accessible options, plant-derived compounds have shown added possible applications in fighting microbial infections. They can re-establish the clinical use of existing antibiotics by improving their effectiveness as a result. Some of the plants and their components that possess antimicrobial activity are described in Table **2**.

Table 2. Plants containing antimicrobial activity.

Scientific Name	Compound	Active Against	References
Berberis vulgaris	*Berberine*	*Bacteria, protozoa*	[27]
Piper nigrum	*Piperine*	*Fungi, Lactobacillus, Micrococcus, E. coli, E. Faecalis*	[28]
Citrus paradisa	*Terpenoid*	*Fungi*	
Camellia sinensis	*Catechin*	*Shigella, S.mutans,Vibrio.*	
Rhamnus purshiana	*Tannins*	*Bacteria, fungi, viruses*	[29]
Glycyrrhiza glabra	*Glabrol*	*S. aureus, M. Tuberculosis*	
Matricaria chamomilla	*Anthemic acid*	*S.typhimurium, M. tuberculosis, S. aureus*	[30]
Laurus nobilis	*Essential oil*	*Bacteria, fungi*	
Piper betel	*Catechols, eugenol*	*General*	
Syzygium aromaticum	*Eugenol*	*General*	[31]
Hydrastis Canadensis	*Berberine, hydrastine*	*Bacteria, Giardia duodenale, Trypanosomes*	
Barosma setulina	*Essential oil*	*General*	
Ranunculus bulbosus	*Protoanemonin*	*General*	
Vaccinium spp.	*Fructose*	*Bacteria*	[32]
Allium cepa	*Allicin*	*Bacteria, Candida*	
Pimenta dioica	*Eugenol*	*General*	
Larrea tridentata	*Nordihydroguaiaretic acid*	*Skin bacteria*	
Eucalyptus globulus	*Tannin*	*Bacteria, viruses*	[33]

(Table 2) cont.....

Scientific Name	Compound	Active Against	References
Quercus rubra *Allium cepa*	*Tannins* *Quercetin*	*Bacteria*	[34, 35]
Thymus vulgaris	*Caffeic acid, Thymol* *Tannins*	*Viruses, bacteria, fungi*	
Gloriosa superba	*Colchicine*	*General*	
Hydrastis Canadensis	*Berberine, hydrastine*	*Giardia duodenale, Bacteria,* *trypanosomes*	
Centella asiatica	*Asiatocoside*	*M. leprae*	
Vicia faba	*Fabatin*	*Bacteria*	
Schinus terebinthifolius	*Terebinthone*	*General*	
Capsicum annuum	*Capsaicin*	*Bacteria*	
Cinnamomum verum	*Essential oils, others*	*General*	
Syzygium aromaticum	*Eugenol*	*General*	
Vaccinium spp.	*Fructose*	*Bacteria*	[36, 37]
Cannabis sativa	*β-Resercyclic acid*	*viruses and Bacteria*	
Carica papaya	*Mix of organic acids, alkaloids* *terpenoids*	*General*	
Onobrychis viciifolia	*Tannins*	*Ruminal bacteria*	
Allium sativum	*Allicin, ajoene*	*General*	[38]
Lawsonia inermis	*Gallic acid*	*S. aureus*	
Humulus lupulus	*Lupulone, humulone*	*General*	
Rabdosia trichocarpa	*Trichorabdal A*	*Helicobacter pylori*	
Melissa officinalis	*Tannins*	*Viruses*	
Aloysia triphylla	*Essential oil, Terpenoid*	*M. tuberculosis, Ascaris, S.* *aureus, E. coli*	
Arnica montana	*Helanins*	*General*	
Olea europaea	*Hexanal*	*General*	
Mahonia aquifolia	*Berberine*	*Trypanosomes, general*	
Tabebuia	*Sesquiterpenes*	*Fungi, Plasmodium*	

METHODS FOR DETERMINING A PLANT'S ANTIBACTERIAL POTENTIAL

Most approaches for assessing antimicrobial activity have been used directly or with alterations. The methods utilized to assess antimicrobial activity can be categorized as shown in Fig. (**2**). An *in vitro* test is utilized for initial evidence of antimicrobial activity. The endpoint experiments give qualitative data about active concentration. In this way, a microorganism is tested for a random period, and the effects reveal the inhibitory potential for the stated time. The descriptive screening methods provide quantitative evidence about the growth dynamics, where intermittent sampling is done to definechanges in viable cell numbers. In applied tests, the antimicrobial is administrated to the actual organism, and the efficacy is evaluated [39].

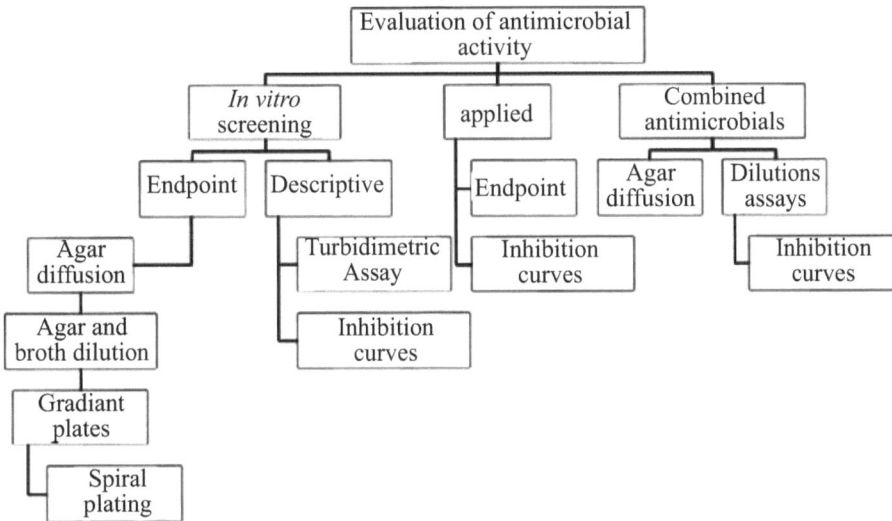

Fig. (2). Methods used for assessment of antimicrobial potential.

Anticancer Activity

Generally, the term cancer is used for diseases associated with anaplastic cell proliferation and malignant tumor formation. This disease is one of the leading causes of death worldwide. Moreover, according to WHO, about 1 in 6 deaths is due to cancer, with an estimated 9.6 million deaths in 2018. Cancer is the second leading cause of death globally. It primarily arises when mutations occur in oncogenes and tumor suppressor genes [40]. Such mutations promote aberrant protein expression and, as a result, the transition of healthy cells into cancerous cancer cells. These cancer cells are characterized by unchecked cell proliferation and the ability to avoid immunity detection and evade apoptosis [41].

Almost 10% to 15% of cancer is entirely hereditary [42]. However, we could not estimate the exact cause of specific cancer in most cases. Some factors are considered responsible for increasing its risks, such as alcohol and tobacco, infectious agents, environmental pollutants, and lifestyles. Although massive attempts have been made to find a solution for this disease, the number of reported cases has been steadily increasing [43].

The main reason for this dismal scenario is that cancer prevention has received less attention than cancer therapy. There is a famous saying by Hippocrates, *i.e.*, "Let food be thy medicine and medicine be thy food."He highlighted the fact that humans have been using plants since immemorial times. This phrase is further supported by the fact that different countries have a different rate of disease incidence that is clearly because of the different types of food products they eat, *i.e.*, in contrast to well-developed Western countries, Southeast Asian populations have a relatively low risk of cancer, including prostate, breast, gastrointestinal and colon cancer [6].

Studies also revealed that some cancers are more common in specific communities than others, such as breast, prostate, colon, and lung cancers in the West, whereas head, neck, and cervix cancers are common in Indian cultures, and in Japan, stomach cancer is the most common one. This difference in cancer incidence might be due to genetic variation between these populations, but when studied, those who migrated from one region to another had developed the risk of cancer of their adopted country; this change occurs within two generations [44].

Since genes only contribute to 10-15% of cancer cases, other external and internal factors also contribute to its incidence. Epidemiologists suggest that those populations that tend to take a plant-rich diet, *i.e.*, seeds, whole grain, fruits, nuts, vegetables and legumes, and animal fat and meat, have a noticeable difference/reduced cancer incidence rate. Such studies about the anti-cancer properties of dietary plants have resulted in large-scale research in this area over the last two decades to identify the compounds responsible for such properties

Early discovery and surgical removal are the key tactics against this illness so far, as are chemotherapy, radiation, and attempts to develop gene therapy, but regrettably, in the majority of instances, these therapies simply result in disease regression. Nowadays, another tremendous challenge the pharmaceuticals face is the increase in tumors' resistance to therapeutic agents [45].

Plants have long been used in the treatment of cancer, and they have thus been the principal source of traditional medications that are effective in the treatment of cancer. Following the discovery of new proteins with important regulatory effects on the tumor cell cycle progression, research into isolating molecules from plants and other living organisms confirmed that plants are essential sources of synthetic inhibitors with the potential to develop anticancer drug production.

More than half of all anticancer medications are currently generated from natural resources such as plants, microbes, and sea creatures. They deliver an outstanding contribution toward modern health care systems. About 70% of anticancer compounds have been originated from natural products. Sufficient studies confirmed that these therapeutic properties of plants are due to their secondary metabolites, prominently phenols and alkaloids [46]. At present, four classed plant-based anticancer compounds are in the market, the camptothecin derivatives (camptothecin and irinotecan), the epipodophyllotoxins (teniposide and etoposide), the taxanes, and the vinca alkaloids (vincristine, vindesine, and vinblastine) [47].

Scientists start to analyze natural organisms as a source of anticancer compounds in the 1950s. Nowadays, it is claimed that this strategy was one of the most successful ones in discovering therapeutic compounds. Plants are the most important source of anticancer compounds among all-natural organisms as they have a complex defense system. They defend themselves against microorganisms by producing toxic compounds, and as human cells and fungi are similar at the biochemical level, these chemicals also have an inhibitory action on human cells, including cancer cells. Some chemicals/compounds are more hazardous to cancer cells than healthy/normal cells and are used as chemotherapy drugs [48].

Anticancer compounds isolated from the herbal source have a very long history, and many of these chemical compounds are still a part of the modern system of medicine, *i.e.*, etoposide, vinblastine, vincristine, irinotecan, topotecan, and camptothecin derivatives. Many traditional systems are still being used, *i.e.*, Ayurveda, Chinese medicine, and Kampo medicine. Many formulations of these traditional systems are still considered effective as a complementary and alternative medicine for cancer worldwide. According to recent data from the World Health Organization (WHO), countries with well-developed medicine systems believe traditional systems to be effective in cancer treatment [49].

Several plants are used for treating cancer, *i.e.*, *Anacardium occidentale* in hepatoma, *Erythrina suberosa* in sarcoma, *Nigella sativa* in Lewis lung carcinoma, *Asparagus racemosus* in human epidermoid carcinoma, *Boswellia serrata* in nasopharynx carcinoma, *Gynandropsis pentaphylla* in hepatoma, *Euphorbia hirta* in Freund virus leukemia, *Alstonia scholaries* on forestomach carcinoma, *picrorhiza kurroa* in hepatic cancers and *Withania somnifera* in various tumors [44]. Plants with anticancer activity are also reported in Table **3**.

Table 3. Plant with anticancer activity.

Plants	Suppressed Cancerous Cell Lines/cancer Models	References
Ageratum conyzoides	*Lung cancer (A-549), colon carcinoma (HT-29), human glioma carcinoma (U-251), prostate cancer (DU-145), mouse leukaemia (P-388),gastric carcinoma (SGC-7901),breast cancer (MDA-M--231),hepatic carcinoma (REL 7402),*	[50]

(Table 3) cont.....

Plants	Suppressed Cancerous Cell Lines/cancer Models	References
Azadirachta indica	*Lung cancer (U937), skin melanoma (B16),leukaemia (HL-60, THP1), prostate cancer (PC-3)*	[51]
Cannabis sativa	*Breast cancer (MCF-7, EFM-19, MDA-MB-231),brain/spine tumour (U87, U373), skin cancer (PDV.C57, HaCa4),*	[52]
Curcuma longa	*Breast cancer (BT-20,T-47D, SK-BR3 and MCF-7), leukaemia (HL60)*	[53]
Zingiber officinale	*Breast cancer (MCF-7 and MDA-MB-231), ovarian cancer (SK-O--3), lung cancer (A549), colorectal adenocarcinoma (HCT15),colon cancer (HCT 116, HT 29),melanoma (SK-MEL-2)*	[54]
Dillenia indica	*Lung cancer (U937), promyelocytic leukaemia (HL60, K562)*	[55]
Aloe vera	*Liver cancer (HepG2), breast cancer (MCF-7), cervical cancer*	[56]
Dillenia pentagyna	*lymphoma T-cell*	[57]
Potentilla fulgens	*(MCF-7)Breast cancer, (U-87) human glioblastoma cancer*	[58]
Blumea balsamifera	*Breast cancer (MCF-7), myeloid leukemia (K562), hepatocellular carcinoma (McA-RH7777), epidermal carcinoma of the mouth (KB),lung cancer (NCI-H187),*	[59]
Ocimum sanctum	*Lung cancer (A549),(HFS) human fibrosarcoma cells*	[60]
Mirabilis jalapa	*Human laryngeal carcinoma (Hep-2), (MCF-7) breast cancer*	[61]
Catharanthus roseus	*Lung cancer (NCI-H69/P)*	[62]
Litchi chinensis	*Breast cancer (MCF-7), colorectal cancer (Colo320DM and SW480), leukaemia (U937,K562 and HL-60),*	[63]
Podophyllum hexandrum	*Lung cancer, neuroblastoma, hepatoma, testicular cancer*	[64]
Xanthium strumarium	*Cervical cancer*	[65]

ACTIVITY GUIDED SEPARATION AND CHARACTERIZATION OF PHYTOCHEMICALS

The ethnopharmacological data-driven phytochemical inquiry can be portrayed as an operational strategy for developing new compounds with potential therapeutic leads. Plants formulations used by local communities to treat numerous ailments describe the foundation of the chemical entities, but no evidence is present of their nature [66]. An outline of the methodologies included in the selected approach is shown in Fig. (**3**)

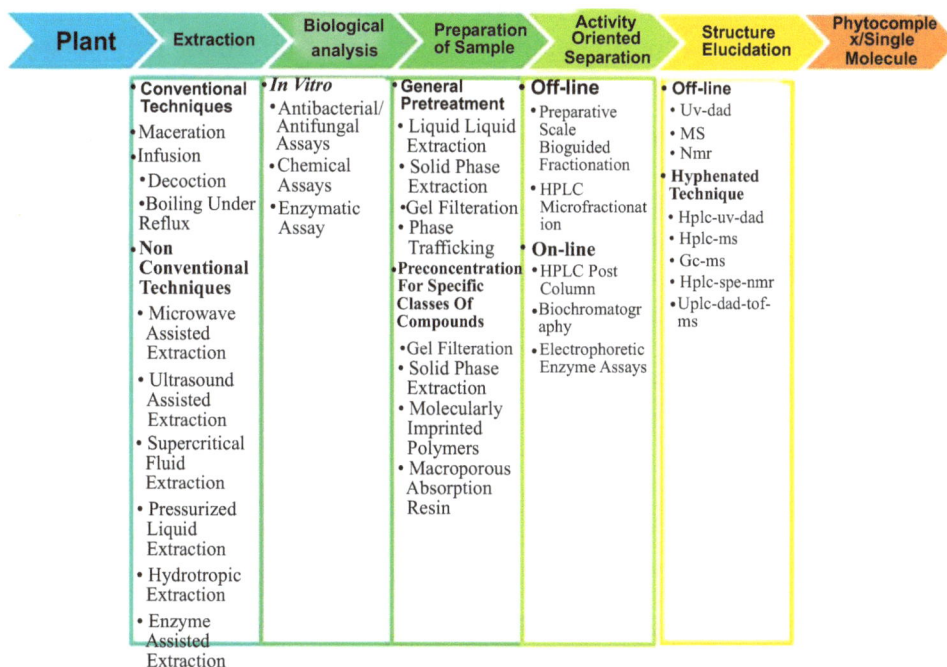

Plant	Extraction	Biological analysis	Preparation of Sample	Activity Oriented Separation	Structure Elucidation	Phytocomplex/Single Molecule
	• **Conventional Techniques** • Maceration • Infusion • Decoction • Boiling Under Reflux • **Non Conventional Techniques** • Microwave Assisted Extraction • Ultrasound Assisted Extraction • Supercritical Fluid Extraction • Pressurized Liquid Extraction • Hydrotropic Extraction • Enzyme Assisted Extraction	• *In Vitro* • Antibacterial/ Antifungal Assays • Chemical Assays • Enzymatic Assay	• **General Pretreatment** • Liquid Liquid Extraction • Solid Phase Extraction • Gel Filteration • Phase Trafficking • **Preconcentration For Specific Classes Of Compounds** • Gel Filteration • Solid Phase Extraction • Molecularly Imprinted Polymers • Macroporous Absorption Resin	• **Off-line** • Preparative Scale Bioguided Fractionation • HPLC Microfractionation • **On-line** • HPLC Post Column • Biochromatography • Electrophoretic Enzyme Assays	• **Off-line** • Uv-dad • MS • Nmr • **Hyphenated Technique** • Hplc-uv-dad • Hplc-ms • Gc-ms • Hplc-spe-nmr • Uplc-dad-tof-ms	

Fig. (3). Methodologies involved in activity guided isolation and characterization of bioactive components.

Extraction Techniques

Before initiation of the drug discovery process, extraction is the first step. Various techniques have been used to isolate active compounds from plants based on polarity or isolation of common secondary metabolites, *i.e.*, alkaloids, saponins, *etc* [67]. Traditionally, techniques such as preservation, infusion, decoction, and boiling in reflux have been employed, but in recent decades, many unique approaches have been established. *i.e.* Extraction with the help of a microwave (Mihigo, Ndombele, Masesane, & Sichilongo), Extraction using ultrasound assistance (UAE), Extraction of supercritical fluids(SFE), Extraction of pressurised fluids (PLE).

Sample Preparation

In order to identify metabolites, a pre-treatment on crude extracts is required. These procedures include:

Common liquid–liquid partition

Solid phase extraction (SPE)

SephadexLH-20 gel filtering

When no information on the chemical composition of the crude extracts is available, an initial purification based on information, such as lipophilic/hydrophilic and/or acidic/essential characteristics, could be carried out.

Biological Screening

Screening of biological activity is necessary in order to validate their traditional uses. As a result, biological assessments are conducted on partially purified or crude extracts, which are carefully selected based on their alleged bioactivity. *In vitro* bioassays are faster and require only a tiny amount of chemicals. Even though they may not be related to clinical conditions, they are precise, sensitive, and extensively used; furthermore, most of these are microplate-based and could be conducted in whole or semi-automation [68].

The complication of the bioassay must be delineated by laboratory facilities, antibacterial and antifungal assays that are "simple to use" are frequently used as "on/off" tests to infer the presence or absence of active compounds. If the IC50 readings are lesser than 100 g/ml, a crude extract is normally deemed exceptional. Chemical and Enzymatic assays based on spectrophotometric measurements can be used to determine whether or not substances with specific activity exist. The substance must be purified to extract the bioactive components once a biological activity has been identified [69].

Chromatographic Techniques

The use of a combination of analytical and spectroscopic methods, referred to as "hyphenated techniques," to generate chemical as well as bioactive data in a sequential manner is becoming increasingly important in the study of phytopharmaceutical effects. These methods are currently used to identify recognizable components on the internet (dereplication) quickly. HPLC, in particular, is widely used for natural product profiling. HPLC can be used in conjunction with simple detectors for monitoring chromatographic traces, such as UV, Electron Capture Detector (ECD), Evaporative Light Scattering Detector (ELSD), or sensors for hyphenated systems to generate multidimensional data for online identification (*e.g.*, UV-diode array (DAD), MS, nuclear magnetic resonance) (NMR)) [68].

In the case of complex mixtures, ultra-high pressure liquid chromatography (UHPLC) is utilized to do high-resolution metabolite profiling. Associated with other analytical methods, it allows rapid analysis, thus allowing greater separation efficiency, sensitivity, resolution, and lower solvent consumption. Attention has also been paid to the development of mass spectroscopy gas chromatography [70].

MS is used for fingerprinting volatile compounds, and LC-MS is used to detect and identify activated metabolites. By quickly identifying known compounds based on structural data derived from their fragmentation outline generated by collision-induced separation (CID) in MS-MS analysis, LC-MS avoids the repetitive separation of known compounds and focuses on the directed isolation of composites producing specific fragmentations. Various chromatographic methods can be combined to improve plant samples' "chromatographic fingerprint." Multiple detections yield a 2D fingerprint analysis, which allows for the accumulation of further chemical data on the overall chemical composition. PCA is a well-known chemometric spectroscopic technique for determining the uniqueness of known or novel compounds. HPLC-NMR can offer big matching data or perhaps a compound's entire structural data. If a molecule's parallel LC peak is visually resolved, HPLC-NMR can immediately provide a detailed structural description of that molecule in an extract [71].

NANO-ELICITATION

In recent years, much attention has been paid to nanotechnology with various field applications in human and animal health. Pharmaceutical applications and nanomaterials as drug delivery systems have been deeply investigated, but few reports are available regarding the use of nanoelicitirs to improve the yield of secondary metabolites in plant cells [72]. Nanoparticles (NPs) or nanomaterials have unique physicochemical properties to enter and interact with plant cells and intracellular organelles. Although many studies focused on revealing outcomes of NPs application on growth, development, physiology, and biochemistry of plants, limited reports are available on the use of nanoelicitors and their impact on the specialized metabolism of plants [73]. Nanoparticles are evolving like extra effective replacements for conventional and biotic elicitors. The reason behind their increased use is their customizable and physicochemical possessions and the reactions they produce. Nanoelicitors act as stress factors, thus effects primary and secondary metabolism in plants resulting in enhanced production of various classes of plant metabolites [74]. Plants' secondary metabolites play an important role in adaptation to the adverse environment and have unlimited applications in developing different commercially-valued products. Plants produce different types of secondary metabolites, including terpenoids, tannins, phenolics, and alkaloids. Under unfavorable conditions, these metabolites play a vital role by interacting with other biotic and non- biotic agents. Elicitors have the potential to induce physiological and biochemical responses in target plants and hence activate defense mechanisms. The use of signaling molecules as elicitors has developed an efficient technique for the enhanced production of pharmaceutically active metabolites in plants [75].

Nanoparticles as Nano-elicitors

The use of nanoparticles as nano-elicitors is significant due to their ability to modify the metabolic profile of plants. Nanoparticles introduce modifications either by getting attached to the cell surface or by internalization and communication with the cell protoplasm. The prospective influence of nanoparticles has been observed on specific metabolic processes, which differ based on the NPs physical makeup like their shape, size, aspect-ratio, concentration, type of cell culture, total time of exposure, duration of culture, and the species of plants in consideration [76]. Though other struggles are pretty necessary to illuminate the mechanism, some researchers proposed that nanoparticles encourage the production of reactive-oxygen-species (ROS) and associated signaling molecules that facilitate the regulation of plant's transcription and secondary metabolism [77]. Reactive-oxygen-species as well as calcium-ions (Ca2+) are significant second messengers facilitating the up-regulation of transcriptional regulators of secondary metabolites in the plant [78].

Classes of Nanomaterials

Different types of nanoparticles are based on the extent, morphological, physical, and chemical assets such as carbon-based, Ceramic-nanoparticles, metallic-nanoparticles, semiconductor NPs, polymeric-nanoparticles, and lipid-based nanoparticles [79]. Inorganic solids containing oxides, carbides, carbonates, and phosphates are ceramic nanoparticles. They are identified to be used as a drug delivery system to diagnose different infections, including glaucoma and tumor [80]. Organic nanoparticles are known as polymeric nanoparticles. Their structure design is similar to a nano-capsular (matrix-like) or nano-sphere (core-shell structure). Due to their highly biodegradable and biocompatible properties are used in drug delivery for diagnostic purposes [81]. Nanoparticles with diameters ranging from 10-100 nm and have spherical shapes are known as lipid nanoparticles. They are used as a drug carrier in the biomedical arena and cancer therapy due to the release of RNA ([82]. Semiconductor nanoparticles have dual properties that are both metal and non-metal properties. These nanoparticles have vast applications in electronic devices, photo-optics, water splitting, and photocatalysis. Based on the utilization of nanomaterials as nanoelicitors, they can be divided into three major categories: metallic nanoparticles, metal oxide nanoparticles, and carbon nanotubes.

Metallic Nanoparticles

Metallic nanoparticles (MNPs) have a metal core surrounded by a shell comprised of organic substances, inorganic substances, or metallic oxides. Our everyday life is based on several appliances of metallic nano-fragments. MNPs have functional groups that can be modified to bind substances such as ligands, drugs, and antibiotics [83]. The size of these MNPs is too tiny and measured in nanometers ranging from 10-100 nm. MNPs have distinctive characteristics, *i.e.*, ocular properties and surface Plasmon resonance. In different types of chemical reactions, MNPs acts as a catalyst. During MNPs preparation, their size and shape can easily be controlled [84]. Metallic nanoparticles (MNPs), when applied to different plant species, caused genetic variation, altered production of secondary metabolites, and mass propagation of plants being investigated. Different metallic nanoparticles (Au, Ag, Cu, Co, and Zn) have been applied to elicit plant secondary metabolites in various species. The effects of MNPs on the production of secondary metabolites depend on the concentration of nanoparticles used, their origin, exposure time, and their characteristics like shape, size, and even species of plants under investigation [85, 86].

Silver nanoparticles are the most studied nanoelicitors that have been employed in all types of systems like cell culture, hairy root culture, and callus culture. Their role as nano elicitors was highlighted [87] to enhance the production of plants' secondary metabolites. AgNPs have enhanced artemisinin content production from 1.67 mg/g dry wt to 2.86 mg/g dry wt in the hairy root culture of *A. Annua*. Similarly, exposure to different concentrations of AgNPs caused an enhanced production of various components of essential oils such as E-caryophyllene, geraniol, and citronellyl formate in seedlings of *Pelargonium graveolens*. In another study, atropine production was observed to be increased in the hairy root culture of *Datura metel* when exposed to 200 mg L−1 concentration of AgNPs [88].

Similarly, the application of AgNPs to the hairy root culture of *Cucumis anguria* increased the production of flavonoids and phenolic compounds [85]. Silver nanoparticles were also found to elicit secondary metabolites' production, specially quercetin and glycyrrhizin, in *in vitro* culture of *Glycyrrhiza glabr* seedlings [89]. Exposure to AgNPs can also exhibit an inhibitory effect on SMs production as in *Calendula officinalis* a strong decrease in carotenoid content was observed while saponin production was enhanced twice [90]. Thus, it can be concluded that AgNPs act as nanoelicitors, and their effects vary depending on the type of plant under study and the type of secondary metabolites. In another study, the effects of bimetallic nanoparticles, that is, gold and silver nanoparticles with plant hormone naphthaleneacetic acid, were investigated to enhance the production of bioactive metabolites in *Prunella vulgaris* L callus culture. Silver nanoparticles without NAA resulted in higher biomass production while AgAu (1:3) and AuNPs enhanced the production of proline, superoxide dismutase, and peroxidase enzymes and accumulation of phenols and flavonoids [91].

Copper nanoparticles have various applications in agricultural science and plant nanotechnology. Their use has indicated improved plant growth and secondary metabolite production. A significant increase in essential oil contents was observed after treatment with CuNPs in *Mentha longifolia*. Similar effects were also indicated by treatment with cobalt nanoparticles [92]. Treatment of CuNPs can activate plants' defense mechanisms and metabolic perturbations, thus resulting in enhanced production of total phenolic contents. However, sometimes destructive results can also be observed, as in the case of cucumber plants growing in a hydroponic environment resulting in decreased production of secondary metabolites [93].

Metallic Oxides NPs

This type of nanoparticle plays a vital role in different fields of science such as chemistry, physics, and material sciences. The metal constituents can construct a considerable diversity of oxide composite. Metallic oxide nanoparticles have electronic structure characteristics such as metallic, semiconductor, or insulator. Metallic oxide nanoparticles have technological applications in which oxides are used in microelectronic circuit construction, sensors, piezoelectric devices, and fuel cells, used as a coating to prevent corrosion and as a catalyst [94]. Ionic or mixed ionic/electronic conductivity can contain oxide materials, and it is tentatively well established that both can be affected by the nano-composite of the solid. Different methods are used to synthesize metallic-oxide NPs like co-precipitation, microemulsion technique, sol-gel processing, solvothermal approach, gas-solid transformation methods, and multiple-pulsed laser deposition.

Nanoparticles in Secondary Metabolite Production

Numerous studies have been conducted to determine the response of secondary metabolite production due to exposure to metal oxide nanoparticles. The most widely used elicitors of secondary metabolites are zinc oxide (ZnO), copper oxide (CuO), cadmium oxide (CdO), titanium oxide (TiO_2), cesium oxide (CeO_2), and aluminum oxide (Al_2O_3) NPs. Various contradictory reports are available regarding their potential to enhance the production of secondary metabolites. Copper being incorporated in many enzymes and proteins is considered an essential micronutrient. It plays a vital role in plant nutrition and health [95]. Copper oxide nanoparticles (CuONPs) are considered elicitors for producing beneficial bioactive metabolites in bioreactor systems, but previous findings indicate different effects of CuONPs on the production of different secondary metabolites [96]. It is suggested to optimize the concentration of CuONPs with respect to the targeted secondary metabolites.

Zinc oxide nanoparticles (ZnONPs) have unique optical properties, bandwidth, catalytic, and high surface area to volume ratio properties. In various plant species and cell culture studies, ZnONPs positively influenced secondary metabolite production. In the callus culture of *Echinacea purpurea*, total flavonoid contents were enhanced due to the application of ZnONPs [97]. Similarly, anthocyanin and phenolic contents in potato plants were affected by applying different concentrations of ZnONPs [98].

Iron oxide nanoparticles (Fe$_3$O$_4$NPs) are readily available nanoparticles due to their simple and cost-effective synthesis. Fe$_3$O$_4$NPs are considered more efficacious than other MONPs to elicit the production of secondary metabolites as their application can enhance the production of bioactive compounds and the growth of plants. Enhanced production of flavonoids, phenolics, and fresh and dry weight of root culture of C. Intybus has been recorded [99]. Although many studies indicated the beneficial effect of FeONPs on the stimulation of secondary metabolites, it is necessary to optimize Fe$_3$O$_4$NPs concentrations and duration of treatment prior to avoiding unwanted toxic effects. Studies on root culture of *Hyoscyamus reticulatus* indicated a concentration and time associative stimulation of the tropane alkaloids and alkaloids biosynthesis in Fe$_3$O$_4$NPs-treated cultures after 24 and 48 h of exposure, respectively. Contrarily, prolonged exposure time (*i.e.*, 96 h) resulted in a 2-fold decrease in tropane production [100].

Titanium dioxide nanoparticles (TiO$_2$ NPs) are mostly used in skincare, cosmetics, antibacterial products, paints, and wastewater treatment. TiO$_2$ NPs can influence plants in beneficial and adverse ways. The impact of TiO$_2$ NPs is concentration-dependent and varies from plant to plant. Similarly, the size of nanoparticles and treatment methods also affect their elicitation potential. Moreover, a concentration of 200 mg L−1 TiO2NPs exerts a significant rise in essential oil content was also observed. It has been noticed that oxidative damages induced by drought stress can be controlled by applying TiO$_2$NPs at optimum quantity. In the context of climate change and rising temperature, this impact should be further investigated [101]. Cerium oxide nanoparticles (CeO$_2$NPs) are very stable due to their limited dissolution in soil. Cerium (Ce) is not necessary for plants, but it stimulates plants physiological processes and growth. The use of CeO$_2$NPs as nanoelicitors resulted in a significant increase in carotenoid contents of *Solanum Lycopersicum* hydroponic [102]. Cadmium oxide (CdO), aluminium oxide (Al$_2$O$_3$), and manganese dioxide (Mn$_2$O$_3$) nanoparticles are the least studied nanomaterials concerning elicit production of secondary metabolites. Limited available data indicates the positive influence of Al$_2$O$_3$NPs phenolic content of tobacco cell culture in a dose-dependent manner with enhanced biomass production [103].

Carbon Nanotubes and Elicitation of Secondary Metabolites

Carbon-based nanomaterials have vast applications in agriculture and biotechnology due to their influence on primary and secondary metabolisms of various plant species. Carbon nanotubes are made up of pieces that are graphene in nature, trundled in the tubes. They are 100 times sturdier than steel, so used for physical fortification. Carbon-based nanomaterials are categorized into single-walled carbon nanotubes (SWCNTs), multi-walled carbon nanotubes (MWCNTs), and derivatives of fullerene (Buckminsterfullerene or Buckyball). Allotropes of carbon having sixty or more carbon particles are arranged in a hollow cage-like construction and are known as fullerenes. They play an essential role in electric conduction, assembly, high gift, and electron affinity. The CNTs have been used to influence diverse biological functions and hence are considered plant growth regulators as they positively influence seeds' germination and seedling performance [104]. Enhanced production of secondary metabolites (total phenolics, flavonoids, caffeic acid, and rosamarinic acid) was also confirmed when *in vitro Satureja khuzestanica* calli were treated with MWCNTs [59]. Based on available literature, CNTs may serve as a good candidate for elicitors in plant tissue culture, thus influencing plant growth, primary and secondary metabolism, and providing defense to plants when exposed to adverse conditions.

MECHANISM OF ELICITATION OF SPECIALIZED METABOLITES BY NANO-MATERIALS

NPs, due to their extremely small size, can easily penetrate and interact with several constituents of plant cells and tissues. They possess a high surface to-volume ratio and surface reactive properties that make their various microscopic, and spectroscopic techniques have been employed to detect and confirm penetration of nanoparticles in plant cells. NPs are taken up by plants through foliar spray, through the soil, and by utilizing artificially prepared nutrient media. Numerous studies have been carried out to clarify the possible effect of nanoparticles on specialized metabolism. Eruption of reactive oxygen species ROS is the primary mechanism for nanoparticles facilitating particular metabolite elicitation in plants [77]. Size, concentration, time of exposure, plant types, tissue form, cell cycle phase, in addition to physiological limits of nanoparticles, can affect the production of secondary metabolites. Membrane destabilization can also help penetrate NPs into plant cells, but there is a need to study transporters' precise mechanism and role in incorporating nanoparticles. It is suggested that a ROS burst is an initial reaction and occurs by the intrusion with the cell after internalization. After bursting, it activates antioxidant machinery, transmembrane K+/H+ions channel initiation, and up-regulation and initiation of MAPKs [105].

Possible explanation of nano-particle-mediated elicitation of selected metabolites is also sketched in Fig. (**4**).

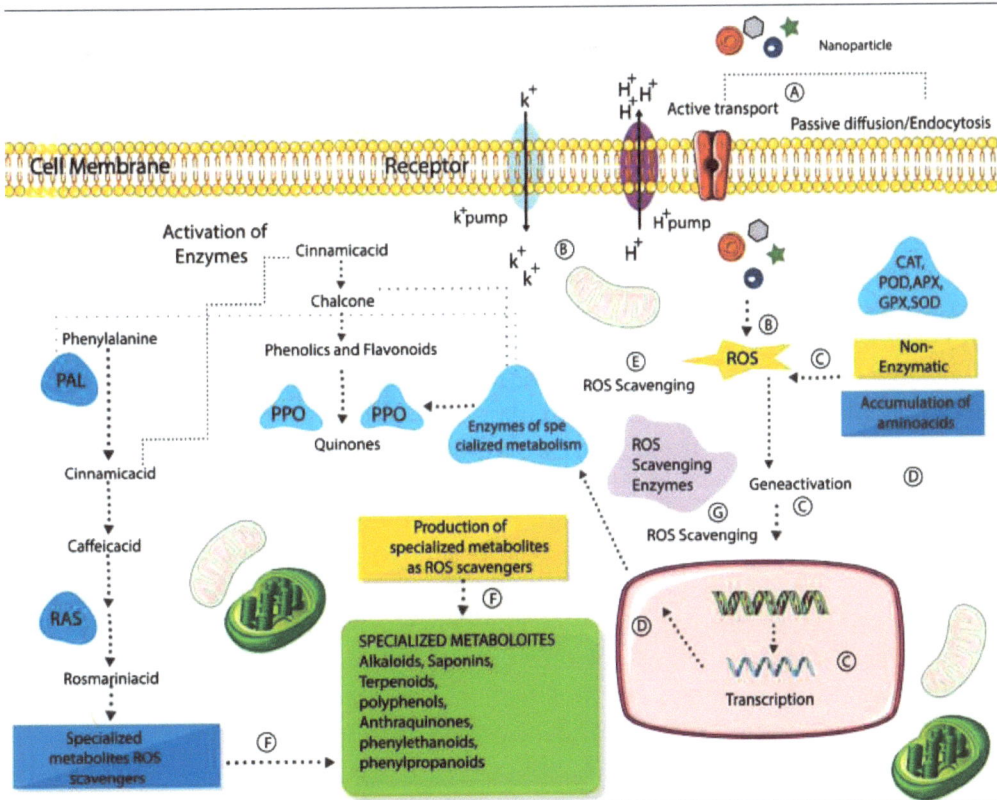

Fig. (4). Promising explanation of nano-particle-mediated elicitation of particular metabolites (**A.** Entry of nanoparticles in plant cells and release of elicitors. **B.** Generation of ROS. **C.** ROS scavengers initiation **D.** Initiation of enzyme production. **E.** Metabolic trails initiation **F.** Fusion of particular metabolites. **G.** Antioxidants and ROS scavengers regulating cellular functions).

ADVERSE EFFECTS OF NANO-ELICITORS ON PLANT CELL METABOLISM

Nanomaterials have many advantages in stimulating the production and accumulation of medicinally and commercially important plant secondary metabolites, but their adverse effects have also been documented. As nanomaterials are used to elicit secondary metabolites in plants, there is a need to judge the toxicity of NPs. Quantum dots have high surface reactivity with ultrafine size and cause undesirable cellular fluctuations within the intracellular compartment and metabolic pathways. The cytotoxicity of nanomaterials depends on many factors like age, health condition, type of plant species, and type of culture being exposed to the nanomaterials. Similarly, nanomaterials' properties also play a significant role in the induction of phytotoxicity, including the type of nanoparticles, their size, concentration, and application methods. Mechanisms of phytotoxicity caused by nanomaterials are also linked with the initiation of oxidative stress and ROS generation, and when the level of ROS exceeds the capability of the defensive mechanism of the plant to cope with the stress, the damage is caused to plant cells. Cell damage may be caused as a result of lipid peroxidation, protein oxidation, DNA damage, or damage caused to the cell membrane resulting in leakage of electrolytes and finally, cell death. In a study, tobacco cell suspension cultures were exposed to Al_2O_3-NPs to study the mechanism of cytotoxicity, and it was found to be linked with enhanced expression of two enzymes, namely dehydrogenase and oxidoreductase [103]. Thus it was concluded that cellular changes occur by the activity of redox signals; these catalysts are energetic for hunting ROS. NPs produce ROS in cell cultures, and bushy roots pledge a cascade of gene/transcript/catalyst stimulations. Cellular shield machinery is triggered by oxidative stress, including ROS scavengers, accretion of carbon-based osmolytes, gesticulating particles, phosphorylation of MAPKs, and activating particular metabolite production. In the case where the level of NPs extends from its ideal limit, the cellular threshold of ROS surpasses primarily due to deoxyribonucleic acid injury, polypeptide/fatty acid peroxidation, film injury, electrolyte drip, and cell decease. Several nanoparticles are, therefore, covered with biological molecules, *i.e* citrate, chitosan, curcumin, and vanillin, or carbon-based stabilizers that act as elicitors such as chitosan [106] that can decrease the exit of surplus ions, thus reducing the toxicity of NPs and controlling NP-mediated elicitation in a significant way. Thus the efficiency of nanoparticles can be increased by coating them with stabilizers and biological molecules that reduce their cytotoxicity, and thus injury to the cellular and subcellular mechanisms can be avoided.

As the literature indicates the size of nanomaterials, their concentration and time of exposure may also cause phytotoxicity; therefore, before utilizing nanomaterials for stimulation of secondary metabolite accumulation, they should be optimized. At optimized conditions, NPs can benefit plant growth and secondary metabolite production.

CONCLUSION

Bioproduction of secondary metabolites is enhanced through nature by fortifying nanostructures. The efficacy of these natural entities varies from antimicrobial to anticancerous effects. Therapeutically important bioactive compounds have been identified, characterized and purified by using different strategies. The current chapter highlights the use of different nanostructures for accelerating the production of phytodrugs in plant bodies by employing different types of nano-oxides as well as the effect of varying concentrations of these nanoforms on pharmacologically important phytoconstituents.

REFERENCES

[1]   Jia Wang , Hui Yang , Zhong-Wen Lin , Han-Dong Sun . Flavonoids from Bidens pilosa var. radiata. Phytochemistry 1997; 46(7): 1275-8.
[http://dx.doi.org/10.1016/S0031-9422(97)80026-X]

[2]   Chaudhary S, Magar GT, Sah SN, Parajuli S. Ethnic plants of Tharu community of Eastern Nepal. Int J Appl Sci Biotechnol 2020; 8(2): 223-30.
[http://dx.doi.org/10.3126/ijasbt.v8i2.28325]

[3]   Pengelly A. The constituents of medicinal plants Cabi 2021.
[http://dx.doi.org/10.1079/9781789243079.0000]

[4]   Jimenez-Lopez C, Fraga-Corral M, Carpena M, *et al.* Agriculture waste valorisation as a source of antioxidant phenolic compounds within a circular and sustainable bioeconomy. Food Funct 2020; 11(6): 4853-77.
[http://dx.doi.org/10.1039/D0FO00937G] [PMID: 32463400]

[5]   Ndiege ML, Kengara F, Maiyoh GK. Characterization of phenolic compounds from leaf extract of bidens pilosa linn. var. Radiata South Asian Res J Nat Prod 2021; 4: 44-58.

[6]   Muniyandi K, George E, Sathyanarayanan S, *et al.* Phenolics, tannins, flavonoids and anthocyanins contents influenced antioxidant and anticancer activities of rubus fruits from western ghats, India. Food Sci Hum Wellness 2019; 8(1): 73-81.
[http://dx.doi.org/10.1016/j.fshw.2019.03.005]

[7]   Maarfia S, Zellagui A. Study of essential oils and phenolic compounds their changes and anticancer activity in some species belonging to Asteraceae and Lamiaceae families. Oum-El-Bouaghi 2019; pp. 1-212.

[8]   Ullah A, Munir S, Badshah SL, *et al.* Important flavonoids and their role as a therapeutic agent. Molecules 2020; 25(22): 5243.
[http://dx.doi.org/10.3390/molecules25225243] [PMID: 33187049]

[9]   Kumari M, Jain S. Tannins: An antinutrient with positive effect to manage diabetes. Res J Recent Sci 2012; 1: 1-8.

[10]   Cushnie TPT, Cushnie B, Lamb AJ. Alkaloids: An overview of their antibacterial, antibiotic-enhancing and antivirulence activities. Int J Antimicrob Agents 2014; 44(5): 377-86.
[http://dx.doi.org/10.1016/j.ijantimicag.2014.06.001] [PMID: 25130096]

[11]   Ashtiania F, Sefidkonb F. Tropane alkaloids of atropa belladonna l. and atropa acuminata royle ex miers plants. J Med Plants Res 2011; 5(29): 6515-22.

[12]   Jain S, Sinha A, Bhakuni DS. The biosynthesis of β-carboline and quinolizidine alkaloids of alangium lamarckii. Phytochemistry 2002; 60(8): 853-9.
[http://dx.doi.org/10.1016/S0031-9422(02)00057-2] [PMID: 12150812]

[13]   Hodaj E, Tsiftsoglou O, Abazi S, Hadjipavlou-Litina D, Lazari D. Lignans and indole alkaloids from the seeds of *centaurea vlachorum* hartvig *(asteraccae),* growing wild in albania and their biological activity. Nat Prod Res 2017; 31(10): 1195-200.
[http://dx.doi.org/10.1080/14786419.2016.1226823] [PMID: 27609488]

[14]   Corea G, Fattorusso E, Lanzotti V, Capasso R, Izzo AA. Antispasmodic saponins from bulbs of red onion, allium cepa l. var. tropea. J Agric Food Chem 2005; 53(4): 935-40.
[http://dx.doi.org/10.1021/jf048404o] [PMID: 15713001]

[15]   Man S, Gao W, Zhang Y, Huang L, Liu C. Chemical study and medical application of saponins as anti-cancer agents. Fitoterapia 2010; 81(7): 703-14.

[http://dx.doi.org/10.1016/j.fitote.2010.06.004] [PMID: 20550961]

[16]   Swain SS, Paidesetty SK, Padhy RN. Development of antibacterial conjugates using sulfamethoxazole with monocyclic terpenes: A systematic medicinal chemistry based computational approach. Comput Methods Programs Biomed 2017; 140: 185-94.
[http://dx.doi.org/10.1016/j.cmpb.2016.12.013] [PMID: 28254074]

[17]   Deng GF, Xu XR, Zhang Y, Li D, Gan RY, Li HB. Phenolic compounds and bioactivities of pigmented rice. Crit Rev Food Sci Nutr 2013; 53(3): 296-306.
[http://dx.doi.org/10.1080/10408398.2010.529624] [PMID: 23216001]

[18]   Parcheta M, Świsłocka R, Orzechowska S, Akimowicz M, Choińska R, Lewandowski W. Recent developments in effective antioxidants: The structure and antioxidant properties. Materials 2021; 14(8): 1984.
[http://dx.doi.org/10.3390/ma14081984] [PMID: 33921014]

[19]   Mishra K, Ojha H, Chaudhury NK. Estimation of antiradical properties of antioxidants using DPPH assay: A critical review and results. Food Chem 2012; 130(4): 1036-43.
[http://dx.doi.org/10.1016/j.foodchem.2011.07.127]

[20]   Vinson JA, Zubik L, Bose P, Samman N, Proch J. Dried fruits: Excellent *in vitro* and *in vivo* antioxidants. J Am Coll Nutr 2005; 24(1): 44-50.
[http://dx.doi.org/10.1080/07315724.2005.10719442] [PMID: 15670984]

[21]   Beeharry N, Lowe JE, Hernandez AR, *et al.* Linoleic acid and antioxidants protect against DNA damage and apoptosis induced by palmitic acid. Mutat Res 2003; 530(1-2): 27-33.
[http://dx.doi.org/10.1016/S0027-5107(03)00134-9] [PMID: 14563528]

[22]   Decker EA. The role of phenolics, conjugated linoleic acid, carnosine, and pyrroloquinoline quinone as nonessential dietary antioxidants. Nutr Rev 1995; 53(3): 49-58.
[http://dx.doi.org/10.1111/j.1753-4887.1995.tb01502.x] [PMID: 7770184]

[23]   Panda SK, Padhi L, Leyssen P, Liu M, Neyts J, Luyten W. Antimicrobial, anthelmintic, and antiviral activity of plants traditionally used for treating infectious disease in the similipal biosphere reserve, odisha, India. Front Pharmacol 2017; 8: 658.
[http://dx.doi.org/10.3389/fphar.2017.00658] [PMID: 29109684]

[24]   Savluchinske Feio S, Gigante B, Carlos Roseiro J, Marcelo-Curto MJ. Antimicrobial activity of diterpene resin acid derivatives. J Microbiol Methods 1999; 35(3): 201-6.
[http://dx.doi.org/10.1016/S0167-7012(98)00117-1] [PMID: 10333071]

[25]   Savluchinske-Feio S, Curto MJM, Gigante B, Roseiro JC. Antimicrobial activity of resin acid derivatives. Appl Microbiol Biotechnol 2006; 72(3): 430-6.
[http://dx.doi.org/10.1007/s00253-006-0517-0] [PMID: 16896605]

[26]   Farjana A, Zerin N, Kabir MS. Antimicrobial activity of medicinal plant leaf extracts against pathogenic bacteria. Asian Pac J Trop Dis 2014; 4: S920-3.
[http://dx.doi.org/10.1016/S2222-1808(14)60758-1]

[27]   Behravan M, Hossein Panahi A, Naghizadeh A, Ziaee M, Mahdavi R, Mirzapour A. Facile green synthesis of silver nanoparticles using berberis vulgaris leaf and root aqueous extract and its antibacterial activity. Int J Biol Macromol 2019; 124: 148-54.
[http://dx.doi.org/10.1016/j.ijbiomac.2018.11.101] [PMID: 30447360]

[28]   Zarai Z, Boujelbene E, Ben Salem N, Gargouri Y, Sayari A. Antioxidant and antimicrobial activities of various solvent extracts, piperine and piperic acid from Piper nigrum. Lebensm Wiss Technol 2013; 50(2): 634-41.
[http://dx.doi.org/10.1016/j.lwt.2012.07.036]

[29]   Cowan MM. Plant products as antimicrobial agents. Clin Microbiol Rev 1999; 12(4): 564-82.
[http://dx.doi.org/10.1128/CMR.12.4.564] [PMID: 10515903]

[30]   Savoia D. Plant-derived antimicrobial compounds: Alternatives to antibiotics. Future Microbiol 2012;

7(8): 979-90.
[http://dx.doi.org/10.2217/fmb.12.68] [PMID: 22913356]

[31]   Gyawali R, Ibrahim SA. Natural products as antimicrobial agents. Food Control 2014; 46: 412-29.
[http://dx.doi.org/10.1016/j.foodcont.2014.05.047]

[32]   Davidson PM, Taylor TM. Chemical peservatives and natural antimicrobial compounds. Food microbiology: Fundamentals and frontiers 2007; 713-45.

[33]   Tiku AR. Antimicrobial compounds and their role in plant defense Molecular aspects of plant-pathogen interaction. Singapore: Springer 2018; pp. 283-307.

[34]   Mansfield JW. Antimicrobial compounds and resistance Mechanisms of resistance to plant diseases. Dordrecht: Springer 2000; pp. 325-70.
[http://dx.doi.org/10.1007/978-94-011-3937-3_10]

[35]   Hayashi MA, Bizerra FC, Junior PI, Eds. Antimicrobial compounds from natural sources. Frontiers E-books 2014.
[http://dx.doi.org/10.3389/978-2-88919-259-5]

[36]   González-Lamothe R, Mitchell G, Gattuso M, Diarra M, Malouin F, Bouarab K. Plant antimicrobial agents and their effects on plant and human pathogens. Int J Mol Sci 2009; 10(8): 3400-19.
[http://dx.doi.org/10.3390/ijms10083400] [PMID: 20111686]

[37]   Jimenez-Lopez C, Fraga-Corral M, Carpena M, *et al.* Agriculture waste valorisation as a source of antioxidant phenolic compounds within a circular and sustainable bioeconomy. Food Funct 2020; 11(6): 4853-77.
[http://dx.doi.org/10.1039/D0FO00937G] [PMID: 32463400]

[38]   Aiyegoro OA, Okoh AI. Use of bioactive plant products in combination with standard antibiotics: Implications in antimicrobial chemotherapy. J Med Plants Res 2009; 3: 1147-52.

[39]   Di Ventura B, Lemerle C, Michalodimitrakis K, Serrano L. From *in vivo* to *in silico* biology and back. Nature 2006; 443(7111): 527-33.
[http://dx.doi.org/10.1038/nature05127] [PMID: 17024084]

[40]   Rajabi S, Maresca M, Yumashev AV, Choopani R, Hajimehdipoor H. The most competent plant-derived natural products for targeting apoptosis in cancer therapy. Biomolecules 2021; 11(4): 534.
[http://dx.doi.org/10.3390/biom11040534] [PMID: 33916780]

[41]   Koper K, Wileński S, Koper A. Advancements in cancer chemotherapy. Phys Sci Rev 2021; 1-22.
[http://dx.doi.org/10.1515/9783110662306-002]

[42]   Anand P, Kunnumakara AB, Sundaram C, *et al.* Cancer is a preventable disease that requires major lifestyle changes. Pharm Res 2008; 25(9): 2097-116.
[http://dx.doi.org/10.1007/s11095-008-9661-9] [PMID: 18626751]

[43]   Ashkenazi A, Holland P, Eckhardt SG. Ligand-based targeting of apoptosis in cancer: The potential of recombinant human apoptosis ligand 2/Tumor necrosis factor-related apoptosis-inducing ligand *(rhApo2L/TRAIL)*. J Clin Oncol 2008; 26(21): 3621-30.
[http://dx.doi.org/10.1200/JCO.2007.15.7198] [PMID: 18640940]

[44]   Iqbal J, Abbasi BA, Mahmood T, *et al.* Plant-derived anticancer agents: A green anticancer approach. Asian Pac J Trop Biomed 2017; 7(12): 1129-50.
[http://dx.doi.org/10.1016/j.apjtb.2017.10.016]

[45]   Sanmartín C, Plano D, Sharma AK, Palop JA. Selenium compounds, apoptosis and other types of cell death: An overview for cancer therapy. Int J Mol Sci 2012; 13(8): 9649-72.
[http://dx.doi.org/10.3390/ijms13089649] [PMID: 22949823]

[46]   NAIR PP. Diet, nutrient intake, and metabolism in populations at high and low risk for colon cancer. Am J Clin Nutr 1984; 40(4): 880-6.
[http://dx.doi.org/10.1093/ajcn/40.4.880] [PMID: 6385689]

[47]   Nakanishi C, Toi M. Nuclear factor-κB inhibitors as sensitizers to anticancer drugs. Nat Rev Cancer 2005; 5(4): 297-309.
        [http://dx.doi.org/10.1038/nrc1588] [PMID: 15803156]

[48]   Jain R, Jain SK. Screening of *in vitro* cytotoxic activity of some medicinal plants used traditionally to treat cancer in chhattisgarh state, India. Asian Pac J Trop Biomed 2011; 1(2): S147-50.
        [http://dx.doi.org/10.1016/S2221-1691(11)60144-5]

[49]   Lee P, Zhang R, Li V, *et al.* Enhancement of anticancer efficacy using modified lipophilic nanoparticle drug encapsulation. Int J Nanomedicine 2012; 7: 731-7.
        [PMID: 22359452]

[50]   Tan NH, Akindahunsi AA, Zeng GZ, Zhang YM, Adebayo AH. Anticancer and antiradical scavenging activity of *Ageratum conyzoides* L. *(Asteraceae)*. Pharmacogn Mag 2010; 6(21): 62-6.
        [http://dx.doi.org/10.4103/0973-1296.59968] [PMID: 20548938]

[51]   Amer H, Helmy WA, Taie HA. *In vitro* antitumor and antiviral activities of seeds and leaves Neem *(Azadirachta indica)* extracts. Int J Acad Res 2010; 2: 47-51.

[52]   Shebaby W, Saliba J, Faour WH, *et al. In vivo* and *in vitro* anti-inflammatory activity evaluation of Lebanese Cannabis sativa L. ssp. indica *(Lam.)*. J Ethnopharmacol 2021; 270: 113743.
        [http://dx.doi.org/10.1016/j.jep.2020.113743] [PMID: 33359187]

[53]   Akram M, Shahab-Uddin AA, Usmanghani KH, Hannan AB, Mohiuddin E, Asif M. Curcuma longa and curcumin: A review article. Rom J Biol Plant Biol 2010; 55: 65-70.

[54]   Jeena K, Liju VB, Kuttan R. Antitumor and cytotoxic activity of ginger essential oil *(Zingiber officinale Roscoe)*. Int J Pharm Pharm Sci 2015; 7: 341-4.

[55]   Gandhi D, Mehta P. Dillenia indica linn. and dillenia pentagyna roxb.: Pharmacognostic, phytochemical and therapeutic aspects. J Appl Pharm Sci 2013; 3: 134-42.

[56]   El-Shemy H, Aboul-Soud M, Nassr-Allah A, Aboul-Enein K, Kabash A, Yagi A. Antitumor properties and modulation of antioxidant enzymes' activity by aloe vera leaf active principles isolated *via* supercritical carbon dioxide extraction. Curr Med Chem 2010; 17(2): 129-38.
        [http://dx.doi.org/10.2174/092986710790112620] [PMID: 19941474]

[57]   Rosangkima G, Prasad SB. Antitumour activity of some plants from meghalaya and mizoram against murine ascites dalton's lymphoma. Indian J Exp Biol 2004; 42(10): 981-8.
        [PMID: 15511001]

[58]   Rosangkima G, Rongpi T, Prasad SB. Ethno-medicinal value of some anticancer medicinal plants from north-east India: An *in vivo* screening in murine tumor model. Sci Vis 2010; 10: 123-32.

[59]   Pang Y, Wang D, Fan Z, *et al.* Blumea balsamifera--a phytochemical and pharmacological review. Molecules 2014; 19(7): 9453-77.
        [http://dx.doi.org/10.3390/molecules19079453] [PMID: 24995927]

[60]   Karthikeyan K, Gunasekaran P, Ramamurthy N, Govindasamy S. Anticancer activity of ocimum sanctum. Pharm Biol 1999; 37(4): 285-90.
        [http://dx.doi.org/10.1076/phbi.37.4.285.5801]

[61]   Rumzhum NN, Rahman MM, Islam MS, Chowdhury SA, Sultana R, Parvin MN. Cytotoxicity and antioxidant activity of extractives from mirabilis jalapa. Stamford j pharm sci 2008.

[62]   Chanda S, Nagani K. *In vitro* and *in vivo* methods for anticancer activity evaluation and some Indian medicinal plants possessing anticancer properties: An overview. J Pharmacogn Phytochem 2013; 2: 140-52.

[63]   Emanuele S, Lauricella M, Calvaruso G, D'Anneo A, Giuliano M. Litchi chinensis as a functional food and a source of antitumor compounds: An overview and a description of biochemical pathways. Nutrients 2017; 9(9): 992.
        [http://dx.doi.org/10.3390/nu9090992] [PMID: 28885570]

[64]   Giri A, Lakshmi Narasu M. Production of podophyllotoxin from podophyllum hexandrum: A potential natural product for clinically useful anticancer drugs. Cytotechnology 2000; 34(1/2): 17-26.
[http://dx.doi.org/10.1023/A:1008138230896] [PMID: 19003377]

[65]   Piloto-Ferrer J, Sánchez-Lamar Á, Francisco M, *et al.* Xanthium strumarium´s xanthatins induces mitotic arrest and apoptosis in CT26WT colon carcinoma cells. Phytomedicine 2019; 57: 236-44.
[http://dx.doi.org/10.1016/j.phymed.2018.12.019] [PMID: 30797985]

[66]   Vijayshri S, Yadava RN. Isolation and characterization of new potential allelochemical from Bidens biternata *(Lour.).* Merrill & Sherff J Chem Pharm Res 2015; 7: 175-9.

[67]   Tobinaga S, Sharma M, Aalbersberg W, *et al.* Isolation and identification of a potent antimalarial and antibacterial polyacetylene from bidens pilosa. Planta Med 2009; 75(6): 624-8.
[http://dx.doi.org/10.1055/s-0029-1185377] [PMID: 19263339]

[68]   Kumirska J, Czerwicka M, Kaczyński Z, *et al.* Application of spectroscopic methods for structural analysis of chitin and chitosan. Mar Drugs 2010; 8(5): 1567-636.
[http://dx.doi.org/10.3390/md8051567] [PMID: 20559489]

[69]   Khan AL, Hussain J, Hamayun M, *et al.* Secondary metabolites from inula britannica l. and their biological activities. Molecules 2010; 15(3): 1562-77.
[http://dx.doi.org/10.3390/molecules15031562] [PMID: 20336001]

[70]   Brandão MGL, Krettli AU, Soares LSR, Nery CGC, Marinuzzi HC. Antimalarial activity of extracts and fractions from bidens pilosa and other bidens species *(asteraceae)* correlated with the presence of acetylene and flavonoid compounds. J Ethnopharmacol 1997; 57(2): 131-8.
[http://dx.doi.org/10.1016/S0378-8741(97)00060-3] [PMID: 9254115]

[71]   Grata E, Guillarme D, Glauser G, *et al.* Metabolite profiling of plant extracts by ultra-high-pressure liquid chromatography at elevated temperature coupled to time-of-flight mass spectrometry. J Chromatogr A 2009; 1216(30): 5660-8.
[http://dx.doi.org/10.1016/j.chroma.2009.05.069] [PMID: 19539942]

[72]   AÇIKGÖZ MA, Kara ŞM, Aygün AH, ÖZCAN MM, Ay EB. Effects of methyl jasmonate and salicylic acid on the production of camphor and phenolic compounds in cell suspension culture of endemic turkish yarrow *(achillea gypsicola)* species. Turk J Agric For 2019; 43: 351-9.
[http://dx.doi.org/10.3906/tar-1809-54]

[73]   Lala S. Nanoparticles as elicitors and harvesters of economically important secondary metabolites in higher plants: A review. IET Nanobiotechnol 2021; 15(1): 28-57.
[http://dx.doi.org/10.1049/nbt2.12005] [PMID: 34694730]

[74]   Anjum S, Anjum I, Hano C, Kousar S. Advances in nanomaterials as novel elicitors of pharmacologically active plant specialized metabolites: Current status and future outlooks. RSC Advances 2019; 9(69): 40404-23.
[http://dx.doi.org/10.1039/C9RA08457F] [PMID: 35542657]

[75]   Hatami M, Naghdi Badi H, Ghorbanpour M. Nano-elicitation of secondary pharmaceutical metabolites in plant cells: A review. Faslnamah-i Giyahan-i Daruyi 2019; 3(71): 6-36.
[http://dx.doi.org/10.29252/jmp.3.71.6]

[76]   Servin A, Elmer W, Mukherjee A, *et al.* A review of the use of engineered nanomaterials to suppress plant disease and enhance crop yield. J Nanopart Res 2015; 17(2): 92.
[http://dx.doi.org/10.1007/s11051-015-2907-7]

[77]   Marslin G, Sheeba CJ, Franklin G. Nanoparticles alter secondary metabolism in plants *via* ros burst. Front Plant Sci 2017; 8: 832.
[http://dx.doi.org/10.3389/fpls.2017.00832] [PMID: 28580002]

[78]   Sudha PN, Sangeetha K, Vijayalakshmi K, Barhoum A. Nanomaterials history, classification, unique properties, production and market.Elsevier. Emerging applications of nanoparticles and architecture nanostructures 2018, pp. 341-84.

[http://dx.doi.org/10.1016/B978-0-323-51254-1.00012-9]

[79]   Khan I, Saeed K, Khan I. Nanoparticles: Properties, applications and toxicities. Arab J Chem 2019; 12(7): 908-31.
[http://dx.doi.org/10.1016/j.arabjc.2017.05.011]

[80]   Thomas S, Harshita BSP, Mishra P, Talegaonkar S. Ceramic nanoparticles: Fabrication methods and applications in drug delivery. Curr Pharm Des 2015; 21(42): 6165-88.
[http://dx.doi.org/10.2174/1381612821666151027153246] [PMID: 26503144]

[81]   Crucho CIC, Barros MT. Polymeric nanoparticles: A study on the preparation variables and characterization methods. Mater Sci Eng C 2017; 80: 771-84.
[http://dx.doi.org/10.1016/j.msec.2017.06.004] [PMID: 28866227]

[82]   Cartaxo AL. Nanoparticles types and properties–understanding these promising devices in the biomedical area. Int. J. Nanomed 2018; pp. 1-8.

[83]   Prasad SR, Elango K, Damayanthi D, Saranya JS. Formulation and evaluation of azathioprine loaded silver nanopartilces for the treatment of rheumatoid arthritis. Ame J Biopharma Pharma Sci 2013; 3: 28-32.

[84]   Kumar H, Venkatesh N, Bhowmik H, Kuila A. Metallic nanoparticle: A review. Biomed J Sci Tech Res 2018; 4: 3765-75.

[85]   Chung IM, Rajakumar G, Thiruvengadam M. Effect of silver nanoparticles on phenolic compounds production and biological activities in hairy root cultures of *cucumis anguria*. Acta Biol Hung 2018; 69(1): 97-109.
[http://dx.doi.org/10.1556/018.68.2018.1.8] [PMID: 29575919]

[86]   Golkar P, Moradi M, Garousi GA. Elicitation of Stevia glycosides using salicylic acid and silver nanoparticles under callus culture. Sugar Tech 2019; 21(4): 569-77.
[http://dx.doi.org/10.1007/s12355-018-0655-6]

[87]   Zhang Y, Chen Y, Zhang H, Zhang B, Liu J. Potent antibacterial activity of a novel silver nanoparticle-halloysite nanotube nanocomposite powder. J Inorg Biochem 2013; 118: 59-64.
[http://dx.doi.org/10.1016/j.jinorgbio.2012.07.025] [PMID: 23123339]

[88]   Logeswari P, Silambarasan S, Abraham J. Synthesis of silver nanoparticles using plants extract and analysis of their antimicrobial property. J Saudi Chem Soc 2015; 19(3): 311-7.
[http://dx.doi.org/10.1016/j.jscs.2012.04.007]

[89]   Tahoori F, Majd A, Nejadsattari T, Ofoghi H, Iranbakhsh A. Qualitative and quantitative study of quercetin and glycyrrhizin in *in vitro* culture of Liquorice *(Glycyrrhiza glabra L.)* and elicitation with AgNO3. Not Bot Horti Agrobot Cluj-Napoca 2018; 47(1): 143-51.
[http://dx.doi.org/10.15835/nbha47111275]

[90]   Ghanati F, Bakhtiarian S. Effect of methyl jasmonate and silver nanoparticles on production of secondary metabolites by Calendula officinalis L *(Asteraceae)*. Trop J Pharm Res 2014; 13(11): 1783-9.
[http://dx.doi.org/10.4314/tjpr.v13i11.2]

[91]   Fazal H, Abbasi BH, Ahmad N, Ali M. Elicitation of medicinally important antioxidant secondary metabolites with silver and gold nanoparticles in callus cultures of prunella vulgaris L. Appl Biochem Biotechnol 2016; 180(6): 1076-92.
[http://dx.doi.org/10.1007/s12010-016-2153-1] [PMID: 27287999]

[92]   Talankova-Sereda TE, Liapina KV, Shkopinskij EA, *et al.* The Influence of Cu и Co Nanoparticles on growth characteristics and biochemical structure of mentha longifolia *in vitro*. InNanophysics, Nanophotonics, Surface Studies, and Applications. Cham: Springer 2016; pp. 427-36.

[93]   Zhao L, Huang Y, Hu J, Zhou H, Adeleye AS, Keller AA. 1H NMR and GC-MS based metabolomics reveal defense and detoxification mechanism of cucumber plant under nano-Cu stress. Environ Sci Technol 2016; 50(4): 2000-10.

[http://dx.doi.org/10.1021/acs.est.5b05011] [PMID: 26751164]

[94] Rodriguez JA, Liu G, Jirsak T, *et al.* Activation of gold on titania: Adsorption and reaction of SO(2) on Au/TiO(2)(110). J Am Chem Soc 2002; 124(18): 5242-50.
[http://dx.doi.org/10.1021/ja020115y] [PMID: 11982389]

[95] Kasana RC, Panwar NR, Kaul RK, Kumar P. Biosynthesis and effects of copper nanoparticles on plants. Environ Chem Lett 2017; 15(2): 233-40.
[http://dx.doi.org/10.1007/s10311-017-0615-5]

[96] Ain N, Haq I, Abbasi BH, Javed R, Zia M. Influence of PVP/PEG impregnated CuO NPs on physiological and biochemical characteristics of *Trigonella foenum graecum* L. IET Nanobiotechnol 2018; 12(3): 349-56.
[http://dx.doi.org/10.1049/iet-nbt.2017.0102]

[97] Karimi N, Behbahani M, Dini G, Razmjou A. Enhancing the secondary metabolite and anticancer activity of *echinacea purpurea* callus extracts by treatment with biosynthesized zno nanoparticles. Adv nanosci nanotechnol 2018; 9(4): 045009.
[http://dx.doi.org/10.1088/2043-6254/aaf1af]

[98] Raigond P, Raigond B, Kaundal B, Singh B, Joshi A, Dutt S. Effect of zinc nanoparticles on antioxidative system of potato plants. J Environ Biol 2017; 38(3): 435-9.
[http://dx.doi.org/10.22438/jeb/38/3/MS-209]

[99] Mohebodini M, Fathi R, Mehri N. Optimization of hairy root induction in chicory *(Cichorium intybus L.)* and effects of nanoparticles on secondary metabolites accumulation. Iran J Genet Plant Breed 2017; 6: 60-8.

[100] Moharrami F, Hosseini B, Sharafi A, Farjaminezhad M. Enhanced production of hyoscyamine and scopolamine from genetically transformed root culture of *Hyoscyamus reticulatus* L. elicited by iron oxide nanoparticles *in vitro*. Cell Dev Biol Plant 2017; 53(2): 104-11.
[http://dx.doi.org/10.1007/s11627-017-9802-0] [PMID: 28553065]

[101] Kamalizadeh M, Bihamta M, Zarei A. Drought stress and TiO_2 nanoparticles affect the composition of different active compounds in the Moldavian dragonhead plant. Acta Physiol Plant 2019; 41(2): 21.
[http://dx.doi.org/10.1007/s11738-019-2814-0]

[102] Hussain I, Singh NB, Singh A, Singh H, Singh SC, Yadav V. Exogenous application of phytosynthesized nanoceria to alleviate ferulic acid stress in Solanum lycopersicum. Sci Hortic (Amsterdam) 2017; 214: 158-64.
[http://dx.doi.org/10.1016/j.scienta.2016.11.032]

[103] Poborilova Z, Ohlsson AB, Berglund T, Vildova A, Provaznik I, Babula P. DNA hypomethylation concomitant with the overproduction of ROS induced by naphthoquinone juglone on tobacco BY-2 suspension cells. Environ Exp Bot 2015; 113: 28-39.
[http://dx.doi.org/10.1016/j.envexpbot.2015.01.005]

[104] Joshi A, Kaur S, Dharamvir K, Nayyar H, Verma G. Multi-walled carbon nanotubes applied through seed-priming influence early germination, root hair, growth and yield of bread wheat (*Triticum aestivum* L.). J Sci Food Agric 2018; 98(8): 3148-60.
[http://dx.doi.org/10.1002/jsfa.8818] [PMID: 29220088]

[105] Chen L, Yang J, Li X, *et al.* Carbon nanoparticles enhance potassium uptake *via* upregulating potassium channel expression and imitating biological ion channels in BY-2 cells. J Nanobiotechnology 2020; 18(1): 21.
[http://dx.doi.org/10.1186/s12951-020-0581-0] [PMID: 31992314]

[106] Shah M, Jan H, Drouet S, *et al.* Chitosan elicitation impacts flavonolignan biosynthesis in silybum marianum (l.) gaertn cell suspension and enhances antioxidant and anti-inflammatory activities of cell extracts. Molecules 2021; 26(4): 791.
[http://dx.doi.org/10.3390/molecules26040791] [PMID: 33546424]

CHAPTER 17

Nanocarriers: Promising Vehicles for Controlled Bioactive Drug Delivery in Current Medical System

Ajam C. Shaikh[1,*], **Ashfaq A. Shah**[2] and **Amit Gupta**[2]

[1] *MAEER's MIT College of Railway Engineering and Research, Barshi-413401, Solapur, Maharashtra, India*

[2] *Department of Life Sciences, Graphic Era (Deemed to be) University, Dehradun-248001, Uttarakhand, India*

Abstract: Nanomaterials have been widely employed in the medical profession in recent decades, thanks to the rapid development of nanotechnology. Their distinctive physical and chemical qualities, such as minimal size, functionalized surface characteristics, stable interactions with ligands, high carrier capacity, and ease of binding with both hydrophilic and hydrophobic substances have made them ideal platforms for the target-specific and controlled delivery of micro-and macromolecules in disease therapy and have revealed an excellent potential pertaining to clinical entities with the goal of fine-tuning bioavailability, bioefficacy, and pharmacokinetics. The absorption, post-administration stability as well as bioavailability of bioactive drugs and other medicinal substances are the key issues. Some critical medications have low gastrointestinal absorption and permeability in their active form, are inactivated by pH and temperature fluctuations and cause catastrophic off-target and undesirable side effects. Certain investigations have also indicated that active efflux mechanisms affect the absorption of some presently integrated compounds with structural alterations across the intestinal wall. Furthermore, intestinal bacteria and/or enzymes break down fragile structures of active substances into a variety of metabolites, each of which has different bioactivity than the original chemical compound. Nanocarrier-mediated distribution improved their solubilization potential, changed absorption paths, and reduced metabolic breakdown by gut bacteria and enzymes. Combining nanobiotechnology with current therapeutic techniques has shown to be effective in bringing innovative and previously rejected bioactive substances to the market for treating a myriad of diseases and disorders. As a result, we predict that nanotechnology will play a larger role in illness detection and treatment in the future, perhaps helping to overcome bottlenecks in current medical approaches. This chapter focuses on a comprehensive discussion of strategies and applications of nanoengineered delivery systems along with the pharmacokinetic properties and drug-delivery mechanism of

* **Corresponding author Ajam C. Shaikh:**MAEER's MIT College of Railway Engineering and Research, Barshi-413401, Solapur, Maharashtra, India; E-mail: shaikhajam@gmail.com

Zulqurnain Khan, Azra Yasmin & Naila Safdar (Eds.)

these nanocarriers. Probably associated drawbacks, challenges, future advancements, and scopes of nanocarriers in clinical care are also highlighted.

Keywords: Controlled release, Dendrimers, Liposomes, Nanocapsules, Nanogels, Nanocrystals, Nanosuspensions, Nanowires, Polymeric micelles, Quantum dots.

INTRODUCTION

Despite the fact that drugs, medicines, and novel bioactive molecules of synthetic or natural origin are beneficial to human health in diseased conditions, the main issues related to these disease-modifying agents of synthetic or natural origin are related to their post-administration instability, off-target, and unwanted side effects. Any disease-modifying agent's therapeutic outcome is determined by how well its pharmacokinetic profile improves after therapy. In the modern period, nanotechnology is tackling all of these difficulties by bridging the physical and biological sciences through the use of nanostructures and nanophases in a variety of sectors of study such as nanomedicine [1, 2]. Nanomedicine and nano-delivery systems are a relatively novel but rapidly growing discipline in which microscopic materials are utilized either as diagnostic tools or for the delivery of therapeutic drugs to specific sites in a controlled manner.

Nanotechnology proved to have a variety of benefits in treating chronic human diseases as it permits the exact delivery of drugs to some particular regions. Medication delivery, chemotherapy, biosensors, and tissue engineering are all using nanoparticles right now in biomedicine. It also covers the use of nano-dimensional materials in live cells, *e.g.*, nano-robots, nano-sensors for diagnostic and sensory applications, and the actuation of materials in living cells. Nanomaterials with diameters ranging from 1 to 100 nanometers, affect the frontiers of nanomedicine, ranging from biosensors to microfluidics, microarray testing to tissue engineering, and drug delivery [2, 3]. In nanotechnology, to produce nanomedicines, a curative agent at the nanoscale level has been utilized. Nanoparticles are typically tiny nanospheres composed of materials created at the atomic or molecular level [3, 4]. Because of nanoscale structural dimensions, they may travel more easily throughout the human body than larger materials and can easily permeate through the tissue system thereby allowing for facile drug absorption by cells to bring out a significant activity at the desired area. The structural, mechanical, chemical, magnetic, biological, and electrical characteristics of nanoscale particles are all different.

Nanomedicines have attracted considerable attention in recent years due to their ability to encapsulate pharmaceuticals or bind bioactive chemicals to nanostructures and deliver them to specific tissues more accurately in a controlled fashion. Furthermore, nanostructures also facilitate the transport of water-insoluble medications to their target area, as well as reducing drug decomposition in the gastrointestinal tract. Because the nanostructures have standard absorptive endocytosis absorption mechanisms as well as enhanced oral bioavailability, they are substantially more rapidly absorbed by cells than large particles ranging in size from 1 to 10 m. Nanostructures last a long time in the circulatory system, allowing combination drugs to be administered at precisely the right amount. As a result, they have fewer negative effects and cause fewer plasma fluctuations [4, 5]. The positive zeta potential, as well as the hydrophobicity of these nanoscopic particles, aids their absorption from the gastrointestinal tract. Other mechanisms found to be supportive in enhancing the absorption of bioactive molecule include electrostatic communication between positively charged nano-vehicle surfaces and negatively charged mucin, amplified transcytosis, and receptor-mediated endocytosis, interaction with junction proteins to modulate tight junctions, the microfold cells mediated phagocytosis of nanoparticles, and chylomicron aided absorption by enterocytes intervened by lipases for lipid nanocarriers. Also, there are many processes by which these nanosystems release their protected bioactive components once, inside the body *e.g.* desorption of adsorbed/surface-bound various components, matrix erosion, enzymatic degradation, matrix diffusion, dissolution, or a combination of either of the processes [5, 6].

A promising approach to nanotechnology in medicine is the synthesis of a variety of nanosized carriers in the size range of 10 to 100 nm (Fig. **1**). The uptake of loaded molecules across the gastrointestinal mucosa is influenced by the particle size, surface characteristics, and shape of the nanoparticles. Polymer nanocarriers (nanocapsules, polymeric micelles, *etc.*), molecular complexes (Cyclodextrins Inclusion Complexes and Phytosomes), lipid-containing nanocarriers (nanostructured lipid capsules (NLC), solid-lipid nanocapsules (SLN), lipid nanospheres, micro- and nanoemulsions, micelles) are all examples of nanosystems for enhanced, controlled, and site-specific delivery of bioactive compounds [6, 7]. Because of the hydrophobicity as well as hydrophilicity inside the polymeric system, polymer-based nanocarriers may accept a wide range of medicinal compounds. Natural or manufactured biodegradable polymers are commonly used as the carrier material in polymer-based nanocarriers.

Natural ones either polysaccharides or proteins are recommended since they have a lower level of toxicity. Plant-based polysaccharides such as pectin, gum arabic, alginate, starch, and its derivatives, cellulose and its derivatives, and animal-based polysaccharides such as xanthan gum, chitosan, *etc.* are used to formulate polymeric nanoparticles. Polyglycolic acid, polylactic acid, poly-cyanoacrylate alkyl esters, polyvinyl alcohol, polylactic-glycolic acid, *etc.* are examples of synthetic polymers. Polysaccharide nanoparticles having distinct characteristics are considered ideal carriers for hydrophilic drug delivery. Polysaccharide nanoparticles are natural biomaterials that are biodegradable, safe, non-toxic, and stable under various conditions. Polysaccharides are plentiful in nature and need little processing costs. In polymeric nanoparticles, bioactive medicines, and plant secondary metabolites are embedded, dissolved, or adsorbed [7, 8]. The nanocarriers release drug molecules ensuring that none of the molecules are released until they reach systemic circulation, bypassing different physiological obstacles that may obstruct drug metabolism. After reaching the apical membrane of intestinal epithelial cells, the majority of the loaded nanoparticles take entry into enterocytes *via* transcellular transport. Small particles *e.g.*, 100 to 400 nm, are ingested by enterocytes through clathrin and caveolae-mediated endocytosis apart from taken up by specialized Peyer's patches (M cells) and GALT follicles in the gastrointestinal tract. The endolysosomal breakdown is prevented by covering nanoparticles with cationic chitosan. To guarantee improved medication absorption, nanocarrier micelles modify membrane permeability, and mucoadhesion inside the GI tract [8, 9]. Different types of nanomaterials employed in drug delivery applications are shown in Fig. (**1**)

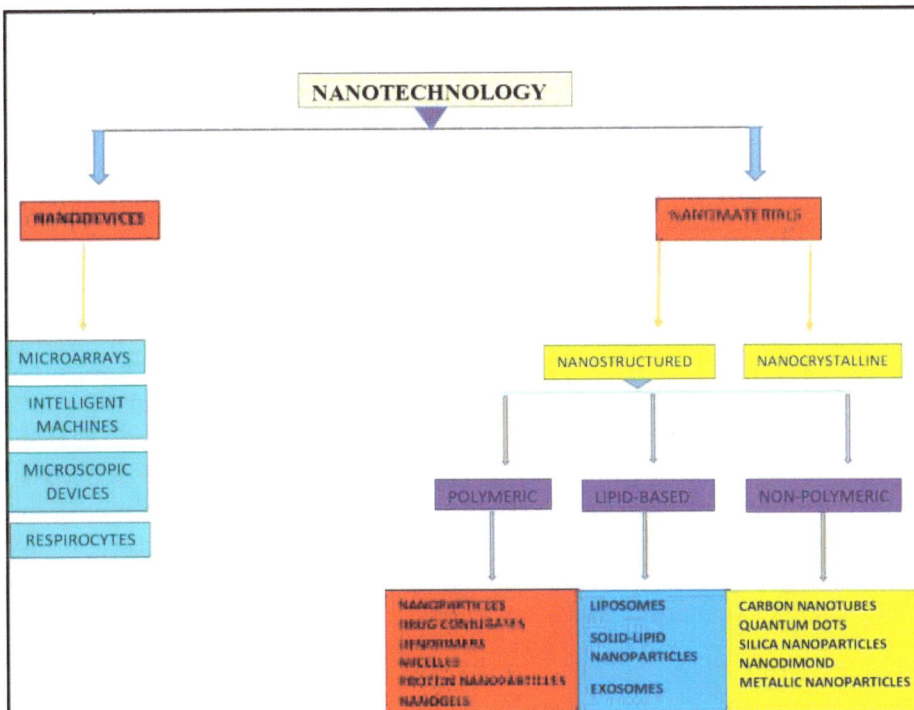

Fig. (1). Medicinal applications of nanotechnology.

NANOMATERIALS AS NANOCARRIERS

Various impasses have been observed for conventional site-specific drug delivery systems due to the harsh microenvironment of diseased tissues of the living system, poor availability of required doses, adverse side effects, low therapeutic indices, and non-specific targeting [10]. But nanotechnology offers a smart solution by devising a nano-sized system, in which encapsulated drug is provided through a nano-vehicle to the targeted site with reasonable biocompatibility. It is being protected from degradation in a hostile physiological environment with low toxicity to healthy cells. In order to qualify as an ideal nanocarrier, it should possess certain ideal characteristics like suitable preparation and purification methods, non-reactivity with other drugs, requisite mechanical strength, stability, particle size, shape, high drug payload, encapsulation efficiency, enhanced *in vivo* residence time, good biocompatibility, and low toxicity [8, 11]. Nanomaterials being used as nanocarriers could be of natural or synthetic origin. Some of the representative nanocarriers of natural origin are:

Chitosan

Due to excellent mucoadhesive properties, chitosan-containing nanomaterials are frequently used as site-specific drug delivery vehicles for various types of epithelia, such as buccal [2], nasal [3], eye [4,] and others [5, 6]. Chitosan, alginate, and pectin-based nanocarriers have been screened for the oral administration of the drug for the oral cavity [7], whereas carboxymethyl chitosan is used for intra-nasal carbamazepine (CBZ) release, for the increased amount of the medication in the brain to reduce the systemic drug exposure [8].

Cellulose

Exploiting the presence of a number of hydroxyl groups protruding out of the cellulose structure to form the hydrogen bonds in the cellulose nanocrystals and the bioactive drug, showed promising results in controlled release of the repaglinide (an anti-hyperglycemic—RPG) [9]. Four derivatives of cellulose such as methylcellulose, cationic hydroxyethyl cellulose, hydroxypropyl methylcellulose, and sodium carboxymethyl cellulose have been screened for the release of the controlled drug into the nasal mucosa, and none of them showed detrimental impacts on either tissues or cells [10].

Liposomes

Liposomes are basically vesicles of globular form which are made up of steroids and phospholipids. Being one of the most studied drug delivery nano-vehicle systems, the required stability is imparted to therapeutic compounds for improved bio-distribution. It is being used along with hydrophilic and hydrophobic bioactive drugs. Biocompatibility and biodegradability are added features of these classes of nano-vehicles. The versatility of the carriers is owed to the membrane structure, which is very similar to the cell membranes that assist in the incorporation of drugs into them [11, 12].

Alginate

Simultaneous lowering of serum glucose levels along with enhanced serum insulin levels has been observed in diabetic rats due to the administration of insulin-appended alginate nanoparticles in which nicotinamide is used as a permeation agent [13]. Apart from this, alginate-based nanocarriers are being used for the release of a drug called venlafaxine (VLF) through the intranasal route for treating depression [14], loading Cisplatin (carcinogen drug) aiming at the non-small lung cancerous cells [15], Synthesis of alginate nanoparticles having chitosan coating to increase the daptomycin permeation in the ocular epithelium intended to achieve an antibacterial effect [16].

The nanomaterials of synthetic origin are also reported as nano-carriers for site-specific drug delivery. These include;

Carbon-based Nanomaterials (CBNs)

CBNs are widely used as nanocarriers owing to their distinct physicochemical properties and structural dimensions. A family of graphene, carbon nanotubes, fullerenes, mesoporous carbon, and nano-diamonds were employed for the targeted delivery of drugs for malignant cancer [17]. These scaffolds are also utilized for cell culture growth as well as diagnostic devices for *in-vivo* as well as *in-vitro* tumor imaging along with cellular dynamics [18]. A dual-ligan--functionalized nano-diamond, such as Cetuximab-NDs-cisplatin bio-conjugated material has been found to suppress the growth of HepG2 cells [19].

Metal-Organic Framework (MOFs)

MOFs are a class of molecules successfully employed as a nanocarrier for steady drug release of anticancer drugs [20], metabolic labeling of molecules, antimicrobial agents [21], hormones, and antiglaucoma medications [22]. ZIFs (Zeolite Imidazolate Frameworks) - a subcategory of MOFs have been successfully employed for drug delivery. ZIF-8 is one of the members of this family used for a system wherein pH-dependent drug release is warranted. This unique mechanism is owed to its property of being unstable under acidic conditions. Similarly, ZIF-8 has been practiced for the delivery of 5-fluorouracil (5-FU), Camptothecin, and Ceftazidime as a pH-responsive nano-vehicle for the targeted, controlled, and nontoxic drug delivery [23]. The other members of the MOF family are also effective and were utilized for transporting the drug at the required site, such as MIL-101, NU-1000, UiO-68, PCN-333 [24], *etc.*

Nanocapsules

Polymeric nanocapsules have attracted the interest of the scientific fraternity for drug delivery, owing to their unique core-shell microstructure. Nanocapsules consist of vesicular systems with an inner liquid core composed of an oily or an aqueous core containing a thin wall of polymer. This offers an increased drug-loading capacity simultaneously reducing the polymeric matrix of the nanoparticle. The encapsulated drug is protected from the hostile tissue environment along with the untimely burst release of the drug due to pH, enzymes, temperature, and various other reasons due to the thin polymeric shell. The versatility of the nanocapsules lies in their ability to tailor the size, shape, and quantity as well. They can be made to suit the desired complex application with specific biochemical, optical, magnetic, and electrical properties, overcoming physiological barriers and the hostile environment of the diseased tissues [25]. Nanocapsules are employed not only to encapsulate proteins, peptides, hormones, metabolites, *etc.* but also for a variety of biomedical applications such as, anti-cancer therapy, immunotherapy, and anti-inflammation therapy [26]. Nanocapsules are among the most sought-after candidates for drug delivery due to the enhanced permeability and solubility of the drug. The enhanced solubility is attributed to the very high surface-to-volume ratio. Nanocapsules were reported to utilize for the targeted delivery of protein for treating tumor cells [27].

Dendrimers

Dendrimers were first introduced in 1985, and since then they have attracted the interest of the scientific community. A well-defined size and globular shape with a hyper-branch offer a unique possibility of tethering a drug molecule and guided delivery of the same at the desired site. A polymerization approach has been adopted in order to construct dendrimers. Monomers are used as building blocks. The tri-body structure of dendrimer consists of a core, branches resembling trees, protruding out the core, and the terminal functional group. This class of molecules is either synthesized by a divergent or convergent approach [28]. PAMAM-type dendrimers having carboxyl or hydroxyl functional groups are used for ocular delivery of bioactive drugs, as a result of its enhanced retention of pilocarpine within the eyes [29]. A multifold cytotoxicity enhancement has been observed for the conjugates Paclitaxel with PAMAM G4dendrimer having hydroxyl groups at the terminus and bis(PEG) polymer. Paclitaxel is known for poor solubility but in this case, not only cytotoxicity but also solubility has been found to be increased

[30]. An investigation has been carried out on doxycycline conjugated PAMAM dendrimer for the potential cellular binding, migration of T47D breast cancer cells and BT-549-Luc, and cytotoxicity [31]. Berberine, a cyclic natural alkaloid containing nitrogen is poorly studied due to its wretched pharmacokinetic behavior. The drastic boost in anticancer activity against MCF-7 and MDA-M--468 breast cancer cells has been observed when berberine was conjugated with G4 PAMAM [32]. These conjugates are safe and biocompatible as well. The other applications of dendrimer-based nanocarrier include enhanced permeation of dendrimers for passive targeting of drugs to tumor tissues [33].

Nanogels

Nanogels have great potential as a promising system for carrying an active ingredient for site-specific control release. These nanogels are three-dimensional cross-linked polymeric networks with high drug encapsulation capacity owing to their hydrophilic or amphiphilic macromolecular chains which are able to hold a great amount of water and swell, keeping their original structure intact. The water incorporating capacity by the nanogels can be attributed to the presence of various hydrophilic moieties along the chain, such as $-CONH_2-,-OH$, $-SO_3H$, and $-CONH-$. This ability of nanogels helps in diffusing and exchanging the metabolic and biomolecules in the tissue fluids and organs to maintain the biochemical balance. Nanogels have been reported to tether to the desired active biomolecule, which upon complexation with nucleic acid/genetic material such as lipoplexes or polyplexes, enhances the transfection of nucleic acid in the cells and ensures its stability [34]. Further nanogel consisting of Ethylenediamine (ED) functionalized PGMA was investigated as potent siRNA and pDNA carriers [35]. The Pluronic nanogels bearing pre-coated PEI and conjugated with heparin, containing vascular endothelial growth factor (bFGF) and pDNA encoding VEGF165 genes encapsulated in the heparin-conjugate have been studied for differentiation and proliferation [35, 36]. Arjunglucoside-I capsulated PNIPAAm/VP nanogels were observed to increase the therapeutic efficacy against parasites when compared to PLA. It is also found that nanogelsare effective in lowering hepatotoxicity as well as nephrotoxicity of the drug [37]. In general, nanogels among the other nanocarriers standout as promising non-viral carriers due to high transfection efficiency, high extracellular stability, low toxicity, and immunogenicity.

Polymeric Micelles

Self-assembly of amphiphilic block copolymers (ABCs) in the spherical, colloidal, supramolecular nanostructure is termed polymeric micelles. A unique structure of this category of carriers with a high loading capacity of the inner core offers a means to site-specific drug delivery of the drug molecules which are otherwise poorly water-soluble. The structure of polymeric micelles consists of a hydrophilic outer core and a hydrophobic inner core in an aqueous environment. In the presence of an aqueous medium, spontaneous rearrangement of the amphiphilic molecules in a supramolecular core takes place, wherein the water-insoluble drug can be introduced in a hydrophobic core. The hydrophilic shell enables prolonged blood circulation by inhibiting opsonins from adsorption on the micelle surface. The pharmacokinetic properties of micelles are size, shape, and surface property dependent, whereas on, the other hand, they are independent of the type and property of the loaded drug [38]. PEG is commonly used in hydrophilic segments in micelle for drug delivery owing to its high water solubility, non-toxicity, and neutral nature [39]. Embelin is an alkyl-substituted hydroxyl benzoquinone bioactive drug molecule known for its hepato-protective effects, anti-diabetic and antitumor activity, and anti-inflammatory potential. In spite of its high bio-activity, poor solubility in water is one of the major drawbacks of its restricted use in drug formulations. Hence, PEG conjugated embelin not only increases its solubility but also forms micelles while retaining its antitumor activity [39]. It was also observed that paclitaxel was delivered using PEG5K-embelin2 micelles yielding excellent antitumor activity as compared to Taxol in murine models of breast and prostate cancers [40]. Adriamycin (ADR)-conjugated PEG–poly(aspartic acid) block copolymers (PEG-P[Asp(ADR)]) show excellent *in vivo* antitumor activity by forming the micelles upon exposure to an aqueous medium [41]. Interestingly Vitamin B_{12} tethered at hydrophilic chains of amphiphilic graft copolymer dextran-g-polyethyleneoxidecetyl ether which is also known as DEX-g-PEO-C16, increases drug permeability 2-3 times when the said drug incorporated in the micelles. It is observed that enhancement in the drug permeability is due to receptor-mediated intestinal absorption through a complex formation [42]. It has been noticed that polymeric micelles are being accumulated at the many diseased tissues such as myocardial infarction tissues, in solid tumors which suggests that micelles are having a great potential therapeutic activity and efficiency on many other diseases [43].

Ceramic Nanoparticles

Ceramic nanoparticles are comprised of inorganic materials such as zirconia, alumina, silica, titania or a combination of these materials with metals, metal oxides, and metal sulfides. These nanoparticles possess properties like low electrical and thermal conductivity, high stiffness, high elastic modulus, and corrosion resistivity. Ceramics nanoparticles are reported to be more stable, easy to manufacture, customizable, possessing long degradation time and bioavailability. Apart from these properties variable size, shape, and porosity make them potential candidates as a nanocarrier. Due to the structural uniqueness, these particles protect the encapsulated molecules such as drug molecules, enzymes, and proteins from denaturalization as a result of external pH and temperature. In addition, ceramic nanoparticles can be decorated with various organic functional groups to enhance their therapeutic usage and controlled, prolonged drug release [44]. Based on the architectural differences, the other members of this nanoscopic ceramic family are Ceramic nano-scaffold and Nano-clay [45]. Silica nanoparticles have been covalently linked to the cationic surface to enhance its effective binding, condensation, as well as protection of plasmid DNA in DNA transfection. These are also reported to be used in delivering proteins and genes [46]. Silica functionalized iron oxide nanoparticles were successfully employed to deliver haloalkane dehalogenase. 5-Fluorouracil along with Amoxicillin conjugated iron impregnated hydroxyapatite nanoparticles were found to be ameliorating bioactivity along with extended drug release [47]. Superparamagnetic iron hydroxyapatite nanoparticles are found to be biocompatible and enhance osteoblastic cell proliferation upon introducing a static magnetic field [48]. Based on the utility of hydroxyapatite vehicles for the transportation of malarial merozoite surface protein-119 (MSP-119) one can conclude that hydroxyapatite nanocarriers can be an excellent immuno-adjuvant that can be exploited as an antigen carrier for the immuno-potentiation [49]. Silica-doped nanoparticles were employed for photodynamic therapy for treating cancer [50]. The double-shelled hollow mesoporous silica spheres utilized as a carrier for hydrophilic and hydrophobic anti-cancer drug transportation vehicles highlighted their biocompatibility of steady release of hydrophilic anticancer drugs *e.g.* irinotecan and high loading capacity of hydrophobic anticancer drug *e.g.* docetaxel along with enhanced anticancer activity [51]. The other systems wherein ceramic nanocarriers have been used are widely reported in the literature.

Nanocrystals and Nanosuspensions

Nanocrystals often known as nanosuspensions, nanoscopic biphasic, colloidal dispersion of aggregates in an aqueous medium of a very large number of molecules stabilized either with a thin coating of surfactants or polymers or both. The minute surfactants quantities are beaded in nanocrystals for electrostatic surface stabilization. The average particle size of the nano-suspensions may range between 200 and 600 nm. The problems associated with the site-specific drug delivery and timely releases of the desired drugs were overcome by the use of nanocrystals in a cost-effective manner without compromising the therapeutic utility of the drug. Typically, the problems associated with nanocarriers are their bioavailability, stability, solubility, and in turn high doses, *etc.* Nanocrystals offer solutions to all these above-mentioned problems due to the higher ratio of volume to the surface area of the nano-suspensions leading to satisfactory therapeutic concentrations along with low doses by altering pharmacokinetics, and bioavailability subsequently increasing drug safety and efficacy. These nanocrystals can accommodate a large amount of drug with minimum dose volume, as is required in parenteral and ophthalmic drug delivery systems. A bottom-up approach (controlled precipitation/crystallization) or top-down approach (nanosizing) has been adapted to produce nano-suspensions with desired particle size and distribution. The ease of scaling-up of nanocrystals has been widely reported [52]. It has been anticipated that nanocrystals serving as a carrier for drugs like amphotericin B and tacrolimus after oral administration are found to be apotent candidate for targeting the mucosa of the gastrointestinal tract while aiming the mononuclear phagocytic system (MPS) cells for the treatment of leishmaniasis and fungal mycobacterial infections. It has been observed that paclitaxel nano-suspensions increase the safety profile many folds than the commercial Taxol injections [53]. With curcumin nano-suspension, excellent cytotoxicity is reported in Hela and MCF-7 cells, in comparison to curcumin solution. The smaller size of the suspension not only enhances the solubility rate but also keeps the crystalline nature intact thereby improving the physical stability of curcumin. Bioavailability has been improved from 5.2% to 82.3% for oral administration of gonadotropin inhibitor Danzaon in nano-suspension form, whereas oral absorption was found to have increased in case of a nano-suspension of amphotericin in comparison with its regular commercial formulation [54]. A successful formulation of cross-linked polymer nano-suspensions of dexamethasone has been adopted for increased anti-inflammatory activity in the model of rabbit eye irritation [55]. In order to check the possibility of curing HIV, indinavir is loaded as a nano-suspension in bone marrow-derived macrophages. The virus-infected cells in lymph nodes plasma, spleen, *etc.* were drastically reduced when the formulation was injected into HIV-1-challenged humanized mice thereby protecting CD4 (+) T-cell .The drug diffusion in the skin has been found to be increased when drug nanoparticles are used in the creams [53].

Nanowires

Assemblies of atoms in a wire fashion known as nanowires have gained the attention of the scientific fraternity owing to their versatility, bioavailability, ease of handling for drug delivery and structural resemblance to an extracellular matrix such as collagen fibers. Increased surface area, wide prospects of functionalization of outer surfaces, and very good mechanical performance make them, particularly an ideal candidate for biomedical devices, and tissue engineering motifs apart from controlled release of drugs to ailing tissues in a fiber diameter-dependent manner [56]. Nanofibers have displayed a high capability of encapsulation and drug loading. Various techniques for synthesizing nanowires such as phase separation, electrospinning [57], template synthesis [58], melt-blown [59], and self-assembly [60] can be employed. Electrospinning technique is widely used due to its ability of large scale production and its effectiveness of fabrication of nanowires of natural polymers and synthetic polymers [57]. Nanowires have potential in the biomedical field for local chemotherapy and for drug delivery in cancer cells [61]. A hydrophobic, anti-fungal agent-itraconazole has been carried as a topical delivery system by non-biodegradable polyurethane-based nanofibers [62]. Similarly, silver nanowires (AgNPs) were incorporated into zein nanofibers resulting in excellent bactericidal activity for *S. aureus* and *E. coli* [63]. Silver nanowires are potent for internalization into A549 and MRC-5 cells devoid of any cytotoxicity, hence furnishing an application of silver nanowires being a biocompatible carrier for lung cancer therapies [64].The iron oxide surface decorated with doxorubicin by a linker which is pH-sensitive, resulted in complete cell death (approx. 90%) when it is applied to cancer cells [65]. Further study showed that nanofibers are safe to a cell which is concluded by *in vitro* cell adhesion and proliferation for the duration of 3 days. Additionally, within the 8 to 72 h, the mPEG-PLA nanowire showed excellent zero-order drug release profiles [66]. Owing to the anisotropy of the nanowires, the superparamagnetic iron oxide nanoparticles allow for deeper tumor-targeting along with higher drug loading [67].

Quantum Dots

Quantum dots are comprised of clumps of atoms having an outer shell made up of a variety of materials. Quantum dots, popularly known as semiconductor nanocrystals as well, having special optical properties (*e.g.*, tunable emission, brightness, and photostability), make them a potent candidate for applications in cellular imaging, and diagnosis and drug delivery by dissolving, dispersing, adsorption, and coupling, *etc*. Owing to their outstanding optical properties these inorganic moieties *i.e.*, QDs glow or fluoresce brightly when shined with laser, which is otherwise not visible. The physical and chemical properties such as rate of dissolution, saturation solubility, surface hydrophilicity, hydrophobicity along with biological responses of the drugs will be altered due to the carrier [68]. The potentiality of the quantum dots lies in their ability to fictionalize with biomolecules such as nucleic acids, cells, and proteins. Light emitted by these quantum dots at a variable wavelength ranging from UV to IR makes cellular or subcellular structures visible. Their reduced toxicity to the organic dyes can be attributed to the inert coating present on the inner surface, thereby reducing the toxicity and resulting in increased efficacy and improved therapeutic properties, hence a potential candidate as a diagnostic tool and nanocarrier for controlled drug release [69].

THERAPEUTIC NANOPARTICLES TARGETED DELIVERY APPLICATIONS

Nanostructured and nanocrystalline nanomaterials are the two basic types of nanomaterials. Polymer-based, non-polymeric, and lipid-based nanoparticles are the three types of nanostructured materials. Dendrimers, nanoparticles, micelles, nanogels, protein nanoparticles, and drug conjugates are all polymer-based nanoparticles. Carbon nanotubes, nanodiamonds, metallic nanoparticles, quantum dots, and silica-based nanoparticles are examples of non-polymeric nanoparticles. Liposomes and solid lipid nanoparticles are two types of lipid-based nanoparticles All have been already discussed as above. Polymer- or lipid-based nanoparticles make up the bulk of clinically approved therapeutic nanoparticles thus far. Nanocrystalline particles, which are generated by the combination of medicinal substances in crystalline form, are utilized in some clinical applications in addition to polymer-based, non-polymeric, or lipid-based nano-structured particles. We described the many types of therapeutically employed nanoparticles, their therapeutic specificity, and current administration techniques in complex pathophysiological diseases in this section (Table **1**).

Table 1. Some approved and under clinical trials therapeutic nanoparticles.

Nanostructure	Nanoparticle Formulation	Active Drug	Indications	Whether Approved or Not	Refs.
Liposome	Nanoliposomes	Irinotecan	Colorectal & Pancreatic Cancer	Yes (FDA 2015) (Europe 2016)	[70]
Lipid-based	Lipid nanoparticles	Transthyretin targeted siRNA	Amyloidosis (Transthyretin-mediated)	Yes (FDA 2018)	[71]
Liposome	Sphingomyelin & cholesterol	Vincristine sulfate	Lymphoid leukemia (Acute)	Yes (FDA 2012)	[72]
Polymer-based	Glatiramer (l-glutamic acid polymer with l-alanine, l-lysine, and l-tyrosine)	MAB	Multiple sclerosis	Yes (FDA 2015)	[73]
Liposome	1-Palmitoyl-2-oleoyl-sn-glycero-3-phosphocholine and 1,2-Dioleoyl-sn-glycero-3-phospho-l-serine liposomes	Mifamurtide	Non-metastasizing osteosarcoma	Yes (Europe 2009)	[74]
Liposome	Distearoylphosphatidylcholine, Distearoylphosphatidylglycerol, Cholesterol	Daunorubicin Cytarabine	Acute myeloid leukemia	Yes (FDA 2017)	[75]
Nanocrystal	Hydroxyapatite	-	Osteoinductive bone graft substitute	Yes (FDA 2009)	[76]
Metallic nanoparticle	Superparamagnetic iron oxide nanoparticle (SPION) covered with dextran	-	Anemia in chronic kidney disease	Yes (FDA 2009) (Europe 2012)	[77]

(Table 1) cont.....

Nanostructure	Nanoparticle Formulation	Active Drug	Indications	Whether Approved or Not	Refs.
Nanocrystal	Paliperidone palmitate	Paliperidone	Schizophrenia	Yes (FDA 2009) (Europe 2011)	[78]
Metallic nanoparticle	Nanoparticles of ferric oxide core-carboxymaltose shell	-	Iron deficiency anemia in chronic kidney disease	Yes (FDA 2013)	[79]
Protein-drug conjugate	Maytansine derivative, DM1	Trastuzumab	HER2+ breast cancer	Yes (FDA 2013)	[80]
Protein-drug conjugate	Glycopegylated coagulation factor IX	Factor IX	Hemophilia	Yes (FDA 2017)	[81]
Metallic nanoparticle	Nanoparticles of superparamagnetic iron oxide coated with amino silane	-	Glioblastoma, prostate, pancreatic cancer	Yes (Europe 2009)	[82]
Nanocrystal	Dantrolene sodium	Dantrolene	Malignant hyperthermia	Yes (FDA 2014)	[83]
Protein-drug conjugate	PEGylated factor VIII	Factor VIII	Hemophilia	Yes (FDA 2015)	[84]
Protein-drug conjugate	Albumin	Paclitaxel	Lung cancer, Pancreatic cancer	Yes (FDA 2012) (Europe 2008)	[85]
Protein-drug conjugate	PEGylated uricase	Pegloticase	Gout	Yes (FDA 2010) (Europe 2013)	[86]
	PEGylated interferon β-1a	Interferon β-1a	Multiple sclerosis	Yes (FDA 2014) (Europe 2014)	[87]
Dendrimer	Polyethylene glycol (PEG)-platinum	α-cyclodextrin	Cancer	clinical/phase I	[88]
Micelle	PEG-polyaspartate polymeric micelle	paclitaxel	Cancer	clinical/phase III	[89]
Carbon nanotube	PEGylated single walled CNT	Cisplatin	Cancer	Pre -clinical phase	[90]
Metallic nanoparticle	Iron oxide magnetic silica-gold nanoparticles	doxorubicin	Cancer	clinical/phase I	[91]
Silica based nanoparticle	Transferrin mesoporous silica	lactobionic acid, doxorubicin	Cancer	clinical/phase 0	[92]
Nanodiamonds	PEGylated nanodiamonds	irinotecan, curcumin	Cancer, Autoimmune diseases	pre-clinical	[93]

CHALLENGES IN THE MANUFACTURING OF NANOMEDICINES

The treatments involving nanocarriers have a high added value than the standard treatments. It is crucial to consider cost-management approaches for the manufacturing process. The high cost of scaled conversion and the manufacturing method itself are the key hurdles to large-scale nanomedicine production. Mass production, the analytical procedures for in-process quality control, and exceptionally precise finished products need time and so are expensive [8, 16]. Nanomedicines must be produced using a process that is both reliable and uniform in quality for every batch. The lack of uniformity in production techniques and quality control testing is another concern. Quality control testing at in-process and end-product stages, regardless of the technology utilized to create most nanomedicines, demands the employment of sophisticated equipment and analytical techniques. By substantiating the selection of analytical methods, and accurately interpreting the results, the establishment of guidelines and standard reference materials are critical in supporting the registration of nanomedicines. Most analytical methods recommended for in-process quality control and final products do not have standard protocols or descriptions in pharmacopeias, making innovative nanomedicines difficult to characterize. The particle size of nanomedicines is a critical quality attribute (CQA) of nanomedicine. At the nanoscale, relatively small variations in large size distribution or a particle size might affect a substance's bioavailability. Furthermore, due to the small particle size, particle shape,and form, special analytical processes are required . Other CQAs for nanomedicines include surface coating, morphology, and charge. Since shape has been shown to alter cell uptake rates, uptake processes, level of uptake and intracellular distribution, and hence toxicity profile. According to a study, the cellular absorption of triangle-shaped nanoparticles was the highest, whereas star-shaped nanoparticles had the lowest [92]. To confirm nanomedicine functionality, manufacturing process repeatability, the functional performance of the systems and its stability, specific analytical techniques such as SEM or TEM (morphological evaluation and size estimation), AFM (morphological evaluation, size, and shape), zeta potential analyzer, laser granulometry and light scattering (particle size and size deviation) are required. Some of the procedures listed are extremely time-consuming, labor-intensive, and require expensive equipment, making them impractical to use in traditional manufacturing lines [94].

Another problem with nanomedicines is the requirement to register them. The Food and Drug Administration (FDA), on the other hand, sees four major challenges in registering nanomedicines: the first is the regulatory status's inadequacy; the second is the potentiality of these drugs for unidentified risks in the post-market; the third is ensuring the development of new risk-benefit measures followed by assessments of clinical trials and further research protocols to determine the clinical efficacy and the fourth is consumer labeling of nanomedicine products. The FDA has released nanomedicine recommendations that stress the need for precise nanocarrier characterization and how these features affect system safety, efficacy, and quality. Few studies have been done to show

how these features affect patient safety. As a result, the FDA's remark is only somewhat useful to a potential nanomedicine developer. Predicting the link between nanocarrier behavior *in vitro* and *in vivo* is one of the most difficult tasks. Tissue accumulation cellular interactions, transit, and biocompatibility are the pivotal elements that need to be investigated with the help of *in vivo* models, and the expenses of these trials are not small. The accumulation capacity of nanocarriers and their release profile over time must be idealized due to the likelihood of nanocarrier persistence in the circulation or deposition in tissues and organs, as well as a general lack of knowledge of long-term consequences. Finally, nanotechnology is a relatively young science, whereas the process of selling a revolutionary drug takes many years. Nanomedicine's long-term effects on animals are unknown, thus demanding proper pharmacovigilance studies before and after commercialization [95].

FUTURE PROSPECTIVE FOR CONTROLLED SITE-SPECIFIC TARGET DELIVERY OF DRUGS

Future research trends in the application of edible nanoparticles for the bioactive phytochemicals administration and delivery will be directly linked to innovative manufacturing techniques that incorporate multiple structural designs, more specifically for the numerous types of bioactive compounds. To attain this aim, the advantages of two or more types of biomaterials will be merged, which will improve nanoparticles' characteristics and the advantages of two or more types of biomaterials. The loading of bioactive compounds on nanoparticles and their efficacy are largely benefited and improved by the effective delayed-release and *in situ* delivery control. Target delivery is one of the expected qualities that have yet to be completely developed, and as a result, its future implementation possibilities remain an issue. In this context, materials and industrial procedures which would foster the creation of edible nanoparticles with enhanced properties are constantly being explored. Improved specificity and flexibility of the nanoparticles would pave the way for orally administering bioactive chemicals *via* edible nanoparticles and it will be a beneficial strategy. Due to precision tissue-targeted administration, this will allow us to make use of phytochemicals' true ability to prevent and treatment of specific pathophysiological conditions [95]. The bioactivity of the drug is being influenced by gut metabolism after oral administration needs a modification of current definitions of "bioaccessibility" and "bioavailability," with a focus on upper gastrointestinal absorption. Indeed, understanding the properties of this physiological process, as well as the derivatives that are being produced during gastrointestinal digestion, are crucial for the design and development of nanoparticle systems. With this goal in mind, choosing nanoparticles that improve absorption in the gastric and small intestine (duodenum) may no longer be the gold standard, because excellent bioefficacy of the additional compounds synthesized, *e.g.* in the large intestine as a result of the local microbiota's metabolism, making their absorption in the upper gastrointestinal undesirable. Furthermore, any changes in nanoparticle structure produced by digestive conditions should be considered when evaluating the influence of nanoparticles and/or nanovesicles on phytochemical bioefficacy. Indeed, altering the structure of nanoparticles might affect their efficiency being as a vehicle of bioactive nutrients and non-nutrients, thus leading to incorrect findings. As a result, assuming that dilution has a substantial influence on the stability of nanoemulsions, liposomes, and micelles when combined with digesting fluids, is important. Furthermore, gastrointestinal fluid pH and ions conditions, as well as enzymatic actions in these compartments, may jeopardize nanoparticle stability and development as bioactive drug carriers [96]. As a result of these constraints, it's critical to evaluate nanoparticle behaviour in complex matrices and biological systems in order to offer reliable information on their practical utility.

In spite of the rational value of nanotechnology in enhancing the biocompatibility, bioavailability, and reduced toxicity of bioactive molecules, nanoparticles are produced utilizing physicochemical methods, which involve the use of costly and dangerous chemicals, particularly in the case of metallic biomaterials, limiting their precise application *in vivo*. These limits also include the environmental effect of residues formed during nanoparticle synthesis, which necessitates the use of green methods that do not involve the use of harmful chemicals to manufacture metal nanoparticles. In this sense, employing environmentally friendly and biocompatible chemicals to synthesize nanoparticles should help to reduce the negative effects of these procedures. The interactions between carriers and loaded molecules with gut commensals, and enzymes as well as the ramifications of these interactions on stability and bioefficacy are yet to be explored in the final intestinal stage, despite the fact that they appear to pose additional obstacles [97]. Indeed, applying what has been learned about the utilization of nanoparticles as vehicles for bioactive compounds will allow researchers to decipher and exploit the reciprocal interactions between these molecules, enzymes, and gut microbiota. Another issue that has to be answered is the extent to which nanoparticles alter the pharmacokinetics of bioactive phytochemicals, which might lead to novel uses, such as internalization speed accelerators. This issue must be clarified in order to properly construct the sampling time for determining the pharmacokinetics, bioavailability, and bioactivity of the bioactive substances under research. In this frame, it becomes obvious that the interaction between encapsulated compounds, matrix material, and body enzymes\proteins in complex biological systems, such as mammals warrants additional investigation in terms of metabolism and bioefficacy [98].

CONCLUSION

Since its inception, nanotechnology has been hailed as a promising method for preventing and treating a variety of human health problems. In fact, the industry has used this technique to look for novel options for drug administration based on the alteration of bioactive substances' solubility. In recent years, significant progress has been made in evaluating the input that nanomaterials and nanoparticles give in a variety of pathophysiological circumstances. Encapsulating disease-modifying drugs, phytochemicals, and novel bioactive peptides within the right nanoparticles improves their bioavailability significantly by rendering them safe from decomposition during storage and gastrointestinal path. This facilitates cellular uptake by augmenting retention time with the intestinal wall, not only increasing their mucus penetration but also intestinal permeation. It is also observed that it improves their solubility in aqueous media, prolongs residence time within the body circulation, controls their rate of release, and delivers them at a specific site. In this framework, this chapter considers the extent to which nanoparticles might boost the therapeutic index of the drug while avoiding toxicity and delivering these bioactive substances to target tissues impacted by certain pathophysiological and chemical conditions.

REFERENCES

[1]   Patra JK, Das G, Fraceto LF, *et al.* Nano based drug delivery systems: Recent developments and future prospects. J Nanobiotechnology 2018; 16(1): 71.
[http://dx.doi.org/10.1186/s12951-018-0392-8] [PMID: 30231877]

[2]   Portero A, Remuñán-López C, Criado MT, Alonso MJ. Reacetylated chitosan microspheres for controlled delivery of anti-microbial agents to the gastric mucosa. J Microencapsul 2002; 19(6): 797-809.
[http://dx.doi.org/10.1080/0265204021000022761] [PMID: 12569028]

[3]   Fernández-Urrusuno R, Calvo P, Remuñán-López C, Vila-Jato JL, José Alonso M. Enhancement of nasal absorption of insulin using chitosan nanoparticles. Pharm Res 1999; 16(10): 1576-81.
[http://dx.doi.org/10.1023/A:1018908705446] [PMID: 10554100]

[4]   De Campos AM, Sánchez A, Alonso MJ. Chitosan nanoparticles: A new vehicle for the improvement of the delivery of drugs to the ocular surface. Application to cyclosporin A. Int J Pharm 2001; 224(1-2): 159-68.
[http://dx.doi.org/10.1016/S0378-5173(01)00760-8] [PMID: 11472825]

[5]   Al-Qadi S, Grenha A, Carrión-Recio D, Seijo B, Remuñán-López C. Microencapsulated chitosan nanoparticles for pulmonary protein delivery: *In vivo* evaluation of insulin-loaded formulations. J Control Release 2012; 157(3): 383-90.
[http://dx.doi.org/10.1016/j.jconrel.2011.08.008] [PMID: 21864592]

[6]   Silva M, Calado R, Marto J, Bettencourt A, Almeida A, Gonçalves L. Chitosan Nanoparticles as a mucoadhesive drug delivery system for ocular administration. Mar Drugs 2017; 15(12): 370-86.
[http://dx.doi.org/10.3390/md15120370] [PMID: 29194378]

[7]   Pistone S, Goycoolea FM, Young A, Smistad G, Hiorth M. Formulation of polysaccharide-based nanoparticles for local administration into the oral cavity. Eur J Pharm Sci 2017; 96: 381-9.
[http://dx.doi.org/10.1016/j.ejps.2016.10.012] [PMID: 27721043]

[8]　Liu S, Yang S, Ho PC. Intranasal administration of carbamazepine-loaded carboxymethyl chitosan nanoparticles for drug delivery to the brain. Asian J Pharm 2018; 13(1): 72-81.
[http://dx.doi.org/10.1016/j.ajps.2017.09.001] [PMID: 32104380]

[9]　Abo-Elseoud WS, Hassan ML, Sabaa MW, Basha M, Hassan EA, Fadel SM. Chitosan nanoparticles/cellulose nanocrystals nanocomposites as a carrier system for the controlled release of repaglinide. Int J Biol Macromol 2018; 111: 604-13.
[http://dx.doi.org/10.1016/j.ijbiomac.2018.01.044] [PMID: 29325745]

[10]　Hansen K, Kim G, Desai KGH, *et al.* Feasibility investigation of cellulose polymers for mucoadhesive nasal drug delivery applications. Mol Pharm 2015; 12(8): 2732-41.
[http://dx.doi.org/10.1021/acs.molpharmaceut.5b00264] [PMID: 26097994]

[11]　Bozzuto G, Molinari A. Liposomes as nanomedical devices. Int J Nanomedicine 2015; 10: 975-99.
[http://dx.doi.org/10.2147/IJN.S68861] [PMID: 25678787]

[12]　Akbarzadeh A, Rezaei-Sadabady R, Davaran S, *et al.* Liposome: classification, preparation, and applications. Nanoscale Res Lett 2013; 8(1): 102-11.
[http://dx.doi.org/10.1186/1556-276X-8-102] [PMID: 23432972]

[13]　Patil NH, Devarajan PV. Insulin: Loaded alginic acid nanoparticles for sublingual delivery. Drug Deliv 2016; 23(2): 429-36.
[http://dx.doi.org/10.3109/10717544.2014.916769] [PMID: 24901208]

[14]　Haque S, Md S, Sahni JK, Ali J, Baboota S. Development and evaluation of brain targeted intranasal alginate nanoparticles for treatment of depression. J Psychiatr Res 2014; 48(1): 1-12.
[http://dx.doi.org/10.1016/j.jpsychires.2013.10.011] [PMID: 24231512]

[15]　Román JV, Galán MA, del Valle EMM. Preparation and preliminary evaluation of alginate crosslinked microcapsules as potential drug delivery system (dds) for human lung cancer therapy. Biomed Phys Eng Express 2016; 2(3): 035015.
[http://dx.doi.org/10.1088/2057-1976/2/3/035015]

[16]　Costa JR, Silva NC, Sarmento B, Pintado M. Potential chitosan-coated alginate nanoparticles for ocular delivery of daptomycin. Eur J Clin Microbiol Infect Dis 2015; 34(6): 1255-62.
[http://dx.doi.org/10.1007/s10096-015-2344-7] [PMID: 25754770]

[17]　Beg S, Rahman M, Jain A, *et al.* Emergence in the functionalized carbon nanotubes as smart nanocarriers for drug delivery applications.William Andrew Publishing 2018; pp. 105-33.
[http://dx.doi.org/10.1016/B978-0-12-813691-1.00004-X]

[18]　Liang F, Chen B. A review on biomedical applications of single-walled carbon nanotubes. Curr Med Chem 2010; 17(1): 10-24.
[http://dx.doi.org/10.2174/092986710789957742] [PMID: 19941481]

[19]　Li D, Chen X, Wang H, *et al.* Cetuximab-conjugated nanodiamonds drug delivery system for enhanced targeting therapy and 3D Raman imaging. J Biophotonics 2017; 10(12): 1636-46.
[http://dx.doi.org/10.1002/jbio.201700011] [PMID: 28635183]

[20]　Cai W, Gao H, Chu C, *et al.* Engineering phototheranostic nanoscale metal–organic frameworks for multimodal imaging-guided cancer therapy. ACS Appl Mater Interfaces 2017; 9(3): 2040-51.
[http://dx.doi.org/10.1021/acsami.6b11579] [PMID: 28032505]

[21]　Mao D, Hu F, Kenry , *et al.* Metal–organic-framework-assisted *in vivo* bacterial metabolic labeling and precise antibacterial therapy. Adv Mater 2018; 30(18): 1706831.
[http://dx.doi.org/10.1002/adma.201706831] [PMID: 29504163]

[22]　Simon-Yarza T, Mielcarek A, Couvreur P, Serre C. Nanoparticles of metal–organic frameworks: On the road to *in vivo* efficacy in biomedicine. Adv Mater 2018; 30(37): 1707365.
[http://dx.doi.org/10.1002/adma.201707365] [PMID: 29876985]

[23]　Sun CY, Qin C, Wang XL, *et al.* Zeolitic imidazolate framework-8 as efficient pH-sensitive drug

delivery vehicle. Dalton Trans 2012; 41(23): 6906-9.
[http://dx.doi.org/10.1039/c2dt30357d] [PMID: 22580798]

[24]   Zhuang J, Kuo CH, Chou LY, Liu DY, Weerapana E, Tsung CK. Optimized metal-organic-framework nanospheres for drug delivery: Evaluation of small-molecule encapsulation. ACS Nano 2014; 8(3): 2812-9.
[http://dx.doi.org/10.1021/nn406590q] [PMID: 24506773]

[25]   Raffin Pohlmann A, Weiss V, Mertins O, Pesce da Silveira N, Stanisçuaski Guterres S. Spray-dried indomethacin-loaded polyester nanocapsules and nanospheres: Development, stability evaluation and nanostructure models. Eur J Pharm Sci 2002; 16(4-5): 305-12.
[http://dx.doi.org/10.1016/S0928-0987(02)00127-6] [PMID: 12208461]

[26]   Baran ET, Özer N, Hasirci V. Poly(hydroxybutyrate-co-hydroxyvalerate) nanocapsules as enzyme carriers for cancer therapy: An *in vitro* study. J Microencapsul 2002; 19(3): 363-76.
[http://dx.doi.org/10.1080/02652040110105355] [PMID: 12022502]

[27]   Rawat M, Singh D, Saraf S, Saraf S. Nanocarriers: Promising vehicle for bioactive drugs. Biol Pharm Bull 2006; 29(9): 1790-8.
[http://dx.doi.org/10.1248/bpb.29.1790] [PMID: 16946487]

[28]   Abbasi E, Aval SF, Akbarzadeh A, *et al.* Dendrimers: Synthesis, applications, and properties. Nanoscale Res Lett 2014; 9(1): 247.
[http://dx.doi.org/10.1186/1556-276X-9-247] [PMID: 24994950]

[29]   Vandamme TF, Brobeck L. Poly *(amidoamine)* dendrimers as ophthalmic vehicles for ocular delivery of pilocarpine nitrate and tropicamide. J Control Release 2005; 102(1): 23-38.
[http://dx.doi.org/10.1016/j.jconrel.2004.09.015] [PMID: 15653131]

[30]   Khandare JJ, Jayant S, Singh A, *et al.* Dendrimer *versus* linear conjugate: Influence of polymeric architecture on the delivery and anticancer effect of paclitaxel. Bioconjug Chem 2006; 17(6): 1464-72.
[http://dx.doi.org/10.1021/bc060240p] [PMID: 17105225]

[31]   Chittasupho C, Anuchapreeda S, Sarisuta N. CXCR4 targeted dendrimer for anti-cancer drug delivery and breast cancer cell migration inhibition. Eur J Pharm Biopharm 2017; 119: 310-21.
[http://dx.doi.org/10.1016/j.ejpb.2017.07.003] [PMID: 28694161]

[32]   Gupta L, Sharma AK, Gothwal A, *et al.* Dendrimer encapsulated and conjugated delivery of berberine: A novel approach mitigating toxicity and improving *in vivo* pharmacokinetics. Int J Pharm 2017; 528(1-2): 88-99.
[http://dx.doi.org/10.1016/j.ijpharm.2017.04.073] [PMID: 28533175]

[33]   Zhong Q, Krishna Rao KSV. Recent advances in stimuli-responsive poly *(amidoamine)* dendrimer nanocarriers for drug delivery. Indian J Adv Chem Sci 2016; 4: 195-207.

[34]   Neamtu I, Rusu AG, Diaconu A, Nita LE, Chiriac AP. Basic concepts and recent advances in nanogels as carriers for medical applications. Drug Deliv 2017; 24(1): 539-57.
[http://dx.doi.org/10.1080/10717544.2016.1276232] [PMID: 28181831]

[35]   Li RQ, Wu W, Song HQ, *et al.* Well-defined reducible cationic nanogels based on functionalized low-molecular-weight PGMA for effective pDNA and siRNA delivery. Acta Biomater 2016; 41: 282-92.
[http://dx.doi.org/10.1016/j.actbio.2016.06.006] [PMID: 27267781]

[36]   Yang HN, Choi JH, Park JS, Jeon SY, Park KD, Park KH. Differentiation of endothelial progenitor cells into endothelial cells by heparin-modified supramolecular pluronic nanogels encapsulating bFGF and complexed with VEGF165 genes. Biomaterials 2014; 35(16): 4716-28.
[http://dx.doi.org/10.1016/j.biomaterials.2014.02.038] [PMID: 24630837]

[37]   Tyagi R, Lala S, Verma AK, *et al.* Targeted delivery of arjunglucoside I using surface hydrophilic and hydrophobic nanocarriers to combat experimental leishmaniasis. J Drug Target 2005; 13(3): 161-71.
[http://dx.doi.org/10.1080/10611860500046732] [PMID: 16036304]

[38]   Qiu L, Zheng C, Jin Y, Zhu K. Polymeric micelles as nanocarriers for drug delivery. Expert Opin Ther

Pat 2007; 17(7): 819-30.
[http://dx.doi.org/10.1517/13543776.17.7.819]

[39]   Li Y, Zhang T, Liu Q, He J. PEG-derivatized dual-functional nanomicelles for improved cancer therapy. Front Pharmacol 2019; 10: 808.
[http://dx.doi.org/10.3389/fphar.2019.00808] [PMID: 31379579]

[40]   Lu J, Huang Y, Zhao W, *et al.* PEG-derivatized embelin as a nanomicellar carrier for delivery of paclitaxel to breast and prostate cancers. Biomaterials 2013; 34(5): 1591-600.
[http://dx.doi.org/10.1016/j.biomaterials.2012.10.073] [PMID: 23182923]

[41]   Kwon G, Suwa S, Yokoyama M, Okano T, Sakurai Y, Kataoka K. Enhanced tumor accumulation and prolonged circulation times of micelle-forming poly *(ethylene oxide-aspartate)* block copolymer-adriamycin conjugates. J Control Release 1994; 29(1-2): 17-23.
[http://dx.doi.org/10.1016/0168-3659(94)90118-X]

[42]   Francis MF, Cristea M, Winnik FM. Exploiting the vitamin B12 pathway to enhance oral drug delivery *via* polymeric micelles. Biomacromolecules 2005; 6(5): 2462-7.
[http://dx.doi.org/10.1021/bm0503165] [PMID: 16153081]

[43]   Lukyanov AN, Hartner WC, Torchilin VP. Increased accumulation of PEG–PE micelles in the area of experimental myocardial infarction in rabbits. J Control Release 2004; 94(1): 187-93.
[http://dx.doi.org/10.1016/j.jconrel.2003.10.008] [PMID: 14684282]

[44]   Lei Y, Sheldon BW, Webster TJ. Nanophase ceramics for improved drug delivery: Current opportunities and challenges. Am Ceram Soc Bull 2010; 89: 24-33.

[45]   Armatas GS, Kanatzidis MG. Mesostructured germanium with cubic pore symmetry. Nature 2006; 441(7097): 1122-5.
[http://dx.doi.org/10.1038/nature04833] [PMID: 16810250]

[46]   Luo D, Han E, Belcheva N, Saltzman WM. A self-assembled, modular DNA delivery system mediated by silica nanoparticles. J Control Release 2004; 95(2): 333-41.
[http://dx.doi.org/10.1016/j.jconrel.2003.11.019] [PMID: 14980781]

[47]   Jain TK, Roy I, De TK, Maitra A. Nanometer silica particles encapsulating active compounds: A novel ceramic drug carrier. J Am Chem Soc 1998; 120(43): 11092-5.
[http://dx.doi.org/10.1021/ja973849x]

[48]   Panseri S, Cunha C, D'Alessandro T, *et al.* Intrinsically superparamagnetic Fe-hydroxyapatite nanoparticles positively influence osteoblast-like cell behaviour. J Nanobiotechnology 2012; 10(1): 32.
[http://dx.doi.org/10.1186/1477-3155-10-32] [PMID: 22828388]

[49]   Sarath Chandra V, Baskar G, Suganthi RV, *et al.* Blood compatibility of iron-doped nanosize hydroxyapatite and its drug release. ACS Appl Mater Interfaces 2012; 4(3): 1200-10.
[http://dx.doi.org/10.1021/am300140q] [PMID: 22316071]

[50]   Roy I, Ohulchanskyy TY, Pudavar HE, *et al.* Ceramic-based nanoparticles entrapping water-insoluble photosensitizing anticancer drugs: A noveldrug-carrier system for photodynamictherapy. J Am Chem Soc 2003; 125(26): 7860-5.
[http://dx.doi.org/10.1021/ja0343095]

[51]   Chen Y, Chen H, Ma M, *et al.* Double mesoporous silica shelled spherical/ellipsoidal nanostructures: Synthesis and hydrophilic/hydrophobic anticancer drug delivery. J Mater Chem 2011; 21(14): 5290-8.
[http://dx.doi.org/10.1039/c0jm04024j]

[52]   Nagarwal RC, Kumar R, Dhanawat M, Das N, Pandit JK. Nanocrystal technology in the delivery of poorly soluble drugs: An overview. Curr Drug Deliv 2011; 8(4): 398-406.
[http://dx.doi.org/10.2174/156720111795767988] [PMID: 21453258]

[53]   Prabhakar C, Krishna KB. A review on nanosuspensions in drug delivery. Int J Pharma Bio Sci 2011; 2: 549-58.

[54]   Gao Y, Li Z, Sun M, *et al.* Preparation and characterization of intravenously injectable curcumin nanosuspension. Drug Deliv 2011; 18(2): 131-42.
[http://dx.doi.org/10.3109/10717544.2010.520353] [PMID: 20939679]

[55]   Shan X, Liu C, Li F, Ouyang C, Gao Q, Zheng K. Nanoparticles vs. nanofibers: A comparison of two drug delivery systems on assessing drug release performance *in vitro.* Des Monomers Polym 2015; 18(7): 678-89.
[http://dx.doi.org/10.1080/15685551.2015.1070500]

[56]   Nguyen TTT, Ghosh C, Hwang SG, Chanunpanich N, Park JS. Porous core/sheath composite nanofibers fabricated by coaxial electrospinning as a potential mat for drug release system. Int J Pharm 2012; 439(1-2): 296-306.
[http://dx.doi.org/10.1016/j.ijpharm.2012.09.019] [PMID: 22989981]

[57]   Huang ZM, Zhang YZ, Kotaki M, Ramakrishna S. A review on polymer nanofibers by electrospinning and their applications in nanocomposites. Compos Sci Technol 2003; 63(15): 2223-53.
[http://dx.doi.org/10.1016/S0266-3538(03)00178-7]

[58]   Che G, Lakshmi BB, Martin CR, Fisher ER, Ruoff RS. Chemical vapor deposition basedsynthesis of carbon nanotubes and nanofibers using a template method. Chem Mater 1998; 10(1): 260-7.
[http://dx.doi.org/10.1021/cm970412f]

[59]   Hassan MA, Yeom BY, Wilkie A, Pourdeyhimi B, Khan SA. Fabrication of nanofiber meltblown membranes and their filtration properties. J Membr Sci 2013; 427: 336-44.
[http://dx.doi.org/10.1016/j.memsci.2012.09.050]

[60]   Hartgerink JD, Beniash E, Stupp SI. Self-assembly and mineralization of peptide-amphiphile nanofibers. Science 2001; 294(5547): 1684-8.
[http://dx.doi.org/10.1126/science.1063187] [PMID: 11721046]

[61]   Liu S, Zhou G, Huang Y, Xie Z, Jing X. 11. Biodegradable electrospun fibers as drug delivery system for local cancer treatment: A rugged path to the bright future: Original research article: Biodegradable electrospun fibers for drug delivery, 2003. J Control Release 2014; 190: 52-3.
[PMID: 25356481]

[62]   Verreck G, Chun I, Rosenblatt J, *et al.* Incorporation of drugs in an amorphous state into electrospun nanofibers composed of a water-insoluble, nonbiodegradable polymer. J Control Release 2003; 92(3): 349-60.
[http://dx.doi.org/10.1016/S0168-3659(03)00342-0] [PMID: 14568415]

[63]   Dashdorj U, Reyes MK, Unnithan AR, *et al.* Fabrication and characterization of electrospun zein/Ag nanocomposite mats for wound dressing applications. Int J Biol Macromol 2015; 80: 1-7.
[http://dx.doi.org/10.1016/j.ijbiomac.2015.06.026] [PMID: 26093320]

[64]   Singh M, Movia D, Mahfoud OK, Volkov Y, Prina-Mello A. Silver nanowires as prospective carriers for drug delivery in cancer treatment: An *in vitro* biocompatibility study on lung adenocarcinoma cells and fibroblasts. Eur J Nanomed 2013; 5(4): 195-204.
[http://dx.doi.org/10.1515/ejnm-2013-0024]

[65]   Martínez-Banderas AI, Aires A, Quintanilla M, *et al.* Iron-based core–shell nanowires for combinatorial drug delivery and photothermal and magnetic therapy. ACS Appl Mater Interfaces 2019; 11(47): 43976-88.
[http://dx.doi.org/10.1021/acsami.9b17512] [PMID. 31682404]

[66]   Shan X, Liu C, Li F, Ouyang C, Gao Q, Zheng K. Nanoparticles vs. nanofibers: A comparison of two drug delivery systems on assessing drug release performance *in vitro.* Des Monomers Polym 2015; 18(7): 678-89.
[http://dx.doi.org/10.1080/15685551.2015.1070500]

[67]   Shen S, Wu Y, Liu Y, Wu D. High drug-loading nanomedicines: Progress, current status, and prospects. Int J Nanomedicine 2017; 12: 4085-109.

[http://dx.doi.org/10.2147/IJN.S132780] [PMID: 28615938]

[68]   Ju J, Chen W. Graphene quantum dots as fluorescence probes for sensing metal ions: Synthesis and applications. Curr Org Chem 2015; 19(12): 1150-62.
[http://dx.doi.org/10.2174/1385272819666150318222547]

[69]   Yao X, Niu X, Ma K, *et al.* Graphene quantum dots-capped magnetic mesoporous silica nanoparticles as a multifunctional platform for controlled drug delivery, magnetic hyperthermia and photothermal therapy. Small 2017; 13(2): 1602225.
[http://dx.doi.org/10.1002/smll.201602225] [PMID: 27735129]

[70]   Zhang H. Onivyde for the therapy of multiple solid tumors. OncoTargets Ther 2016; 9: 3001-7.
[http://dx.doi.org/10.2147/OTT.S105587] [PMID: 27284250]

[71]   Akinc A, Maier MA, Manoharan M, *et al.* The Onpattro story and the clinical translation of nanomedicines containing nucleic acid-based drugs. Nat Nanotechnol 2019; 14(12): 1084-7.
[http://dx.doi.org/10.1038/s41565-019-0591-y] [PMID: 31802031]

[72]   Silverman JA, Deitcher SR. Marqibo® *(vincristine sulfate liposome injection)* improves the pharmacokinetics and pharmacodynamics of vincristine. Cancer Chemother Pharmacol 2013; 71(3): 555-64.
[http://dx.doi.org/10.1007/s00280-012-2042-4] [PMID: 23212117]

[73]   Komlosh A, Weinstein V, Loupe P, *et al.* Physicochemical and biological examination of two glatiramer acetate products. Biomedicines 2019; 7(3): 49.
[http://dx.doi.org/10.3390/biomedicines7030049] [PMID: 31277332]

[74]   Venkatakrishnan K, Liu Y, Noe D, *et al.* Pharmacokinetics and pharmacodynamics of liposomal mifamurtide in adult volunteers with mild or moderate renal impairment. Br J Clin Pharmacol 2014; 77(6): 986-97.
[http://dx.doi.org/10.1111/bcp.12260] [PMID: 24134181]

[75]   Anselmo AC, Mitragotri S. Nanoparticles in the clinic: An update. Bioeng Transl Med 2019; 4(3): e10143.
[http://dx.doi.org/10.1002/btm2.10143] [PMID: 31572799]

[76]   Choi YH, Han HK. Nanomedicines: Current status and future perspectives in aspect of drug delivery and pharmacokinetics. J Pharm Investig 2018; 48(1): 43-60.
[http://dx.doi.org/10.1007/s40005-017-0370-4] [PMID: 30546919]

[77]   Bullivant J, Zhao S, Willenberg B, Kozissnik B, Batich C, Dobson J. Materials characterization of Feraheme/ferumoxytol and preliminary evaluation of its potential for magnetic fluid hyperthermia. Int J Mol Sci 2013; 14(9): 17501-10.
[http://dx.doi.org/10.3390/ijms140917501] [PMID: 24065092]

[78]   Kaduk JA, Dmitrienko AO, Gindhart AM, Blanton TN. Crystal structure of paliperidone palmitate (INVEGA SUSTENNA ®), $C_{39} H_{57} FN_4 O_4$. Powder Diffr 2017; 32(4): 222-7.
[http://dx.doi.org/10.1017/S0885715617000896]

[79]   Lyseng-Williamson KA, Keating GM. Ferric Carboxymaltose. Drugs 2009; 69(6): 739-56.
[http://dx.doi.org/10.2165/00003495-200969060-00007] [PMID: 19405553]

[80]   Kim MT, Chen Y, Marhoul J, Jacobson F. Statistical modeling of the drug load distribution on trastuzumab emtansine *(Kadcyla),* a lysine-linked antibody drug conjugate. Bioconjug Chem 2014; 25(7): 1223-32.
[http://dx.doi.org/10.1021/bc5000109] [PMID: 24873191]

[81]   Farjadian F, Ghasemi A, Gohari O, Roointan A, Karimi M, Hamblin MR. Nanopharmaceuticals and nanomedicines currently on the market: Challenges and opportunities. Nanomedicine 2019; 14(1): 93-126.
[http://dx.doi.org/10.2217/nnm-2018-0120] [PMID: 30451076]

[82]   Weissig V, Pettinger T, Murdock N. Nanopharmaceuticals (part 1): Products on the market. Int J

Nanomedicine 2014; 9: 4357-73.
[http://dx.doi.org/10.2147/IJN.S46900] [PMID: 25258527]

[83]   Patra JK, Das G, Fraceto LF, *et al.* Nano based drug delivery systems: Recent developments and future prospects. J Nanobiotechnology 2018; 16(1): 71.
[http://dx.doi.org/10.1186/s12951-018-0392-8] [PMID: 30231877]

[84]   Chelle P, Yeung CHT, Croteau SE, *et al.* Development and validation of a population-pharmacokinetic model for rurioctacog alfa pegol (adynovate®): A report on behalf of the wapps-hemo investigators ad hoc subgroup. Clin Pharmacokinet 2020; 59(2): 245-56.
[http://dx.doi.org/10.1007/s40262-019-00809-6] [PMID: 31435896]

[85]   Miele E, Spinelli GP, Miele E, Tomao F, Tomao S. Albumin-bound formulation of paclitaxel *(Abraxane ABI-007)* in the treatment of breast cancer. Int J Nanomedicine 2009; 4: 99-105.
[PMID: 19516888]

[86]   Pardridge WM. Drug transport across the blood-brain barrier. J Cereb Blood Flow Metab 2012; 32(11): 1959-72.
[http://dx.doi.org/10.1038/jcbfm.2012.126] [PMID: 22929442]

[87]   Patel HN, Patel PM. Dendrimer applicationsAsian J. Pharm.: A review. Int J Pharma Bio Sci 2013; 4: 454-63.

[88]   Wang X, Wang C, Zhang Q, Cheng Y. Near infrared light-responsive and injectable supramolecular hydrogels for on-demand drug delivery. Chem Commun 2016; 52(5): 978-81.
[http://dx.doi.org/10.1039/C5CC08391E] [PMID: 26588349]

[89]   Sendi P, Proctor RA. *Staphylococcus aureus* as an intracellular pathogen: The role of small colony variants. Trends Microbiol 2009; 17(2): 54-8.
[http://dx.doi.org/10.1016/j.tim.2008.11.004] [PMID: 19162480]

[90]   Bhirde AA, Patel S, Sousa AA, *et al.* Distribution and clearance of PEG-single-walled carbon nanotube cancer drug delivery vehicles in mice. Nanomedicine 2010; 5(10): 1535-46.
[http://dx.doi.org/10.2217/nnm.10.90] [PMID: 21143032]

[91]   Kharlamov AN, Feinstein JA, Cramer JA, Boothroyd JA, Shishkina EV, Shur V. Plasmonic photothermal therapy of atherosclerosis with nanoparticles: Long-term outcomes and safety in NANOM-FIM trial. Future Cardiol 2017; 13(4): 345-63.
[http://dx.doi.org/10.2217/fca-2017-0009] [PMID: 28644056]

[92]   Zhou J, Li M, Lim WQ, *et al.* A transferrin-conjugated hollow nanoplatform for redox-controlled and targeted chemotherapy of tumor with reduced inflammatory reactions. Theranostics 2018; 8(2): 518-32.
[http://dx.doi.org/10.7150/thno.21194] [PMID: 29290824]

[93]   Madamsetty VS, Pal K, Keshavan S, *et al.* Development of multi-drug loaded pegylated nanodiamonds to inhibit tumor growth and metastasis in genetically engineered mouse models of pancreatic cancer. Nanoscale 2019; 11(45): 22006-18.
[http://dx.doi.org/10.1039/C9NR05478B] [PMID: 31710073]

[94]   Wang C, Jin J, Sun Y, Yao J, Zhao G, Liu Y. *In-situ* synthesis and ultrasound enhanced adsorption properties of MoS 2 /graphene quantum dot nanocomposite. Chem Eng J 2017; 327. 774-82.
[http://dx.doi.org/10.1016/j.cej.2017.06.163]

[95]   Ruzycka-Ayoush M, Kowalik P, Kowalczyk A, *et al.* Quantum dots as targeted doxorubicin drug delivery nanosystems in human lung cancer cells. Cancer Nanotechnol 2021; 12: 1-27.

[96]   Peynshaert K, Soenen SJ, Manshian BB, *et al.* Coating of quantum dots strongly defines their effect on lysosomal health and autophagy. Acta Biomater 2017; 48: 195-205.
[http://dx.doi.org/10.1016/j.actbio.2016.10.022] [PMID: 27765679]

[97]   Wang Y, Yang C, Hu R, *et al.* Assembling Mn:ZnSe quantum dots-siRNA nanoplexes for gene silencing in tumor cells. Biomater Sci 2015; 3(1): 192-202.

[http://dx.doi.org/10.1039/C4BM00306C] [PMID: 26214202]

[98]   Demir GM, Ilhan M, Akkol EK, Taştan H, Işık A, Değim İT. Effect of Paclitaxel loaded chitosan nanoparticles and quantum dots on breast cancer. Multi Digit Pub Inst Proc 2017; 1(10): 1074. [http://dx.doi.org/10.3390/proceedings1101074]

CHAPTER 18

Phyto-nanoformulations for the Treatment of Clinical Diseases

Vaishali Ji[1,*], Chandra Kishore[2] and **Krishna Prakash[3]**

[1] *Department of Botany, Patna Science College, Patna, Bihar, India*

[2] *Stem Cell Research Centre, Department of Hematology, Sanjay Gandhi Postgraduate Institute of Medical Sciences, Lucknow, Uttar Pradesh, India*

[3] *ICAR-Indian Agricultural Research Institute (IARI), Hazaribagh, Jharkhand, India*

Abstract: Plant-derived drugs or formulations have always been explored because of their lesser side effects and toxicities compared to synthetic drugs and they have been widely used as traditional and complementary medicines for the management of many diseases including cancer. The major challenges faced were the absorption of the plant-derived drugs, their stability, bioavailability, and transport to the intended sites inside the body. Recent progress in nanotechnology has helped to minimize these limitations and hence phyto-nanoformulations are slowly growing in preclinical trials as well as clinical use. The use of various nanostructures such as nano-micelles, lipid nanoparticles, carbon nanotubes, polymer nanoparticles, and nanoliposomes and various types of drug delivery vehicles such as polybutylcyanoacrylate, polylactic-c--glycolic acid, and lactoferrin has immensely helped in increasing the effectiveness of phytochemical drugs by increasing their stability, better pharmacokinetics and reducing the toxicity and side effects. Phyto-nanoformulations having natural product components such as curcumin, piperine, quercetin, berberine, scutellarin, baicalin, stevioside, silybin, gymnemic acid, naringenin, capsicum oleoresin, emodin, and resveratrol have been shown to improve the condition of patients diagnosed with diseases such as neurodegenerative disorders, diabetes, infections, and cancer. Phyto nanoformulations can also be used to treat disorders of the brain where the blood-brain barrier is impervious to the drugs. These phyto-nanoformulations have been shown to target several molecular cell-signaling and metabolic pathways. This chapter covers the compositions of phyto-nanoformulations and how they have been used to control several diseases.

Keywords: Nano micelles, Neurodegenerative disorder, Nanoemulsions, Pharmacokinetics, Phyto-nanoformulation.

---

* **Corresponding author Vaishali Ji:** Department of Botany, Patna Science College, Patna, Bihar, India; E-mail: vaishaki?oct@gmail.com

Zulqurnain Khan, Azra Yasmin & Naila Safdar (Eds.)

INTRODUCTION

Phytochemicals are secondary metabolites in plants that are largely involved in creating pigmentation and flavor and these metabolites also constitute the phytoimmune system. It is a common observation that plants that have more pigmentation generally contain a higher level of phytochemicals and nutrients such as phenols, steroids, allicin, carotene, terpenes, pyrimidines, and nucleic acids. Plants are a better source of medicinal and nutritive compounds and these compounds can be obtained easily and at a cheaper cost compared to lab synthesized compounds. The toxic side effects of synthetic drugs have forced researchers to look for plant-derived phytochemicals as complementary medicines and natural or semisynthetic medicines. These goals have motivated the agriculture, pharmaceutical, and food industries to focus on phytochemicals for drug leads. Many plant extracts have antioxidant properties that have health benefits such as enhancing vision and vascular health, decreasing the severity or chances of cardiovascular diseases, and preventing diabetes, cancer, and microbial diseases. Garlic and onion extracts have been used in traditional medicines as antimicrobial [1], antitumor [2], antihypertension [3], and cholesterol [4] reduction agents [5 - 8]. Carotenoids from the carrot have been shown to control colon cancer [9], inflammation [10] and oxidation [11]. Nanotechnology has helped to push the limitations of traditional methods of treatment [12]. The diverse nanostructures and vehicle systems have increased the stability and effectiveness and reduced the side effects of the drugs [13]. The nanoformulation of phytochemicals has beneficial effects as these have a greater surface area to volume ratio, delivery of drug to the precise location, controlled and well-regulated release of therapeutic drug, and better bioavailability [14] and hence nanoformulation and nanonization of phytochemicals are a potential replacement for conventional drug delivery system [15]. Phyto-nanoformulations alone or in combination with other drugs have shown a promising proficiency and potential for the treatment of cancer, neurodegenerative disorders, diabetes, metabolic disorders, and microbial infections.

PHYTO-NANOFORMULATION IN CANCER

Polymeric Nanoparticles

Cancer is a group of various pathological abnormalities that lead to a painful death and for poor and developing countries, the compromised health infrastructure presents a further challenge in controlling the disease [16]. Cancer becomes hard to control when it undergoes metastasis or develops drug resistance [17, 18]. The conventional approaches to treat the disease are chemotherapy, radiotherapy, and surgery but these therapies sometimes seem limited in their efficiency especially when cancer cells get resistant to the drugs or have disseminated in different parts of the body [19].

Nanoparticles are structures having a minimum one dimension lesser than a hundred nanometers and they can be grouped into nanospheres and nanocapsules [20]. Drugs are added to the matrix of the nanosphere or embedded into the inner cavity of the reservoir of the nanocapsules. Nanoparticles can be made up of natural or synthetic polymers and are sometimes known as polymeric nanoparticles [21]. The surface of nanoparticles can be modified by PEGylation to enhance their intake into the cells [22]. Chitosan surface coating increases stability and polyethylene glycol suppresses the protein expression in cell lines of breast cancer [23]. P glycoprotein is suppressed in PEG MDA-MB 453 and it has the capability to suppress resistance to the anticancer drugs in breast cancer therapy. The coating of PEG and chitosan on the BioPerine nanoparticles improves the intake inside the cells and fabricated BioPerine nanoparticles have better uptake than free BioPrine [24].

Some of the important nanoformulations that have improved the efficiency of phytochemicals are carbon nanotubes, liposomes, polymeric nanoparticles, lipid nanoparticles, lipid bilayers, dendrimers, micelle, nanocapsules, and functional gold nanoparticles [25].

Liposomes

These are vesicles made up of a lipid bilayer which have an aqueous inner core separated from a bulky outer part and it has a size of 10 nm-1 µm or sometimes higher [26]. The aqueous environment in the core serves as drug or other molecules' carrier. To target the specific sites on the cell, stealth liposome [27] is used whereas to improve the transportation of the drugs across various barriers, transferosomes [28] are used. Some of the advantages of liposomes are biocompatibility, amphiphilic nature, and biodegradability. The surface of the liposomes can be modified and ligated for the targeted delivery of the cargo [29]. The liposomes loaded with betulinic acid have been shown to target the HepG2 cells [30]. Icariin-loaded phytosomes have lesser IC50 compared to the pure form of Icariin [31]. Phytosomes are also useful for targeting ovarian cancer and as an inducer of apoptosis. Liposomal targeting of breast cancer cell lines using curcumin has shown a promising result. The half-life of liposomal curcumin is improved compared to the pure form of curcumin [32].

Solid lipid nanoparticles have solid and lipid cores and particle size ranges from 10-1000nm. Loading and targeting efficiency were improved by using solid lipid nanoparticles [33]. The lower particle size range has been used in biomedical applications as solid lipid core can be stabilized by interfacial surfactants such as phospholipids, sterols, and bile salts. Biocompatible excipients can be used for their preparation and can be used to encapsulate lipophobic and lipophilic drugs. For the target-specific release of the drugs, the surface can be modified with a proper coating of polymer of ligands. Ferulic acid-loaded solid lipid nanoparticles were developed using chitosan coating on the surface [34]. Solid lipid nanoparticles having myricetin loaded in them, increased the toxicity in A549 cells and the expression level of genes involved in necroptosis [35].

Nanomicelles

Nanomicelles are nanoscale vesicular carriers that can self-assemble and have a diameter of less than 100 nm and are spherical in shape. Nanomicelles are amphiphilic and hence they can accommodate hydrophilic as well as lipophilic chemical substances. The liposome can also be modified on its surface by suitable polymers to reach specific targets in the body. The hydrophobic core is made up of fatty acyl chains and the hydrophilic core is made up of hydrophilic polymers. Nanomicelles can accommodate hydrophobic as well as hydrophilic drugs because of their amphiphilic nature [36].

Honokiol-loaded Rebaudioside A nanomicelles were formulated to study their anticancer effect against HuH-7 cells and they have inhibited the ROS generation by inducing the blockage of the cell cycle and induction of cell death [37]. Quercetin Nanomicelles prevent the efflux of Adriamycin and disrupt mitochondrial potential [38]. Dihydroartemisinin-loaded nanomicelles were shown to selectively target the HeLa cells [39]. Curcumin nanomicelles promote apoptosis, regulate T cell activity, and inhibit the process of blood vessel growth [36, 40].

Dendrimers

These are star-shaped polymers having multiple branches and comprise the central core, interior branches, and exterior multiple functional groups and they can be used as drug carriers for chemical compounds and biological molecules such as proteins and amino acids [41]. The functional groups present on the surface of the dendrimers can be modified to carboxyl, amine, or hydroxyl groups, and the dendrimers provide reproducible, reliable pharmacokinetics and superior uptake efficiency which make them a better candidate to be used in disease management. Celastrol dendrimer was formulated to target SW620 colon cancer cells and they have shown enhanced apoptosis [42]. Ursolic acid and folic acid were formulated with polyamidoamine dendrimers for inhibiting the growth of HepG2 cells [43].

Gold Nanoparticles

Gold nanoparticles have a larger surface area, higher stability, better biocompatibility, and lower toxicity, and they can be conjugated with various molecules and they can have a controlled release [44]. Gold forms a chemical bond with amine and thiol groups and hence it can be easily tagged with the desired ligand and can be directed to the specific targets. *Curcuma wenyujin* loaded gold nanoparticles were shown to have better toxicity against MDBA-M--231 and they have decreased the expression of HER2/neu protein [45].

Magnetic Nanoparticles

Magnetic nanoparticles are showing a lot of potential in disease management as they can be used to manipulate the magnetic field. Magnetic nanoparticles are modified with polymers, surfactants, or silica to increase their stability [46]. Magnetic iron oxide nanoparticles are used for delivery of the chemical compounds, targeted delivery, imaging, and binding to iron proteins [47]. Iron oxide magnetic nanoparticles were shown to be highly cytotoxic against prostate cancer cell lines [48]. They also caused increased apoptosis, suppression of necrosis, and increased cell population in the subG1 phase. The dendritic magnetic nanoparticles were used to deliver the curcumin compound efficiently [49] and they had better toxicity against cell lines of nasopharyngeal cancer and MCF7 breast cancer cells. Magnetic nanoparticles can also be used to selectively bind to the folate receptor-expressing cancer cells and induce apoptosis.

Carbon Nanotubes

Carbon can be present in different allotropic forms such as Graphite, Diamond, Fullerene, and Graphene. The carbon nanotubes (CNT) are made up of graphene sheets that are rolled in a cylindrical shape that has a larger surface area, wider length, and a narrow diameter [50] and they can easily accommodate therapeutic drugs [51]. The presence of oxygen-containing functional groups on the surface of the CNT makes the surface easily modifiable. The single-walled CNTs have been used to deliver curcumin and its stability was enhanced by lipid and polymeric coating on the surface of the CNT [52]. Single-walled CNT-Cur has improved intake in the cells of human prostate cancer compared to curcumin only. The coating with lipid and polymer conjugation can also improve the delivery and inhibit drug resistance [53].

MANAGEMENT OF METABOLIC SYNDROMES

Metabolic syndrome (MS) consists of a cluster of disease conditions that occur at the same time and enhance the risk of atherosclerotic cardiovascular diseases, obesity, insulin resistance, systemic hypertension, cerebrovascular accident, and atherogenic dyslipidemia [54]. It is generally associated with increased blood pressure, high blood sugar, excess body fat around the waist, and abnormal cholesterol or triglyceride levels. Many natural nanoformulations have been designed to prevent or diminish the effects of MS. Nanoformulations have been used as promising substitutes for conventional drugs. Some of the important phytochemicals and their nanoformulations used in various MS disorders are included in Table **1**.

Table 1. Phyto-nanoformulations used in various Metabolic Syndrome disorders.

Phytochemicals	Nanoformulations	Disorder
Curcumin	PBLG-PEG-PBLG	Diabetic cardiomyopathy
-	PLGA-PVA polymers	Diabetic cataract
-	SNEDDS	Diabetic neuropathy
-	CNPs@GMs/hydrogel	Diabetic wound
-	PLGA based NPs with Q10	Diabtetes complications
-	Curcumin nanoemulsion	Hypertension and hypercholesterolemia
-	PEGMA-DMAEMA-MAO	Diabetic peripheral neuropathy
Capsicum oleoresin	Nanoemulsion	Obesity
-	Alginate double layer nanoemulsion	Obesity
Naringenin	Alginate coated chitosan core shell	Diabetes
Emodin	PEGMA-DMAEMA-MAMMAM nanomacromolecule	Diabetic neuropathic pain
-	Nanoemodin transferosome	Anti obesity
Berberine	Solid lipid nanoparticle	Diabetes
-	O-hexadecyl dextran	High glucose induced apoptosis in hepatocytes
-	PLGA-PEG-PLGA block co polymer	High LDL cholesterol
Quercetin	(QUE/P)NP	Diabetic nephropathy
-	pH sensitive chitosan alginate core shell	Diabetes
-	PLGA NPs	Diabetes
-	Nano emulsion	Oxaliplatin-induced neuro and hepatotoxicity
-	Quercetin nanorods	Diabetes
Alpha eleostearic	Bitter gourd seed oil nanoemulsion	Diabetes mellitus
Myricitrin	SLNs	Diabetes

(Table 1) cont.....

Phytochemicals	Nanoformulations	Disorder
Gymnemic acid	Nanosuspension	Diabetes
Resveratrol	Nanoliposome	Diabetes mellitus
-	Nanocapsule	Blood pressure regulation
Baicalin	Nanostructured lipid carrier d	Diabetes
Scutellarin	Amphiphilic chitosan derivatives	Diabetic retinopathy
Silybin	PLGA polymers	Systemic hyperglycemia
Stevioside	PLA nanoparticle	Diabetes

To improve the bioactivity and functionality of natural compounds for the management of metabolic syndromes, the compounds have been encapsulated in nanoformulations to increase the absorption as well as efficiency of the compound. Some of the methods used to increase the solubility and bioavailability are nanoemulsion, loading of NPs, nanosizing, use of SEDDs, nanoliposomes, and nanosuspensions. These methods have enhanced the half-life and efficacy of chemotherapeutic drugs.

HERBAL EXTRACTS NANOFORMULATIONS USED IN THE TREATMENT OF NEURODEGENERATIVE DISORDERS

Nanoformulations consisting of curcumin, piperine, *Nigella sativa*, quercetin, resveratrol, and *Ginkgo biloba*, have been shown to improve the conditions in neurodegenerative disorders [55]. Various other formulations are also well known Table **2** that contribute to the treatment of neuro disorders. Nanoformulations-based herbal products have also been shown to pass the blood-brain barrier to reach the inner cores of the brain [56].

Table 2. Nanoformulations used in neurodegenerative disorders.

Product	Method/Nanovehicle	Disease
Nattokinase enzyme (NK)	PNPs Parkinson's Disease	Alzheimer's Disease
Chrysin	SLNs	Alzheimer's Disease
Coenzyme Q10 6-coumarin	Trimethylated chitosan-conjugated PLGA nanoparticle	Alzheimer's Disease
Hesperetin	Nanocrystal	Alzheimer's Disease
HupA	lactoferrin-conjugated N-trimethylated chitosan nanoparticles (Lf-TMC NPs	Alzheimer's Disease
Sialic acid and peptide-B6	Selenium nanoparticles	Alzheimer's Disease

(Table 2) cont.....

Product	Method/Nanovehicle	Disease
Berberine	MWCNTs coated with phospholipid and polysorbate	Alzheimer's Disease
Vitamin E	PEG-based nanospheres	Alzheimer's Disease
Cysteine	Selenium nanoparticles	Alzheimer's Disease
Retinoic acid (RA)	PNPs	Parkinson's Disease

The phytochemical-based nanocarriers can be safe, lesser in toxicity, easily scalable, eco-friendly, and inexpensive. Plant-based nanosystems can also enhance the pharmacokinetic profile and drug bioavailability to the CNS, increase the disaggregation or inhibit the aggregate formation, and increase the penetration efficiency to the brain. There is evidence to support that NPs can restore neurological disorders but further studies are required to analyze the safety concerns in humans and their long-term neurological effects.

PLGA is a potent nanocarrier [57] used in the oral delivery of most plant-derived compounds such as quercetin, ferulic acid, curcumin, resveratrol, silymarin, crocetin, pelargonidin, γ-oryzanol, and thymoquinone. PLGA-loaded nanoformulations can increase the bioavailability and antidiabetic efficacy of the above-mentioned compounds but these nanoformulations are susceptible to hydrolytic degradation while passing through the gastrointestinal tract [58]. Hence, consideration of the PLA/PGA ratio is important while formulating PLGA and different grades of PLGA should be considered and assessed for a better result. Chitosan has shown great promise as a natural product carrier as it improves compliance and pharmacokinetic attributes of the antidiabetic phytochemicals. Pluronic is another important nanocarrier for phytochemicals [59]. Fabrication of nanocarriers offers a better drug loading capacity, lower toxicity, and gradual releases and it can be a useful antidiabetic formulation. The formation of nanocarriers using PLGA, PEG, chitosan, polycaprolactone, Pluronic acid, and polyacrylamide has increased the efficiency of antidiabetic compounds and improved the pharmacokinetic attributes. Galactosylation of PLGA improves the nanocarrier for transepithelial transport, therapeutic efficiency, and oral bioavailability and succinylated chitosan alginate and o-hexadecyl-dextran have been shown to improve the nanocarrier ability for oral delivery in diabetes [60].

Apigenin extracted from *Castanea sativa* leaves has antibiotic properties against multiple types of bacteria and better efficiency when entrapped in PLGA nanoparticles [61]. Apigenin-loaded micelles have shown improved solubility. Artemisinin has antibiotic properties against bacteria and fungus and its solubility was increased by encapsulation in liposomes using phosphatidylcholine and cholesterol [62]. Berberine is used to treat gastrointestinal disorders and as an antibiotic and PLGA nanoparticles were used to entrap the compound and liposomal formulation was prepared to delay the elimination of berberine [63]. Vegetables-derived luteolin was shown to have a cytotoxic effect against antibiotic-resistant bacteria and nanocapsules and nanospheres were used to encapsulate it [64]. Luteolin-loaded MPEG-PCL micelles were prepared with a better encapsulation efficiency and lesser toxicity [65]. Intensive research work is required to develop nano-formulated phytochemicals as an antibiotic agent.

Silver nanoparticles were used to enhance the antimicrobial property of droserone, ramentaceone, chloroplumbagin, and plumbagin [66]. These compounds show synergy with AgNPs and have an increased cytotoxic effect against *Staphylococcus aureus* [67]. Combination agents of AgNPs and naphthoquinones suggest a potential strategy to control antibiotic-resistant bacteria such as *S. aureus* [68].

CONCLUSION

Phyto-nanoformulations have been widely used in preclinical and clinical stages in the diseases such as cancer, metabolic disorders, and neuronal disorders and as antimicrobial agents. The unique characteristics and functionality of nanostructures combined with phytochemicals can have major advantages in medical treatment and it can be more common in use in coming years.

REFERENCES

[1]   Leontiev R, Hohaus N, Jacob C, Gruhlke MCH, Slusarenko AJ. A comparison of the antibacterial and antifungal activities of thiosulfinate analogues of allicin. Sci Rep 2018; 8(1): 6763.
        [http://dx.doi.org/10.1038/s41598-018-25154-9] [PMID: 29712980]

[2]   Petrovic V, Nepal A, Olaisen C, *et al.* Anti-cancer potential of homemade fresh garlic extract is related to increased endoplasmic reticulum stress. Nutrients 2018; 10(4): 450.
        [http://dx.doi.org/10.3390/nu10040450] [PMID: 29621132]

[3]   Ried K. Garlic lowers blood pressure in hypertensive subjects, improves arterial stiffness and gut microbiota: A review and meta-analysis. Exp Ther Med 2020; 19(2): 1472-8.
        [PMID: 32010325]

[4]   Entezari MH, Aslani N, Askari G, Maghsoudi Z, Maracy M. Effect of garlic and lemon juice mixture on lipid profile and some cardiovascular risk factors in people 30-60 years old with moderate hyperlipidaemia: A randomized clinical trial. Int J Prev Med 2016; 7(1): 95.
        [http://dx.doi.org/10.4103/2008-7802.187248] [PMID: 27563431]

[5]   Kim JH. Anti-bacterial action of onion *(Allium cepa L.)* extracts against oral pathogenic bacteria. J Nihon Univ Sch Dent 1997; 39(3): 136-41.
        [http://dx.doi.org/10.2334/josnusd1959.39.136] [PMID: 9354029]

[6]   González-Peña D, Angulo J, Vallejo S, *et al.* High-cholesterol diet enriched with onion affects endothelium-dependent relaxation and nadph oxidase activity in mesenteric microvessels from wistar rats. Nutr Metab 2014; 11(1): 57.
        [http://dx.doi.org/10.1186/1743-7075-11-57] [PMID: 25926860]

[7]   Brüll V, Burak C, Stoffel-Wagner B, *et al.* Effects of a quercetin-rich onion skin extract on 24 h

ambulatory blood pressure and endothelial function in overweight-to-obese patients with (pre-)hypertension: A randomised double-blinded placebo-controlled cross-over trial. Br J Nutr 2015; 114(8): 1263-77.
[http://dx.doi.org/10.1017/S0007114515002950] [PMID: 26328470]

[8]   Nicastro HL, Ross SA, Milner JA. Garlic and onions: Their cancer prevention properties. Cancer Prev Res 2015; 8(3): 181-9.
[http://dx.doi.org/10.1158/1940-6207.CAPR-14-0172] [PMID: 25586902]

[9]   Deding U, Baatrup G, Christensen LP, Kobaek-Larsen M. Carrot intake and risk of colorectal cancer: A prospective cohort study of 57,053 Danes. Nutrients 2020; 12(2): 332.
[http://dx.doi.org/10.3390/nu12020332] [PMID: 32012660]

[10]   Kaulmann A, Bohn T. Carotenoids, inflammation, and oxidative stress—implications of cellular signaling pathways and relation to chronic disease prevention. Nutr Res 2014; 34(11): 907-29.
[http://dx.doi.org/10.1016/j.nutres.2014.07.010] [PMID: 25134454]

[11]   Woodall AA, Lee SWM, Weesie RJ, Jackson MJ, Britton G. Oxidation of carotenoids by free radicals: Relationship between structure and reactivity. Biochim Biophys Acta, Gen Subj 1997; 1336(1): 33-42.
[http://dx.doi.org/10.1016/S0304-4165(97)00006-8] [PMID: 9271248]

[12]   Kishore C, Bhadra P. Targeting brain cancer cells by nanorobot, a promising nanovehicle: New challenges and future perspectives. CNS Neurol Disord Drug Targets 2021; 20(6): 531-9.
[http://dx.doi.org/10.2174/1871527320666210526154801] [PMID: 34042038]

[13]   Lu H, Wang J, Wang T, Zhong J, Bao Y, Hao H. Recent progress on nanostructures for drug delivery applications. J Nanomater 2016; 2016: 1-12.
[http://dx.doi.org/10.1155/2016/5762431]

[14]   Khan I, Saeed K, Khan I. Nanoparticles: Properties, applications and toxicities. Arab J Chem 2019; 12(7): 908-31.
[http://dx.doi.org/10.1016/j.arabjc.2017.05.011]

[15]   Patra JK, Das G, Fraceto LF, *et al.* Nano based drug delivery systems: Recent developments and future prospects. J Nanobiotechnology 2018; 16(1): 71.
[http://dx.doi.org/10.1186/s12951-018-0392-8] [PMID: 30231877]

[16]   Selmouni F, Zidouh A, Belakhel L, *et al.* Tackling cancer burden in low-income and middle-income countries: Morocco as an exemplar. Lancet Oncol 2018; 19(2): e93-e101.
[http://dx.doi.org/10.1016/S1470-2045(17)30727-1] [PMID: 29413484]

[17]   Seyfried TN, Huysentruyt LC. On the origin of cancer metastasis. Crit Rev Oncog 2013; 18(1 - 2): 43-73.
[http://dx.doi.org/10.1615/CritRevOncog.v18.i1-2.40] [PMID: 23237552]

[18]   Kishore C, Sundaram S, Karunagaran D. Vitamin K3 (menadione) suppresses epithelial-mesenchymal-transition and Wnt signaling pathway in human colorectal cancer cells. Chem Biol Interact 2019; 309: 108725.
[http://dx.doi.org/10.1016/j.cbi.2019.108725] [PMID: 31238027]

[19]   Mansoori B, Mohammadi A, Davudian S, Shirjang S, Baradaran B. The different mechanisms of cancer drug resistance: A brief review. Adv Pharm Bull 2017; 7(3): 339-48.
[http://dx.doi.org/10.15171/apb.2017.041] [PMID: 29071215]

[20]   Guterres SS, Alves MP, Pohlmann AR. Polymeric nanoparticles, nanospheres and nanocapsules, for cutaneous applications. Drug Target Insights 2007; 2: 147-57.
[http://dx.doi.org/10.1177/117739280700200002] [PMID: 21901071]

[21]   Jain AK, Thareja S. *In vitro* and *in vivo* characterization of pharmaceutical nanocarriers used for drug delivery. Artif Cells Nanomed Biotechnol 2019; 47(1): 524-39.
[http://dx.doi.org/10.1080/21691401.2018.1561457] [PMID: 30784319]

[22]   Suk JS, Xu Q, Kim N, Hanes J, Ensign LM. Pegylation as a strategy for improving nanoparticle-based

drug and gene delivery. Adv Drug Deliv Rev 2016; 99(Pt A): 28-51.
[http://dx.doi.org/10.1016/j.addr.2015.09.012] [PMID: 26456916]

[23]   Taherian A, Esfandiari N, Rouhani S. Breast cancer drug delivery by novel drug-loaded chitosan-coated magnetic nanoparticles. Cancer Nanotechnol 2021; 12(1): 15.
[http://dx.doi.org/10.1186/s12645-021-00086-8]

[24]   Pillai SC, Borah A, Jindal A, Jacob EM, Yamamoto Y, Kumar DS. BioPerine encapsulated nanoformulation for overcoming drug-resistant breast cancers. Asian J Pharm 2020; 15(6): 701-12.
[http://dx.doi.org/10.1016/j.ajps.2020.04.001] [PMID: 33363626]

[25]   Wang S, Su R, Nie S, *et al.* Application of nanotechnology in improving bioavailability and bioactivity of diet-derived phytochemicals. J Nutr Biochem 2014; 25(4): 363-76.
[http://dx.doi.org/10.1016/j.jnutbio.2013.10.002] [PMID: 24406273]

[26]   Akbarzadeh A, Rezaei-Sadabady R, Davaran S, *et al.* Liposome: Classification, preparation, and applications. Nanoscale Res Lett 2013; 8(1): 102.
[http://dx.doi.org/10.1186/1556-276X-8-102] [PMID: 23432972]

[27]   Immordino ML, Dosio F, Cattel L. Stealth liposomes: Review of the basic science, rationale, and clinical applications, existing and potential. Int J Nanomedicine 2006; 1(3): 297-315.
[PMID: 17717971]

[28]   Fernández-García R, Lalatsa A, Statts L, Bolás-Fernández F, Ballesteros MP, Serrano DR. Transferosomes as nanocarriers for drugs across the skin: Quality by design from lab to industrial scale. Int J Pharm 2020; 573: 118817.
[http://dx.doi.org/10.1016/j.ijpharm.2019.118817] [PMID: 31678520]

[29]   Khan AA, Allemailem KS, Almatroodi SA, Almatroudi A, Rahmani AH. Recent strategies towards the surface modification of liposomes: An innovative approach for different clinical applications. 3 Biotech 2020; 10: 1-5.

[30]   Shu Q, Wu J, Chen Q. Synthesis, characterization of liposomes modified with biosurfactant mel-a loading betulinic acid and its anticancer effect in hepG2 cell. Molecules 2019; 24(21): 3939.
[http://dx.doi.org/10.3390/molecules24213939] [PMID: 31683639]

[31]   Alhakamy NAA, A Fahmy U, Badr-Eldin SM, *et al.* Optimized icariin phytosomes exhibit enhanced cytotoxicity and apoptosis-inducing activities in ovarian cancer cells. Pharmaceutics 2020; 12(4): 1-17.
[http://dx.doi.org/10.3390/pharmaceutics12040346] [PMID: 32290412]

[32]   Feng T, Wei Y, Lee R, Zhao L. Liposomal curcumin and its application in cancer. Int J Nanomedicine 2017; 12: 6027-44.
[http://dx.doi.org/10.2147/IJN.S132434] [PMID: 28860764]

[33]   Scioli Montoto S, Muraca G, Ruiz ME. Solid lipid nanoparticles for drug delivery: Pharmacological and biopharmaceutical aspects. Front Mol Biosci 2020; 7: 587997.
[http://dx.doi.org/10.3389/fmolb.2020.587997] [PMID: 33195435]

[34]   Picone P, Bondi ML, Picone P, *et al.* Ferulic acid inhibits oxidative stress and cell death induced by Ab oligomers: Improved delivery by solid lipid nanoparticles. Free Radic Res 2009; 43(11): 1133-45.
[http://dx.doi.org/10.1080/10715760903214454] [PMID: 19863373]

[35]   Khorsandi L, Mansouri E, Rashno M, Karami MA, Ashtari A. Myricetin loaded solid lipid nanoparticles upregulate mlkl and ripk3 in human lung adenocarcinoma. Int J Pept Res Ther 2020; 26(2): 899-910.
[http://dx.doi.org/10.1007/s10989-019-09895-3]

[36]   Trivedi R, Kompella UB. Nanomicellar formulations for sustained drug delivery: Strategies and underlying principles. Nanomedicine 2010; 5(3): 485-505.
[http://dx.doi.org/10.2217/nnm.10.10] [PMID: 20394539]

[37]   Wang J, Yang H, Li Q, *et al.* Novel nanomicelles based on rebaudioside A: A potential nanoplatform

for oral delivery of honokiol with enhanced oral bioavailability and antitumor activity. Int J Pharm 2020; 590: 119899.
[http://dx.doi.org/10.1016/j.ijpharm.2020.119899] [PMID: 32971177]

[38]   Liu S, Li R, Qian J, *et al.* Combination therapy of doxorubicin and quercetin on multidrug-resistant breast cancer and their sequential delivery by reduction-sensitive hyaluronic acid-based conjugate/d---tocopheryl poly *(ethylene glycol)* 1000 succinate mixed micelles. Mol Pharm 2020; 17(4): 1415-27.
[http://dx.doi.org/10.1021/acs.molpharmaceut.0c00138] [PMID: 32159961]

[39]   Lu Y, Wen Q, Luo J, *et al.* Self-assembled dihydroartemisinin nanoparticles as a platform for cervical cancer chemotherapy. Drug Deliv 2020; 27(1): 876-87.
[http://dx.doi.org/10.1080/10717544.2020.1775725] [PMID: 32516033]

[40]   Li W, Zhou Y, Yang J, Li H, Zhang H, Zheng P. Curcumin induces apoptotic cell death and protective autophagy in human gastric cancer cells. Oncol Rep 2017; 37(6): 3459-66.
[http://dx.doi.org/10.3892/or.2017.5637] [PMID: 28498433]

[41]   Abbasi E, Aval SF, Akbarzadeh A, *et al.* Dendrimers: Synthesis, applications, and properties. Nanoscale Res Lett 2014; 9(1): 247.
[http://dx.doi.org/10.1186/1556-276X-9-247] [PMID: 24994950]

[42]   Ge P, Niu B, Wu Y, *et al.* Enhanced cancer therapy of celastrol *in vitro* and *in vivo* by smart dendrimers delivery with specificity and biosafety. Chem Eng J 2020; 383: 123228.
[http://dx.doi.org/10.1016/j.cej.2019.123228]

[43]   Gao Y, Li Z, Xie X, *et al.* Dendrimeric anticancer prodrugs for targeted delivery of ursolic acid to folate receptor-expressing cancer cells: Synthesis and biological evaluation. Eur J Pharm Sci 2015; 70: 55-63.
[http://dx.doi.org/10.1016/j.ejps.2015.01.007] [PMID: 25638419]

[44]   Dykman LA, Khlebtsov NG. Gold nanoparticles in biology and medicine: Recent advances and prospects. Acta Nat 2011; 3(2): 34-55.
[http://dx.doi.org/10.32607/20758251-2011-3-2-34-55] [PMID: 22649683]

[45]   Khosropanah MH, Dinarvand A, Nezhadhosseini A, *et al.* Analysis of the antiproliferative effects of curcumin and nanocurcumin in MDA-MB231 as a breast cancer cell line. Iran J Pharm Res 2016; 15(1): 231-9.
[PMID: 27610163]

[46]   Akbarzadeh A, Samiei M, Davaran S. Magnetic nanoparticles: Preparation, physical properties, and applications in biomedicine. Nanoscale Res Lett 2012; 7(1): 144.
[http://dx.doi.org/10.1186/1556-276X-7-144] [PMID: 22348683]

[47]   Wu W, Wu Z, Yu T, Jiang C, Kim WS. Recent progress on magnetic iron oxide nanoparticles: Synthesis, surface functional strategies and biomedical applications. Sci Technol Adv Mater 2015; 16(2): 023501.
[http://dx.doi.org/10.1088/1468-6996/16/2/023501] [PMID: 27877761]

[48]   Tse BWC, Cowin GJ, Soekmadji C, *et al.* PSMA-targeting iron oxide magnetic nanoparticles enhance MRI of preclinical prostate cancer. Nanomedicine 2015; 10(3): 375-86.
[http://dx.doi.org/10.2217/nnm.14.122] [PMID: 25407827]

[49]   Montazerabadi A, Beik J, Irajirad R, *et al.* Folate-modified and curcumin-loaded dendritic magnetite nanocarriers for the targeted thermo-chemotherapy of cancer cells. Artif Cells Nanomed Biotechnol 2019; 47(1): 330-40.
[http://dx.doi.org/10.1080/21691401.2018.1557670] [PMID: 30688084]

[50]   Eatemadi A, Daraee H, Karimkhanloo H, *et al.* Carbon nanotubes: Properties, synthesis, purification, and medical applications. Nanoscale Res Lett 2014; 9(1): 393.
[http://dx.doi.org/10.1186/1556-276X-9-393] [PMID: 25170330]

[51]   Elhissi AMA, Ahmed W, Hassan IU, Dhanak VR, D'Emanuele A. Carbon nanotubes in cancer

therapy and drug delivery. J Drug Deliv 2012; 2012: 1-10.
[http://dx.doi.org/10.1155/2012/837327] [PMID: 22028974]

[52]   Li H, Zhang N, Hao Y, *et al.* Formulation of curcumin delivery with functionalized single-walled carbon nanotubes: Characteristics and anticancer effects *in vitro*. Drug Deliv 2014; 21(5): 379-87.
[http://dx.doi.org/10.3109/10717544.2013.848246] [PMID: 24160816]

[53]   Sanginario A, Miccoli B, Demarchi D. Carbon nanotubes as an effective opportunity for cancer diagnosis and treatment. Biosensors 2017; 7(4): 9.
[http://dx.doi.org/10.3390/bios7010009] [PMID: 28212271]

[54]   Bellomo A, Mancinella M, Troisi G, Ettorre E, Marigliano V. Diabetes and metabolic syndrome *(MS)*. Arch Gerontol Geriatr 2007; 44 (1): 61-7.
[http://dx.doi.org/10.1016/j.archger.2007.01.009] [PMID: 17317435]

[55]   Moradi SZ, Momtaz S, Bayrami Z, Farzaei MH, Abdollahi M. Nanoformulations of herbal extracts in treatment of neurodegenerative disorders. Front Bioeng Biotechnol 2020; 8: 238.
[http://dx.doi.org/10.3389/fbioe.2020.00238] [PMID: 32318551]

[56]   Gong Y, Chowdhury P, Nagesh PKB, *et al.* Novel elvitegravir nanoformulation for drug delivery across the blood-brain barrier to achieve HIV-1 suppression in the CNS macrophages. Sci Rep 2020; 10(1): 3835.
[http://dx.doi.org/10.1038/s41598-020-60684-1] [PMID: 32123217]

[57]   Ghitman J, Biru EI, Stan R, Iovu H. Review of hybrid PLGA nanoparticles: Future of smart drug delivery and theranostics medicine. Mater Des 2020; 193: 108805.
[http://dx.doi.org/10.1016/j.matdes.2020.108805]

[58]   Rezvantalab S, Drude NI, Moraveji MK, *et al.* PLGA-based nanoparticles in cancer treatment. Front Pharmacol 2018; 9: 1260.
[http://dx.doi.org/10.3389/fphar.2018.01260] [PMID: 30450050]

[59]   Kim S, Kwon K, Cha J, *et al.* Pluronic-based nanocarrier platform encapsulating two enzymes for cascade reactions. ACS Appl Bio Mater 2020; 3(8): 5126-35.
[http://dx.doi.org/10.1021/acsabm.0c00591] [PMID: 35021689]

[60]   Gupta S, Agarwal A, Gupta NK, Saraogi G, Agrawal H, Agrawal GP. Galactose decorated PLGA nanoparticles for hepatic delivery of acyclovir. Drug Dev Ind Pharm 2013; 39(12): 1866-73.
[http://dx.doi.org/10.3109/03639045.2012.662510] [PMID: 22397550]

[61]   Basile A, Sorbo S, Giordano S, *et al.* Antibacterial and allelopathic activity of extract from *Castanea sativa* leaves. Fitoterapia 2000; 71 (1): S110-6.
[http://dx.doi.org/10.1016/S0367-326X(00)00185-4] [PMID: 10930721]

[62]   Kim WS, Choi WJ, Lee S, *et al.* Anti-inflammatory, antioxidant and antimicrobial effects of artemisinin extracts from Artemisia annua L. Korean J Physiol Pharmacol 2014; 19(1): 21-7.
[http://dx.doi.org/10.4196/kjpp.2015.19.1.21] [PMID: 25605993]

[63]   Chen C, Yu Z, Li Y, Fichna J, Storr M. Effects of berberine in the gastrointestinal tract: A review of actions and therapeutic implications. Am J Chin Med 2014; 42(5): 1053-70.
[http://dx.doi.org/10.1142/S0192415X14500669] [PMID: 25183302]

[64]   Cai W, Xiong Y, Han M, *et al.* Characterization and quantification of luteolin-metal complexes in aqueous extract of lonicerae japonicac flos and huangshan wild chrysanthemum. Int J Anal Chem 2021; 2021: 1-9.
[http://dx.doi.org/10.1155/2021/6677437] [PMID: 33777144]

[65]   Zhou L, Li A, Wang H, Sun W, Zuo S, Li C. Preparation and characterization of luteolin-loaded mpeg-pcl-g-pei micelles for oral candida albicans infection. J Drug Deliv Sci Technol 2021; 63: 102454.
[http://dx.doi.org/10.1016/j.jddst.2021.102454]

[66]   Yin IX, Zhang J, Zhao IS, Mei ML, Li Q, Chu CH. The antibacterial mechanism of silver

nanoparticles and its application in dentistry. Int J Nanomedicine 2020; 15: 2555-62.
[http://dx.doi.org/10.2147/IJN.S246764] [PMID: 32368040]

[67]   Li WR, Xie XB, Shi QS, Duan SS, Ouyang YS, Chen YB. Antibacterial effect of silver nanoparticles on *Staphylococcus aureus*. Biometals 2011; 24(1): 135-41.
[http://dx.doi.org/10.1007/s10534-010-9381-6] [PMID: 20938718]

[68]   Krychowiak M, Kawiak A, Narajczyk M, Borowik A, Królicka A. Silver nanoparticles combined with naphthoquinones as an effective synergistic strategy against *Staphylococcus aureus*. Front Pharmacol 2018; 9: 816.
[http://dx.doi.org/10.3389/fphar.2018.00816] [PMID: 30140226]

SUBJECT INDEX

A

Absorption 10, 68, 93, 105, 163, 217, 218,
        244, 311, 313, 326, 327, 336, 342
    calcium 105
    nutrient 93, 163, 244
Acid(s) 4, 5, 7, 15, 39, 59, 83, 96, 98, 100,
        105, 135, 148, 161, 163, 167, 175, 194,
        213, 214, 217, 232, 234, 237, 245, 273,
        281, 282, 283, 285, 286, 289, 301, 324,
        336, 338, 339, 342, 343
    acetic 96, 98
    amino 4, 15, 59, 83, 105, 167, 194, 245, 339
    angelic 283
    ascorbic 285, 286
    betulinic 163, 338
    caffeic 214, 281, 289, 301
    chloroauric 237
    chlorogenic 281, 282
    cinnamic 273
    citric 5
    coumaric 213
    ellagic 232
    ferulic 273, 281, 343
    folic 339
    fumaric 5
    gallic 213, 214, 281, 289
    gibberellic 245
    ginkgolic 39
    glycolic 217, 336
    gymnemic 336, 342
    jasmonic 7, 163
    lactobionic 324
    phenolic 148, 161, 175, 234, 281, 286
Agrobacterium-mediated transformation 145
Agronomy of medicinal plants 74
AHL lactonase enzyme 7
Alzheimer's disease 192, 208, 342, 343
Amplification refractory mutation system
        (ARMS) 114, 115, 122
Amplified fragment length polymorphism
        (AFLP) 112, 113, 122

Analysis, fingerprint 296
Anti-diabetic 240, 241
    agent 240
    effect 241
    properties 241
Anti-inflammatory 101, 240
    agents 240
    effects 101
Anti-ulcer properties 236
Antimicrobial 68, 75, 235, 238, 287, 288, 289
    activity 68, 75, 235, 287, 288, 289
    effects 238
Antioxidant 39, 70, 218, 285
    defense system 39
    effects 70, 218, 285
Antitumor activity 164, 215, 220, 319
Antitussive effects 283
Applications, sensory 312
Arabidopsis plants 56, 145, 149, 150
Arbuscular mycorrhizal fungi (AMF) 5, 6
Arc discharge process 259
Arthritis pain 196

B

Bacillus cereus 60
Bacteria, lactic acid 97
Bactericidal 210, 236
    activity 210
    effect 236
Bioactive drug carriers 327
Biomaterials 175, 314, 326, 327
    metallic 327
    natural 314
Bioremediating techniques 245
Blood vessel growth 339
Breast cancer therapy 338